HALF A WING, THREE ENGINES AND A PRAYER

HALF A WING, THREE ENGINES AND A PRAYER

B-17s over Germany
Special Edition

by BRIAN D. O'NEILL

McGRAW-HILL

New York San Francisco Washington, D.C. Auckland Bogotá
Caracas Lisbon London Madrid Mexico City Milan
Montreal New Delhi San Juan Singapore
Sydney Tokyo Toronto

Library of Congress Cataloging-in-Publication Data

O'Neill, Brian D.
Half a wing, three engines and a prayer : B-17s over Germany / by Brian O'Neill.
— Special Edition
p. cm.
Includes bibliographical references and index.
ISBN 0-07-134145-5 (alk. paper)
1. World War. 1939-1945,—Aerial operations, American. 2. World War, 1939-1945,—Germany. 3. World War, 1939-1945 Personal narratives, American. I. Title.
States. I. Spence, Charles F. II. Title.
D790.058 1999
940.54'4373—dc21

99-14339
CIP

McGraw-Hill

A Division of The ***McGraw-Hill*** *Companies*

Permission is gratefully acknowledged to quote from the following copyrighted works in this book:

From *Half a Wing, Three Engines and a Prayer: The Diary of an English Air Force Navigator from Schweinfurt to Big Week* by Elmer Logan Brown, Jr., unpublished work. Copyright © 1944 Elmer Logan Brown, Jr.

From the poem "The Lost Pilot" by James Tate in *The Lost Pilot* by James Tate. Copyright © 1966 Yale University Press.

From the screenplay of *Twelve O'Clock High* by Beirne Lay, Jr. and Sy Bartlett. Twentieth Century Fox.

For the quote attributed to Dale W. Rice in Chapter 5, excerpted from *The Schweinfurt-Regensburg Mission*, Copyright © 1983 Martin Middlebrook. Reprinted with the permission of Charles Schribner's Sons, an imprint of Macmillian Publishing Company.

For previously published excerpts of Wilbur Klint's account of the second Schweinfurt mission in Chapter 14, *In My Book You're All Heroes* by Robert E. O'Hearn, Copyright © 1984 Robert O'Hearn.

For the excerpt from Psalm 130 at the beginning of Chapter 21, which is taken from the *New American Bible*, Copyright © 1970, by the Confraternity of Christian Doctrine, Washington, D.C., used by permission of copyright owner. All rights reserved.

2 3 4 5 6 7 8 9 0 DOC/DOC 9 0 3 2 1 0 9

ISBN 0-07-134145-5

The sponsoring editor of this book was Shelley Carr, the editing supervisor was Sally Glover, and the production supervisor was Sherri Souffrance. It was set in BK Hiroshige Book per the NM1 Design by Paul Scozzari of McGraw-Hill's Hightstown desktop composition unit.

Printed and bound by R. R. Donnelley & Sons Company.

This book is printed on recycled, acid-free paper containing a minimum of 50% recycled, de-inked fiber.

This book is for Donald F. O'Neill, a KB-50 pilot cut from the same mold as the men who appear in these pages.

...I would
touch your face as a disinterested
scholar touches an original page.
However frightening, I would
discover you...

—James Tate
"The Lost Pilot"

General Savage, I'd like to tell you something else, I think. I'm a lawyer by trade. I think I'm a good one. And when a good lawyer takes on a client he does it because he believes in the client's case. And that's all that matters. When I came over to England I took on my biggest client. That client is the 918th Bomb Group. I want to see my client win its case. Does that answer what you had in mind?

—Major Harvey Stovall to General Frank Savage in the screenplay from "Twelve O'Clock High" by Beirne Lay, Jr. and Sy Bartlett, Twentieth Century Fox

Contents

Special Edition Acknowledgments

THIS SPECIAL EDITION of *Half A Wing, Three Engines and a Prayer: B-17s Over Germany* owes its existence to the men it is written about perhaps even more than its best-selling predecessor, which was first published in 1989 and then went through a remarkable seven printings in seven years. Fifty years after the events it chronicles, one would think that the eyewitness sources available for this work would decrease rather than increase; but the exact opposite is true. The men of the famous 303rd "Hell's Angels" Bomb Group, who persevered in their mission throughout the entire daylight bombing campaign against Nazi Germany, are today equally dedicated to another mission: preserving the history of their accomplishments so that succeeding generations may know exactly what it took to win this terrible war.

The Hell's Angels of World War II live on in the 303rd Bomb Group (H) Association, one of the most active, dedicated, and open veterans' organizations to be found anywhere. It is because of the continuing work of these veterans, their spouses, and other 303rd family members and associates that I have much of the new material in this Special Edition.

Hal Susskind, past president of the 303rd Association and Editor Emeritus of the *Hells Angels Newsletter*, has been instrumental as editor from December 1985 through May 1998 in making the *Hells Angels Newsletter* both "the glue that holds the Association together" and the best Eighth Air Force unit publication by far. A number of personal accounts in the Special Edition, including Guy Lance's story of his crew's experiences on the November 29, 1943 Bremen mission, and Jack

Fawcett's lead bombardier's view of the January 11, 1944 Oschersleben raid, appear not only with their cooperation but also with that of the *Hells Angels Newsletter*. Happily, the *Hells Angels Newsletter* has a worthy new editor in Eddie Deerfield, another contributor to this book and an author in his own right,* and those interested in the Group's history are sure to welcome the forthcoming publication of the *Hells Angels Newsletter's* back issues in bound volumes.

Harry D. Gobrecht, current president of the 303rd Association and long-term group historian, is another veteran whose work has been indispensable to the Special Edition. Harry's encyclopedic *Might in Flight, Daily Diary of the Eighth Air Force's Hell's Angels 303rd Bombardment Group (H)*, first published in 1993 and now issued in a 1997 second edition, is a 920-page labor of love whose detail is simply amazing. For the first edition of *Half A Wing* much of my motivation in chronicling the group's history, during the August 1943 to February 1944 period when Bob Hullar's crew flew, lay in my strong desire to find out what had happened to the many crews that were lost as Hullar's crew successfully completed their tour. While my own research provided many answers, *Might in Flight* has provided a wealth of new information on this subject that is contained both in the text of the Special Edition and in its new Appendix I. Thank you, Harry, for making this possible, as well as for the other *Might in Flight* materials used herein.

Another 303rd veteran to whom I owe a special debt of gratitude is Dr. Carl J. Flyer, past president of the 303rd Association and a pilot in the Group's 360th Squadron. Dr. Flyer's tour (until he was shot down on his 25th mission on November 29, 1943) included many of the most difficult missions Hullar's crew flew. His point of view on these missions was—literally and figuratively—unique and I am grateful for permission to quote from his book, *Staying Alive*, in the mission narratives. In return, it has been my privilege to add a new chapter to this book, "Sawicki's Sacrifice," in support of Dr. Fyler's decades-long effort to secure a posthumous Medal of Honor for Joseph R. Sawicki, Flyer's tail gunner, whose heroic actions on the crew's last mission have gone unrecognized for far too long. Assisting in this effort is Robert A. Hand, Sr., a 303rd veteran, author, and artist whose vivid painting of Sawicki's last moments is included in the new chapter with many thanks.

It has been my pleasure to add the contributions of Grover C. Mullins, engineer on Lt. Bill Fort's crew, to the chapter entitled "Star Dust," my

* See Appendix II for a listing of other books written by 303rd veterans.

personal favorite in the first edition of *Half A Wing*. Mullins' recollections show why Charles Spencer won the Distinguished Service Cross, the nation's second highest award for valor, without detracting in any way from the central themes of near-death experience and religious revelation which lay at the heart of Spencer's own account. I am also thankful to Mullins for a small but telling incident he has related in the newly revised Oschersleben chapter.

The Special Edition features a complete rewrite of the Oschersleben chapter, describing the 303rd's ordeal by fire on January 11, 1944, when ten out of forty aircraft were shot down, to update the chronology of losses in light of new information in *Might in Flight*. In addition, the revised chapter includes not only the previously mentioned account of Jack B. Fawcett, but also the historically valuable observations of Vern L. Moncur, who was in the center of the action in the lead group's lead squadron. My thanks to Vern Moncur's son Gary, the 303rd's Web master*, for permission to include excerpts from his father's journal in the Special Edition.

Mrs. Beryl Coburn, the widow of Ralph Coburn, bombardier on Don Gamble's crew, kindly gave permission to use Ralph's diary to add to the mission narratives of Don gamble and his navigator, Bill McSween, which figured so prominently in the original edition. Charles S. Schmeltzer, Gamble's left waist gunner, provided a unique photo of *Sky Wolf*, the crew's favorite B-17 and one of the Group's original ships. Dave L. Rogan, the Gamble crew's original "Instructor Pilot," likewise provided me with a long-sought-after photo of him with Gamble's crew.

Other 303rd veterans generously contributed to my expanded treatment of the fiercely contested August 19, 1943 mission to Gilze-Rijen, Holland, the infamous "milk run" that wasn't. I had the pleasure of meeting Howard L. Abney, the tail gunner on Jack Hendry's crew, whose story formed a significant subtheme in the original edition of this book. Abney's account of being wounded on this August 19th raid enhances this chapter considerably, as does the contribution of Richard C. Grimm, who also marked this operation as the last he flew for the 303rd.

Europeans have contributed generously to this book. I am indebted to Major Ivo M. de Jong of the Dutch Army for presenting me with a "hot off the press" copy of his newly published book, *Mission 85, A Milk Run That Turned Sour*, which provides an account of the August 19, 1943 mission from both sides. Major de Jong's work has allowed me to identify the German fighter units that gave the 303rd such a rough time on this raid.

---

* See Appendix II for additional information on 303rd Web sites.

Serge Lebourg, a Frenchman born and raised in Nantes, willingly shared his extensive research into the bombing raids of September 16th and 23rd, 1943 against his home town to help explain what went wrong and why French civilian casualties were so severe.

Many Germans have offered interesting information about the 303rd's adversaries. Dr. Volkmar Wilckens and the *Luftwaffenhelfer* (Luftwaffe helpers) of Schweinfurt, especially Gerhard Belosa, Otto Gruner, Georg Schäfer (whose father owned the ballbearing plant the 303rd bombed on both of the 1943 Schweinfurt missions) provided details of the city's flak defenses during the "Black Thursday" mission of October 14, 1943, and of the effects of the bombing. Rudolf Tyrassek, who was a *Luftwaffe* night fighter trainee at the end of the war, has helped me come as close as I can to learning who was in the Me-410 that almost shot down Elmer Brown's bomber on the mission which inspired this book's title. The hospitality of all these gentlemen, and of the other Germans whom I met during a trip to Schweinfurt in June of this year, cannot be adequately described in words.

Tremendous thanks are due to Brian S. McGuire, a retired U.S. Air Force officer who is currently working as a civilian contractor for the U.S. European Command's Joint Analysis Center (JAC) located at RAF Molesworth, the 303rd's wartime base. He introduced me to the JAC's Commanding Officer, Col. Frances M. Early, USAF (herself the daughter of an Eighth Air Force airman) and made it possible for me to tour the base and view the impressive collection of 303rd artwork, memorabilia, and information he has collected there with the JAC's support. McGuire also facilitated my contacts with Kenneth Davey, a 303rd veteran who flew with Lewis Lyle on the first Schweinfurt mission; and with Loren Zimmer, right waist gunner on James F. Fowler's crew. Both Davey and Zimmer's contributions to this book are greatly appreciated.

My family's long association with three members of Hullar's crew, and with their families, deserves special mention. Bud and Mary Klint, Merlin and Marge Miller, and George and Elfriede Hoyt have become fast friends since 1985; and the Klints' and Millers' assumption that there will always be one or more O'Neills at the next 303rd Reunion has ensured my continuing interest in all the Hell's Angels. I can say of them in particular what I can fairly state for every 303rd Association member I have met in this time: My respect and admiration for you grows with each passing year.

Ms. Shelley Ingram Carr, Aviation Acquisitions Editor of McGraw-Hill's Professional Book Group, has my warm thanks for championing my proposal for the Special Edition. A special word of thanks also goes to

Ms. Jo Stein of McGraw-Hill's art department for her enthusiasm and willingness to entertain my suggestions for the Special Edition's new cover artwork and to editing supervisor Sally Glover, who has been a real pleasure to work with.

Home is where the heart is, and in my case neither this Special Edition nor its predecessor would have come about without my family's active encouragement, sacrifice, and support. Aileen, Lauren, and Richard, thank you for understanding when your Dad lacked sufficient time for you. Kathy, let me thank you most of all.

Brian D. O'Neill
September 30, 1998

Acknowledgments

THIS BOOK IS uniquely dependent on the contributions of others, for in writing it my principal object has been to let the reader see the events in question through the eyes of the participants. Thus, a tremendous amount of the text is taken up in quotations: from diaries, from letters, from official reports, and from present-day recollections. In consequence, I am greatly indebted to all who are quoted in these pages, beginning with the surviving members of Hullar's crew. But I also wish to acknowledge many whose names do not directly appear, starting with the families of Hullar's crew members who have passed away in the long years since the events in this book took place.

Mrs. Jean J. Hullar contributed Bob Hullar's notebook, otherwise unobtainable photographs, and many of her husband's personal records. Without them, Bob Hullar's feelings about the air war would have been impossible to present. Mrs. Jean M. Rice and Mrs. Sarah A. Marson were equally generous with their memories, photographs, and the personal records of their late husbands, as were Dale Rice's brother, The Rev. Charles Rice, and Charles "Pete" Fullem's recently deceased brother, Joseph. The latter sent me a package I will never forget—a small, string-wrapped envelope that contained the entire written remains of Pete Fullem's life, from his birth certificate to the record of his death.

I am most grateful to a number of B-17 veterans whose contributions went far beyond their accounts of the missions Hullar's crew flew. Major General Lewis E. Lyle graciously consented to write the Foreword to this work and greatly facilitated my research into the 303rd Bomb Group's "mission files" at the National Military Records Center, Suitland, Maryland. It was General Lyle who introduced me to the treasure trove of documents there, and who put me in touch with Mr. John Sanda, a B-17 pilot with over 50 missions to his credit. John Sanda spent uncounted hours at Suitland answering my every documentary request.

Colonel Edgar E. Snyder, Jr., and Colonel Louis M. "Mel" Schulstad, both of the 303rd's original cadre, gave me a real feel for the Group during its early days. Ed Snyder's role as the Hullar crew's Squadron Commander also gave me a unique perspective on the air war. To Mel Schulstad I owe a tremendous debt for the invaluable source material he made available to me. Without the 303rd's "Honor Roll," the names of many planes and crews lost during the Hullar crew's tour could not be properly mentioned; without his mission maps, none of the individual mission routes reproduced here would have been possible.

Grover C. Henderson, Jr., contributed an unforgettable mission account and an equally unforgettable portrait of his pilot, Captain Edward "Woodie" Woddrop. Woddrop was killed in a freak accident at the very end of the war, but enough of him shone through the pages of Elmer Brown's diary to make me want to know the man more. To Grover Henderson I owe the rich satisfaction of seeing "Woodie" as he truly was in life.

David P. Shelhamer's strong interest in this work gave me much encouragement, and he spared no effort in answering my many technical questions and inquiries about the missions he flew with Hullar's crew. He died on February 12, 1988, and while I never met him except in telephone conversations and tapes, I knew him as a friend and greatly regret his passing.

I am grateful to Forrest L. Vosler, winner of the Medal of Honor, for his personal account of the day that changed his life and for his suggestion that I contact two other members of his crew: Mr. Edward Ruppel and Mr. William H. Simpkins. They also rank among the real heroes of the air war.

I want to thank Wilbur "Bud" Klint not only for his personal contributions, but for his early contacts with the other members of Hullar's crew, for his thoughtful critiques of draft chapters, and for his invaluable "missionary work" soliciting contributions from other 303rd sources, including The Rev. Charles W. Spencer.

I want to mention especially George F. Hoyt, Norman A. Sampson, Merlin D. Miller, and William D. McSween, Jr. George Hoyt's insistence that this material deserved a book, Norman Sampson's early contributions, Merlin Miller's steady flow of information, and the importance Bill McSween attached to this effort were all critical ingredients in my decision to expand the work from an article into a full-fledged book.

I am very much indebted to Waters Design Associates, Inc., of Manhattan for their superlative work on the mission maps, and for the book's other artwork. Both John Waters and Mr. Robert Kellerman of his staff were most generous with their talent and time. The staff of the National Air and Space Museum Library in Washington D.C. was quite helpful to me in my photographic research there; one gentleman who wishes to remain nameless I want to thank particularly. I also owe much to Mr. Don Linn, fellow aviation writer and friend, who did excellent photo reproduction work for me.

In England, thanks are due Mr. Martin Middlebrook for permission to quote Dale Rice's recollections of the First Schweinfurt briefing in his fine book, *The Schweinfurt-Regensburg Mission*; to Mr. Roger A. Freeman, dean of Eighth Air Force historians, for putting me in contact with Mr. J.P. Foynes, an expert on British air sea rescue efforts during the war; to Mr. Foynes himself, who acted immediately on my request for information by confirming the identity of the RAF launch that rescued Hullar's crew; and to Mr. John T. Gell, who was instrumental in providing me with information about the *Cat O' 9 Tails* and her crew on the second Schweinfurt raid of October 14, 1943.

Finally, I want to thank Elmer Brown for providing the diary that inspired me to undertake this effort, and his son Larry for convincing his father to let someone outside the Brown family read the diary for the first time. None of this would have happened had Larry not persuaded his father to let me look at the diary, nor asked me if I could "do something" with it.

To all whose contributions made this book possible, my heartfelt thanks. I owe each of you a debt these pages can never repay.

Brian D. O'Neill
July 1988

Foreword

BRIAN O'NEILL HAS chosen my favorite subject, "the combat crew," the lowest and the most important element of Bombardment Aviation.

The 303rd Bomb Group, activated just after Pearl Harbor, had the good fortune to be selected as one of the first groups to go into combat with the Eighth Air Force in England. Formed in January 1942 at Boise, Idaho, it was scheduled to move to England in June 1942. This gave it a high priority for personnel and aircraft, at the highest echelons of the Army Air Forces. Squadron Commanders on up were assigned from the most experienced senior people with bombardment and reconnaissance backgrounds.

Unfortunately, this proved to be the early downfall of the organization; with the rapid expansion of the Air Force, these senior people kept moving on to other forming units. A lack of leadership at the Group Commander's level prevented the 303rd from ever becoming combat-ready. Squadrons with senior commanders fought for resources. Those who were to be the combat crews had to fend for themselves. Guidance and control from the top were sorely needed.

The Operational Readiness Test given to the 303rd in May 1942 was a total disaster, and the group was virtually broken up as key people were sent to other units that were scheduled for combat soon. The group was bounced from Boise, Idaho, to Alamogordo, New Mexico, to Munroc Lake, California, to Biggs, Texas, and to Kellogg, Michigan. It was fortunate for the crews who were left that the group's ground echelon remained intact during this period. The ground echelon was made up of highly qualified enlisted men and professionals from the civilian world. They were the salvation of the organization.

Desperate to augment the Eighth Air Force's attempts to get into the air war against Germany, the Air Force shipped the Group's ground echelon to Molesworth, England, in September 1942. The flying echelon, without any readiness inspection, left Kellogg Field for England in October 1942. We began combat flying in November, far from being combat-ready. Many crews, including pilots, had not been formally checked out. This was a very agonizing period for the group, as our original crews had heavy casualties and replacement crews and aircraft were not available for many months. Thank God for that wonderful ground echelon. They hung in there and produced the best record in the Eighth as far as aircraft combat readiness went.

We began to get good replacement crews in the summer of 1943. Hullar's crew was one of the first we received that had been properly trained in the United States. McCormick, this crew's bombardier, taught me to be a bombardier. As the new Group Deputy Commander for Air Operations, I latched onto these people and can vouch for the tremendous contribution they and all the other replacement crews made not only in the 303rd but in all the other bomb groups of the Mighty Eighth.

The Hullar crew's tour spanned the end of the survival period of the 303rd and the beginning of the real bombing phase of the air war. Brian O'Neill has realistically portrayed, through diaries, archival records, and authentic descriptions, what a typical crew thought and did during those days of fierce fighter attacks and flak, when thousands of crews were destined to go down. Those who lived through the experience, and those who made it through their missions, will attest to the accuracy of human behavior under tremendous stress that is portrayed here—and all readers will become engrossed in the story of this crew's fight to survive.

Lewis E. Lyle
Maj. Gen., USAF, Ret.

Introduction

FIVE DECADES AFTER its last combat mission was flown in April 1945, the U.S. Eighth Army Air Force's role in winning the Allied victory over Nazi Germany is well recognized. Historians consider it the air force that was responsible for breaking the back of the German Luftwaffe in the spring of 1944. Together with British Bomber Command, it is also credited by Albert Speer, Nazi Minister of Armaments and Munitions, with creating a "second front" in the West long before D-Day. Finally, there is no question that the Eighth crippled vital enemy war industries, especially during the "oil campaign" against German petroleum resources that ran from May 1944 through March 1945.

Significant though these accomplishments are, they do not account in full for the place the Eighth occupies as one of the most famous military organizations in the annals of armed conflict, for military greatness is measured not only by an army's ability to win battles. When necessary, its soldiers must also show the capacity to sustain great loss and still return to the fight. Here, too, in the half-year running from mid-1943 to early 1944, the Eighth was tested and not found wanting. This was before the ascendancy of the long-range P-51 Mustang escort fighter, when the Eighth's major assets were the ruggedness of the B-17 and the incredible courage and tenacity of its bomber crews. These crews endured bloody reversals and heartbreaking losses attempting to vindicate the concept of "daylight precision bombing" on unescorted deep penetration raids into Germany. Their perseverance was, by any reckoning, a major factor in the Eighth's ultimate triumph. Their collective experience constitutes an epic whose heroic dimensions cannot be denied.

This is the story of one of the B-17 crews that flew during this critical time of the war. Its outline and primary source is the marvelous wartime diary of the crew's navigator, Elmer L. Brown, Jr. Major portions of his diary are reproduced, with minor editing, just as he wrote them during the dark and difficult days of his crew's tour in the Eighth from August 1943 through February 1944. His account of the crew's missions is supplemented by the writings and recollections of five other crewmembers: the pilot, Robert J. Hullar; the copilot, Wilbur Klint; the radioman, George F. Hoyt; the ball turret gunner, Norman A. Sampson; and the tail gunner, Merlin D. Miller. Additional materials have been supplied by other crewmembers or their families: the bombardier, James E. McCormick; the engineer and top turret gunner, Dale W. Rice; the right waist gunner, Charles H. Marson; and the left waist gunner, Charles "Pete" Fullem.

This book also tells part of the history of the group to which the crew was assigned, the famous 303rd "Hell's Angels" Bombardment Group (H). It contains the recollections of many individuals mentioned in Brown's diary and the experiences of others who took part in the same battles with him and the rest of Hullar's crew high in the skies over wartime Germany.

Through the words of these men, supplemented by the official records of the 303rd Bomb Group and other pertinent Eighth Air Force sources, you will learn first hand what it was like to take the famous Flying Fortress into combat during the worst period of America's air war against the Germans.

1

Hullar's Crew

THE WORLD WAR II heavy bomber crew has always been a powerful symbol of the American air war against the Axis Powers. Because it invariably contained a cross section of American youth whose teamwork was a natural reflection of the larger national effort, the bomber crew was a wartime propagandist's dream—the perfect way to depict "America at War" in microcosm. Since the number of men manning a four-engine bomber was usually large—ten is the typical number—the crew has also served as a prime vehicle for postwar novelists to make their own points about the impact of war on the human psyche. John Hersey's *The War Lover* and Jim Shepard's more recent *Paper Doll* are but two examples of the genre.

What was the truth behind the wartime propaganda? How much of what the novelists write is an accurate reflection of "the way it was" for the men who were actually there doing the flying and the fighting? With the help of real combat men, this book attempts to answer these questions.

The men of "Hullar's crew" fit the "All-American Bomber Crew" ideal to perfection. They were all "average citizens" who the U.S. Army Air Forces training program, operating at full speed from early 1942 through mid-1943, brought together and began to mold into a first-class fighting team. The first eight members met in April and May of 1943 at Gowen Field, a B-17 crew training center near Boise, Idaho. Before learning how they were trained, a brief introduction to each man is in order.

The pilot and commander was Robert J. Hullar, a native of Syracuse, New York, born March 14, 1917. Known as "the Skipper," at 26 Bob Hullar was a tall, athletic man whose outgoing, friendly personality combined with strong flying and leadership skills to make an outstanding

Crew portrait. The men of Hullar's crew are: Bottom row, L-R, Sgt. Merlin D. Miller, tail gunner; Sgt. Norman A. Sampson, ball turret gunner; Sgt. Charles "Pete" Fullem, left waist gunner; Sgt. Dale W. Rice, engineer; Sgt. George F. Hoyt, radioman. Top row, L-R, Sgt. Charles H. "Chuck" Marson, right waist gunner; Lt. Robert J. Hullar, pilot; Lt. Elmer L. Brown, Jr., navigator; Lt. Wilbur "Bud" Klint, copilot; and Lt. James E. "Mac" McCormick, bombardier. (Photo courtesy Mrs. Jean M. Rice.)

bomber crew commander. He spent much of his boyhood in Brooklyn, attended Townsend Harris Prep School in the city, and then City College and NYU where, in 1940, he received a Second Lieutenant's commission in the famous Seventh "Silk Stocking" Regiment of the New York National Guard. In early 1941 Hullar went on active duty with the Regiment (renumbered the 207th Anti-Aircraft Regiment) at Camp Stewart near Savannah, Georgia. In May 1942 he joined the Army Air Force, attended various flight schools, and ultimately won his pilot's wings. April 1943 found him in Boise forming the heart of a brand new B-17 bomber crew.

If the crew had a "loner," it was the bombardier, James E. "Mac" McCormick. He was born July 10, 1919, and grew up in Holland, Michigan. McCormick was 20 years old and two years into college when his National Guard unit—the 126th Infantry, 32nd Division—was activated and shipped to Camp Livingston near Alexandria, Louisiana. Here McCormick found himself serving as a company clerk, but he soon tired of it. In 1941 he transferred into the "Ground Air Corps" and went to Keesler Field, the huge AAF training center at Biloxi, Mississippi. At Keesler, McCormick decided to become an aviation cadet. He entered the AAF training pipeline at Montgomery, Alabama, and emerged in early 1943 at Victorville, California, as a rated bombardier. He then joined the Hullar crew in Boise.

The Pilot's Manual for the Flying Fortress strictly enjoined the novice crew commander to "Size up the man who is to be your engineer. This man is supposed to know more about the airplane you are to fly than any other member of the crew...Make him a man upon whom you can rely." These lines reflected a strong Army Air Force view that the bomber crew's engineer should be an enlisted "second in command." The ideal flight engineer was supposed to know the aircraft's nuts and bolts *and* be a leader to the other men. Dale W. Rice, Hullar's engineer, fit the bill in every way.

Born in Brooklyn on December 3, 1920, Rice was a child of the Great Depression who never stopped looking on the bright side of life. As the son of a union lather (the forgotten trade of installing wooden, wire, or rock foundations for interior plaster walls and window "fancy work"), Rice's early years were spent in Islip, Long Island, and the Queens communities of Saint Albans and Queens Village. They were hard times, but, as his elder brother Charles recalls, "We still had a great life during the

"The Skipper." Bob Hullar as a 25-year-old 2nd Lt. with the 207th (7th N.Y.) Anti-Aircraft Regt. at Camp Stewart, Georgia, in 1941 before he transferred to the Army Air Force. (Photo courtesy Mrs. Jean J. Hullar.)

Depression. In spite of our problems, we never missed a meal. There was a great amount of love in our family, and it had its effect on my brother, his outlook on life, and his feelings towards people."

Real adversity came to Rice in 1937, when he had to leave high school at the end of his junior year to help his father and brother make ends meet. Three years later, Rice was a service station mechanic, and with the war going full blast in the summer of 1942, he opted for the Army Air Force because he wanted to become an airplane mechanic. It seemed a great way to make up for his lost education, and for a while, it looked like a good move. Rice went to Airplane Mechanics School at Keesler, but then he wound up headed for the wild blue yonder via gunnery school at Las Vegas. After further flight engineer training at the Seattle Boeing plant, he joined Hullar's crew at Boise.

Loved for his wonderful sense of humor, easy camaraderie, and deadly ability with the top turret's twin fifties, Dale Rice was, according to the crew's ball turret gunner, "the one who always took responsibility for the enlisted men. He was the one who made the crew what it was."

The ball turret gunner was Norman A. "Sammy" Sampson, a shy young man born February 5, 1921, and raised in the farming community of Mason City, Iowa. Sampson was drafted into the Army in August 1942 after three years of high school and a job in an Armour meat packing plant, and at first he didn't realize that he would end up in the Army Air Force. Sampson was sent to Sheppard Field, Texas, where "after a lot of different tests one morning, I was called out and shipped to Radio School. It was all 'Hurry up Air Force!' after that." There were 18 weeks of Radio School at Sioux Falls, South Dakota, and then Sampson went to Las Vegas for gunnery training. He arrived in Boise with the other men and was given the ball turret and the second radio operator's job.

The crew's most "gung ho" member was George F. Hoyt, the radioman. A Southerner from Brunswick, Georgia, he was born July 6, 1924, into a prosperous family that set great store in traditional values. He grew up with a fervent belief in God and country, a strong streak of romanticism, and a deeply ingrained sense of duty. After graduating in June 1942 from a private school called Glynn Academy, Hoyt had a strong desire to join the war effort; the conflict was already outside his doorstep, for this was the "happy time" of the German U-Boat force prowling close inshore along America's virtually undefended coasts. As Hoyt recalls now, "More than once I heard the dishes in my mother's china cabinet rattle as oil tankers exploded right off the beaches of St. Simons Island near my home."

Hoyt had earned a Private Pilot Certificate while in high school, and he felt that the Army Air Force was the best way of striking back. He enlisted in Waycross, Georgia, on August 3, 1942. After basic training in Atlantic City, he went to Radio School in Sioux Falls, gunnery school at Las Vegas, and then on to Boise and Hullar's crew.

Charles H. "Chuck" Marson was the crew's practical joker and arch "character." Born on November 11, 1920, in Mystic, Connecticut, he grew up on a family farm at Boothbay Harbor, Maine. He spent more time hunting in the woods with his rifle than he did in the classroom, and finally ran off to join the Army during his senior year of high school. He was sent to Panama, where he hacked out jungle trails. After a year, Marson "purchased his discharge" for $100 and came home to finish school. But he remained a restless sort, and after Pearl Harbor was one of countless young Americans crowding U.S. Army recruiting offices. He opted for the Army Air Force, and after basic training his shooting skills won him a place in gunnery school at Las Vegas, followed by armorer's school in Denver. He wound up as the crew's right waist gunner, where his thick Maine accent and rough-hewn ways made a lasting impression on everyone.

Charles "Pete" Fullem was the crew's left waist gunner. He was born in Jersey City, New Jersey, on May 31, 1921, into a large Catholic family of three boys and three girls whose parents ran a small candy store. Pete grew up in the city, where he went to St. Mary's Parochial School and attended St. Michael's High School, completing tenth grade. He got work in 1938 with a wallpaper company and was a factory foreman when he enlisted in August 1942. His first duty was Airplane Mechanics School at Keesler, where he completed the 20-week course in December 1942. He then was transferred to a Douglas Aircraft factory school in California, where he trained in both B-24s and A-20s. Afterwards, he was sent to the Flexible Gunnery School at Buckingham Field, Fort Myers, Florida, for a six-week aerial gunnery course. In May 1943, he joined the crew at Boise as left waist gunner and assistant engineer. Fullem is remembered as a very sociable, friendly soul who was "the most happy-go-lucky member of the crew, the one who always had a big smile and a wisecrack."

Merlin D. Miller, the tail gunner, was classic soldier material out of the southern Indiana farm belt where the small town of Sullivan is located. Miller was born there on July 1, 1923, into a family whose breadwinner was an ex-coal miner; his father switched to farming after his grandfather was killed in the mines. As a farm boy, Miller developed an early, intimate familiarity with all kinds of firearms that served him well in hostile German skies. As he describes it, "I had a shotgun, rifle, pistol, or something like that in my hands ever since I was big enough to hold one, so shooting a gun was second nature to me, and I was a fairly decent shot."

Charles "Pete" Fullem with his parents, Mr. and Mrs. Joseph and Georgianna Fullem, outside the family candy store at 4 Coles Street in Jersey City, New Jersey, during the spring of 1943. The store was the scene of much neighborhood excitement when an AP photo appeared in the press amid news of the crew's ditching and rescue from the English Channel on September 6, 1943. This photo was taken while Pete was on leave before traveling to Boise, Idaho, to join Hullar's crew in May 1943. Mr. Fullem died suddenly that July and Pete's emergency leave with the family almost caused him to miss the crew's deployment overseas. While he and his two brothers were in the military, his mother and three sisters literally "minded the store." Note the sign in the center of the shop window, which proudly proclaims: "Answered the Call." (Photo courtesy Mrs. Rita Dispoto, Executrix of the Estate of Joseph J. Fullem.)

His teenage years were a combination of farm chores and high school, enlivened by attendance at local airshows to watch airplanes he couldn't afford a ride in. Following graduation in June 1941, Miller went to Chicago Heights, Illinois, where he worked for a roofing company. Enlistment in September 1942 came naturally: "It was almost a family tradition. My father had enlisted in the Canadian Army in World War I and several of my uncles were also in that war." After gunnery school in Las Vegas and armament school in Denver, Merlin Miller joined Hullar's crew in Boise and wound up as tail gunner and first armorer.

George Hoyt best recalls how the crew first came together: "We met and assembled as a flight crew at Gowen Field, a first phase training base for B-17 crews, during the first week of April 1943. McCormick came to see the enlisted men first, introducing himself and informing us that we would be under Bob Hullar. At this time our copilot was a fellow named Milton Turner, who later checked out as a first pilot and got his own crew, with 'Willie' Klint joining us in the copilot's seat at Walla Walla, which is where we went through second and third phase crew training."

In the early part of first phase training there were other personnel changes as well. As Miller recalls, "Chuck Marson was originally assigned as the tail gunner, but the extreme buffeting in the tail was too much for his stomach. I had no susceptibility to motion sickness and switched places so that Marson became the right waist gunner. Our first left waist gunner also suffered from airsickness—or at least he claimed he did. After a few training flights he was grounded, and we got Pete Fullem in his place."

The purpose of first phase training was threefold. First, the new crew had to become familiar with the Flying Fortress. Then, individual crewmember skills had to be developed. Finally, by working together, the crew began to become a team. The pace was grueling, as Hoyt recalls: "First phase training was a rough crash course. Much of the time we were flying eight hours on and eight off around the clock in old worn out B-17Es, though there were a few newer F models around. We also flew a lot of 11-hour training missions with a 400-gallon 'self-sealing' auxiliary gas tank in the right bomb bay. The fumes back in the radio room were strong, and several aircraft with this arrangement blew up in midair. I can remember 12 planes going down during one especially rough 14-day period.

Polishing basic flying skills was the first priority, as both pilots were still transiting from twin-engine trainers. Initially, an instructor pilot flew with the crew while the pilots practiced "touch and go" landings. As Merlin Miller recalls, "At first, Hullar had a hard time with landings and he let the plane get away from him at times. In the back, we enlisted men placed bets about how many times the tailwheel was likely to bounce with Bob at the controls. But he soon ironed out the problem."

George Hoyt remembers that "in this phase of training, the pilot and copilot were taught another interesting landing procedure, which was done without the rudder steering the plane. During the landing approach, they would take their feet off the rudder pedals, and using only the control column, would turn and steer by 'goosing' the two outboard engines. To turn left, they would advance the right outboard engine and vice versa. A complete landing with the feet off the rudder pedals was required. During landings, the tailwheel also had to be down and its swivel locked. On one occasion we landed with it unlocked, and the swivel's mad, wild gyrations nearly vibrated our teeth out."

Once the pilots reached a certain level of proficiency, the crew concentrated on other skills. McCormick utilized a bombing range for practice runs at 20,000 feet. George Hoyt sent out the required radio reports every half-hour in the air and practiced radio navigation with the Bendix radio compass in the nose. And the gunners shot countless .50-caliber rounds at towed target sleeves and at ground targets on a gunnery range near the Snake River. There were ground school classes too, with much emphasis on aircraft and ship identification, opportunities for the pilots to practice instrument flying in Link trainers, and training sessions for the gunners in turrets that were set up in buildings.

First phase training at Boise eventually ended and Hullar's crew became part of a "replacement group" ordered to Walla Walla, Washington, for second and third phase training. It was here that the crew got their last two members: the navigator, Elmer Brown, and the copilot, Wilbur Klint.

Elmer L. Brown, Jr. was born on Easter Sunday, March 31, 1918, at Golconda, Illinois, but he grew up in St. Louis, Missouri. In high school, Elmer excelled in science and sports, and after graduation pursued technical training. He attended engineering classes at Washington University and at Rankin Trade School in St. Louis, where he learned to design and install air conditioning and heating systems. He worked first as an air conditioning engineer and then as a draftsman with a St. Louis machinery company.

In 1941, Brown sought greater opportunities in Kansas City, Missouri, where he got a civil service job as a topographic draftsman. It was, he recalls, a happy time. "K.C. was a fun city. The popular dance was swing. On a salary of $120 a month, I had a 1940 four-door Packard and three girlfriends that I liked very much. I frequently went dancing, usually at nightclubs. I also played golf and tennis. It was a good year." Kansas City was also the origin of Brown's interest in aviation. He took Civil Aviation Administration courses in night school and would have enlisted in the Aviation Cadets but didn't have the two years of college required. After Pearl Harbor, it magically became possible to qualify if one passed an exam in lieu of college. Elmer took the exam, passed, and enlisted as a cadet on January 24, 1942.

On February 23, 1942, Brown reported to Air Corps Basic Flying School at Higley Field, Arizona, where he received orders almost immediately to join Class 42-I at Santa Ana Army Air Base, California. Looking back on his military training there, Brown recalls very well the feelings of the men who volunteered for military service just after America's entry into the war:

"On Sunday afternoons after weekend leave, we marched around the parade grounds to the beat of a military band playing very patriotic music. This involved thousands of cadets all in step, marching in a wide formation, 16 abreast, each line maintaining as straight an alignment as humanly possible through one complicated turn after another. Awards were given to the marching division that performed best. Each cadet knew that his unit performed only as well as each and every cadet in that division. Consequently, each cadet tried very hard to be perfect. A team spirit developed. And the team was not limited to the marching division. At the sight of the American flag and the sound of that inspiring band music, it was obvious that the spirit in the air was for America."

Brown received his primary pilot training at Rankin Aeronautical Academy in Tulare, California. It was a beautifully landscaped "country club" with bright white buildings whose interiors were always spotlessly clean and perfectly arranged, for white glove inspections could come at any time. Brown thrived in this environment and found the PT-17 Stearman primary trainer a delight to fly, but then he went on with the rest of Class 42-I to Basic Flight Training at Lemoore Army Flying School in Lemoore, California. Here he was confronted with "the noisy, clumsy BT-13" and the check pilot who ended his pilot aspirations.

As he recalls: "There was an instructor pilot who each morning would select a cadet to check on his flying. The cadets named him 'Capt. Maytag' because any cadet that he selected for a check ride usually was washed out. And that happened to me."

Brown was bitterly disappointed by this setback, and was ordered to return to Santa Ana Army Air Base. En route on July 11, 1942, he went with two other "washed out" cadets to a social at the Beverly Hills Hotel USO, and it was here that he encountered a silver lining to the black cloud on his career. By sheer coincidence her last name was Brown, and her first was Peggyann. After seeing a lot of each other, the two began to seriously consider marriage.

Elmer and Peggyann soon discovered that their personal plans took a back seat to the war effort. Elmer received orders to join Bombardier Class WC 43-1 at Roswell Army Flying School, Roswell, New Mexico. He reported there in September 1942 and the couple spent October apart considering their situation. Peggyann joined Elmer in early November, and on the 14th of the month they got married in the base chapel. The sabres of eight cadets formed an arch under which the two walked.

Elmer Brown finished bombardier school and "hit the jackpot" on January 2, 1943. "On that day I was appointed and commissioned in the Army of the United States as a Second Lieutenant in the Air Reserve, was rated as an 'Aircraft Observer (Bombardier),' and was ordered to active duty." Four days later, he and 23 other new bombardiers got orders to

Wartime wedding. Elmer and Peggyann Brown wed at the base chapel of Roswell Army Flying School, Roswell, New Mexico, on November 14, 1942. (Photo courtesy Elmer L. Brown, Jr.)

Army Air Force Navigation School at Hondo, Texas. Elmer and Peggyann boarded a troop train with the other new officers and their wives and arrived two days later.

At Hondo, they had to rent a room and share a bathroom with another young officer couple, like thousands of other war newlyweds. Elmer Brown completed the navigation school course on May 13, 1943, and after two weeks leave, went to Walla Walla, Washington, where Peggyann joined him for the last six weeks before his bomber crew deployed overseas.

When the pair would see each other again was anyone's guess.

The last man to join the crew was Wilbur Klint, known variously as "Willie," "Bud," or "Bill." Born on July 18, 1919, and raised in a strict Presbyterian family, Bud Klint's childhood included six years in the farming community of Plymouth, Indiana. Klint's home was Chicago, however, where he finished high school and graduated from Morgan Park Junior College in 1939. He then took an accounting clerk's position in a candy factory and enlisted April 21, 1942, as a Private in the Army Air Force to avoid the draft and to pursue an early love of the air. Klint had caught the flying bug as a child much as Merlin Miller had—by watching the barnstormers who toured rural Indiana in the 1920s.

Klint's military service began on August 19, 1942, when he boarded a troop train to an Army Air Force classification center in Nashville, Tennessee. The next two weeks were filled with physical checks, shots, and psychology tests, and on Friday, September 4th, "about one-third of the outfit was notified that they had been classified as pilots and would be shipping out the following day." Klint was one of the lucky ones, and "if I hadn't been in that group, I think I would have been ready to 'go over the hill.' Never had I been so certain, as I was then, that I wanted to be a pilot."

Klint moved on to nine weeks of preflight training in San Antonio, Texas, followed by nine weeks of primary flight training at Fred Harmon Primary Flight Training School at Bruce Field, near Ballinger, Texas. He soloed in the PT-19 on December 1, 1942, and underwent nine weeks of basic training at Goodfellow Field, San Angelo, Texas where he mastered the more demanding BT-13, night flying, cross-country navigation, and instrument training. Multiengine training in AT-9s, 10s, and 17s followed at Blackland Army Flying School, Waco, Texas, where Klint won his wings on May 24, 1943.

Klint next drew exactly what he wanted—heavy bomber training at Ephrata, Washington. But he soon found things "severely tied up there" and after two weeks of sightseeing "the Army put an end to our ease by shipping us out—to Walla Walla, Washington. We arrived there June 22, 1943, and learned we were being assigned as copilots to crews that were entering their third and last phase. The accelerated schedule we were entering called for just 17 more days of training at Walla Walla, 10 days of staging, and arrival in England on or about July 20th. In just three weeks time, we copilots were supposed to step from a twin-engine ship of 900 horsepower, to a four-engine B-17 of 5,000 horsepower. Even today, looking back at it, I am sometimes frightened when I think how little I knew about the airplane and about emergency procedures when I went into combat."

In second phase training the emphasis was on formation flying, and the crew flew in progressively larger formations composed of three-plane elements, six-plane squadrons, and 18-plane groups. In the third phase, mock missions and war games were the principal staple.

The chief pilot instructor at Walla Walla during both phases was Colonel Hewitt T. Wheless, one of the few B-17 pilots to escape the Japanese offensive in the Philippines after Pearl Harbor. The climax to training was a mission in which Colonel Wheless led the replacement group on a mock raid against the Boeing factory in Seattle.

As George Hoyt recalls the mission, "The group took off on a flight plan to proceed out over the Pacific climbing in a tight formation to 25,000 feet. We were then to do a 180-degree turn to head due East on an approach that would take us over one of the nation's prime defense plants. The plan was for Lockheed P-38s and Bell P-39s to intercept us in its defense.

"Things did not go according to plan. When we got 'X' miles out into the Pacific, Colonel Wheless ordered the group to descend rapidly down to the deck, and our heading was changed to an East by Northeast one. We were all perplexed as we skimmed along just above the whitecaps of a choppy sea at 175 mph. When we crossed the West Coast at Vancouver, Canada, we made a 90-degree turn towards the South, followed the coast into the Strait of Juan de Fuca, and flew over the water straight into Seattle. We then sliced over the roofs of the Boeing factory at smokestack level, and 'bombed it off the map.'

"The P-38 and P-39 squadrons were completely duped! They were back on the ground, having exhausted their fuel looking for us at 25,000 feet. But some miles beyond the factory, some of them made mock passes at us, and the next morning they tried to get back at us by making mock strafing runs on the flight line while our crew was standing on the apron. They came in at over 300 mph and went below the tops of the B-17 rudders, down the line between the first and second rows of parked planes. At the end of their runs, they pulled up into chandelles and made off."

With the "bombing" of the Boeing plant, the replacement group and the Hullar crew were almost set for their overseas deployment. Bud Klint remembers that "our new B-17 arrived in Walla Walla about July 4th. We were assigned to it and had the pleasant task of selecting a name for it. The name that we picked was *Winnie the Pooh*."

On July 14, 1943, the crew began an 11-day journey overseas, proceeding by way of Kearney, Nebraska; Syracuse, New York; Bangor, Maine; and Gander Lake, Newfoundland, en route to Prestwick, Scotland.

They were headed for the European Theatre of Operations, universally known as "The ETO" and regarded by all as the toughest theatre of the war for American airmen. Hullar's crew was about to learn what it was all about.

2

The ETO

HULLAR'S CREW GOT a rude introduction to the European Theatre of Operations as soon as they landed at Prestwick. "They took our plane," noted Brown in his diary. "Lost *Winnie the Pooh,*" wrote Hullar in his notebook.

"It was," Klint recalls, "certainly a surprise when we landed at Prestwick and were told that *Winnie the Pooh* was to go to combat immediately and we were to go to school.* There was a real scramble as the crew tried to recover all the contraband goodies we had stashed in various parts of the airplane, such as Hershey bars, silk stockings, soap, chewing gum, and other items we heard would be in short supply once we got to England."

The crew's enlisted men were also affected by the loss of the airplane they had named and expected to fly into battle. To Norman Sampson, "Taking the plane away from our crew seemed to be a loss. After all, it had taken us this far."

But Hullar's men took this event in stride and found the pace too quick, and the excitement of experiencing this strange country too exotic for anyone to brood very long. As Elmer Brown recorded, "They shipped us out that night by rail. Caught train at Kilmarnock."

The crew had to reach their interim destination and begin their theater training. On the railway coach, Brown further recorded that they "had first-class sleepers to London. A little compartment for each man."

**Winnie the Pooh*, B-17F 42-3422, was assigned to the 551st Bomb Sqdn. of the 385th Bomb Group based at Great Ashfield. The aircraft was shot down by flak over Regensburg, Germany on February 25, 1944.

But London was not the end of the line. On July 27, 1943, Brown also wrote that they "changed trains for Hemel Hempstead, a school." In the environs of this English village 20 miles outside London at the RAF base of Bovingdon, the crew was to pass two weeks in a combat crew replacement center.

The purpose of school was, as Klint recalls, "to spend time learning from men who had already been through combat about flight tactics, enemy tactics, flak, formation flying, and other things that were supposed to help us complete our tour of duty."

No one on Hullar's crew made any record while at Bovingdon that would show to what degree the crew appreciated the risks the ETO's combat environment held for them, but another B-17 pilot on whose wing Hullar's crew was to fly many of their missions did write down his impressions of theatre training. He was Lt. David P. Shelhamer, a professional photographer from Chicago who rushed into the Army Air Force after Pearl Harbor. He arrived in the ETO about two months ahead of Hullar's crew. The notes Shelhamer made during his theatre training

Element leader. Hullar's crew flew many of their early missions as a wingman to the man pictured on the left, Lt. David P. Shelhamer. To the right is Capt. Richard P. Dubell, with whom Hullar's crew flew their last mission. (Photo courtesy Mrs. Lorraine Shelhamer.)

speak eloquently for every bomber crew that passed into the ETO during this period of the air war.

"As of this writing on May 26th, some of the boys that I'd gone through training with have already gotten three or four missions, and some of them haven't come back. But life seems to be cheap here and you can't worry too much about the other fellow except to try to find out how he got it, so perhaps you can get out of a like situation...Frankly, as far as us getting through, this requires the completion of 25 missions. They may not seem to be very many, but when you consider how much one can be exposed to antiaircraft fire, fighter attacks, it's quite a few trips, I guess...Looking to the future a bit and even before our first mission, I'm sure looking forward to landing after Number 25. This is supposed to be the toughest theater of operations in the world and this number of trips is a long hard haul.

"Well, as far as the rest of the crew [goes], I keep hoping we will all come through in good order, and I'll do all in my power to see that they do all come through, but as yet I have not heard of a case where an entire crew came through their final trip without losing a man or so somewhere along the line. Well, let's just hope that we're the first."

Unfortunately, Shelhamer's crew did not make it through intact as he hoped. His tail gunner was badly hurt on a mission against Hamburg on July 25th, the very day Hullar's men were making their transatlantic flight. The air war was heating up, and to place the Hullar crew's arrival in context, it is necessary to see what the Eighth's leaders were up to.

To win the daylight bombing campaign against Germany, the U.S. Army Air Force generals responsible for directing the Eighth's fortunes had two foes to fight—the Germans and their adversaries in the Allied camp. Three men headed up the Army Air Force's efforts in both struggles. They were General Ira C. Eaker, Commander of VIII Bomber Command and later of the Eighth as a whole; his superior throughout much of the Eighth's air war, General Carl A. "Tooey" Spaatz; and General H.H. "Hap" Arnold, Chief of Staff of all the U.S. Army Air Forces. Their successful representation of AAF interests against the competing demands of the U.S. Navy (which favored wholesale diversion of the Allied effort to the Pacific) and RAF Bomber Command (which sought incorporation of U.S. heavy bomber resources into its night area bombing campaign) was an essential precondition to daylight strategic bombing of Germany.

The Eighth began its existence in Great Britain 11 weeks after Pearl Harbor when, on February 20, 1942, General Eaker and a small cadre of staff officers arrived in England to organize VIII Bomber Command. In the early months, Eaker had to rely heavily on Air Chief Marshall Sir Arthur Harris, Commander of RAF Bomber Command, for help in the acquisition of headquarters facilities and clerical personnel. He got on well with Harris, but his job was a delicate one; the British commander was pressing for the Eighth to join the RAF's night bombing campaign, and Eaker's own forces were woefully late in coming.

It wasn't until July 6, 1942, that the Eighth's first B-17s arrived in England, and it wasn't until the following month, on August 17, 1942, that the Eighth launched its first heavy bomber raid against Occupied Europe. On that day Eaker flew with a tiny force of 12 B-17Es of the 97th Bomb Group that attacked the railroad marshaling yards at Rouen, France, a mere 35 miles from the English Channel.

Though this mission and the Eighth's other early raids were generally successful, the months that followed were filled with frustration as Eaker found his slowly built-up bomber forces being siphoned off to North Africa to take part in the Allied invasion scheduled there in November 1942. In September, Eaker lost his first three bomb groups and a fourth was sent directly to Africa in exchange for four brand-new bomb groups. The first of these arrived in England early in September and the last appeared in late October. Through the balance of 1942, the Eighth was unable to launch a single mission against Nazi Germany proper, and this unhappy fact helped bring about the pivotal incident in the political battle for daylight bombing.

The crisis occurred during the Casablanca Conference on January 20, 1943, when General Eaker met privately with Winston Churchill. Eaker's mission was to convince the Prime Minister not to press a demand that President Roosevelt had already agreed to—that the Eighth abandon daylight bombing and join RAF Bomber Command's nighttime strategy. Sounding a call for "round the clock" bombing, General Eaker was able to persuade the English leader and save his Air Force, but it was now imperative to make daylight bombing of Germany itself a reality. On January 27, 1943, the Eighth launched its first such attack, bombing Wilhelmshaven with 55 B-17s.

Over the next six months the Eighth tested its daylight bombing theories, mounting missions gradually growing in size from under 100 up through 300 heavy bombers. Then came the seven days following July 23, 1943. Now called "Blitz Week," they marked the first sustained aerial offensive mounted by the Eighth against important industrial targets deep in the German Reich. Clear skies were predicted over the Continent and General Eaker was fortified by the recent arrival of two new B-17 groups, giving him a total of 15 bomb groups with well over

300 B-17s. He and his new deputy, Brig. General Frederick L. Anderson, Jr., Commander of VIII Bomber Command, set the wheels in motion for the first raid of Blitz Week.

The July 24th mission saw 309 B-17s directed against targets in Norway. On July 25th, 264 B-17s were sent to bomb Hamburg and Kiel. The next day, the Eighth dispatched 303 B-17s against Hamburg and Hannover. After a day's relief due to bad weather, on July 28th 302 B-17s were sent to attack aircraft plants at Oschersleben and Kassel. July 29th saw the Eighth sending 168 bombers to Kiel, with 81 more going to attack an aircraft factory in Warnemünde. Finally, on July 30th, the Eighth bombed Kassel again, sending a force of 186 Fortresses.

On July 31st the weather was clear, but the Eighth was exhausted. After six missions in seven days, General Eaker's Air Force had lost nearly 1000 men: dead, missing, or wounded, and a total of 105 B-17s—88 in aerial combat and a further 17 damaged beyond effective repair, or "Category E." Eaker's "effective" strength had been reduced from over 330 B-17s prior to Blitz Week to fewer than 200 by its end. Men and machines were sorely in need of a rest; the last day of July and the first 11 days of August were devoted to recuperation and rebuilding.

It was at the very end of this lull, on August 11, 1943, that Elmer Brown wrote in his diary: "We left the school at Bovingdon and reported to the 427th Squadron, 303rd Bomb Group at Molesworth." Hullar's crew had come just in time to take part in the second stage of the Eighth's major offensive against Germany.

Ahead lay a series of missions whose losses would far exceed those of Blitz Week, and whose results would call into question the whole concept of "daylight precision bombing" as a means of winning the air war.

3

Molesworth and the 303rd Bomb Group

MOLESWORTH, HOME OF the 303rd "Hell's Angels" Bomb Group, was a typical Eighth Air Force bomber base. Taking its name from a nearby English village, it was officially known as Station 107 and was situated about 70 miles north of London in the midlands of East Anglia. It was surrounded by hayfields and contained within its spacious confines all the facilities and support units necessary to service the Group's four B-17 squadrons—the 358th, 359th, 360th, and 427th. The Base also served as Headquarters of the Eighth's 41st Combat Bomb Wing, comprising the 303rd, the 379th Bomb Group located at Kimbolton four miles to the south, and the 384th Group at Grafton Underwood, some eight miles to the west.

From the air, Station 107's dominant feature was its triangle of three runways. The field had a long, 7000-foot east-west main runway intersected at its western end by a smaller north-south runway. A third northwest-to-southeast runway intersected the other two and formed the final triangle leg. Ringing these in an irregular pattern were multiple taxiways joining 50 heavy bomber hardstands. The base's heart was a large "technical site" of buildings situated northeast of the runway triangle and adjacent to the center section of the main runway. The site contained the control tower, a huge J-type aircraft hanger, two smaller T-2 hangers, and a complex of smaller structures east of these buildings that included 41st Wing Headquarters, Group HQ, Base Operations, officers' and enlisted mess halls, and the buildings and barracks that belonged to the men of the 427th Squadron. It was here that Hullar's crew began to settle in.

George Hoyt remembers those barracks well. "Some squadrons had steel Nissan huts, but most of our barracks were long, low, drab-looking

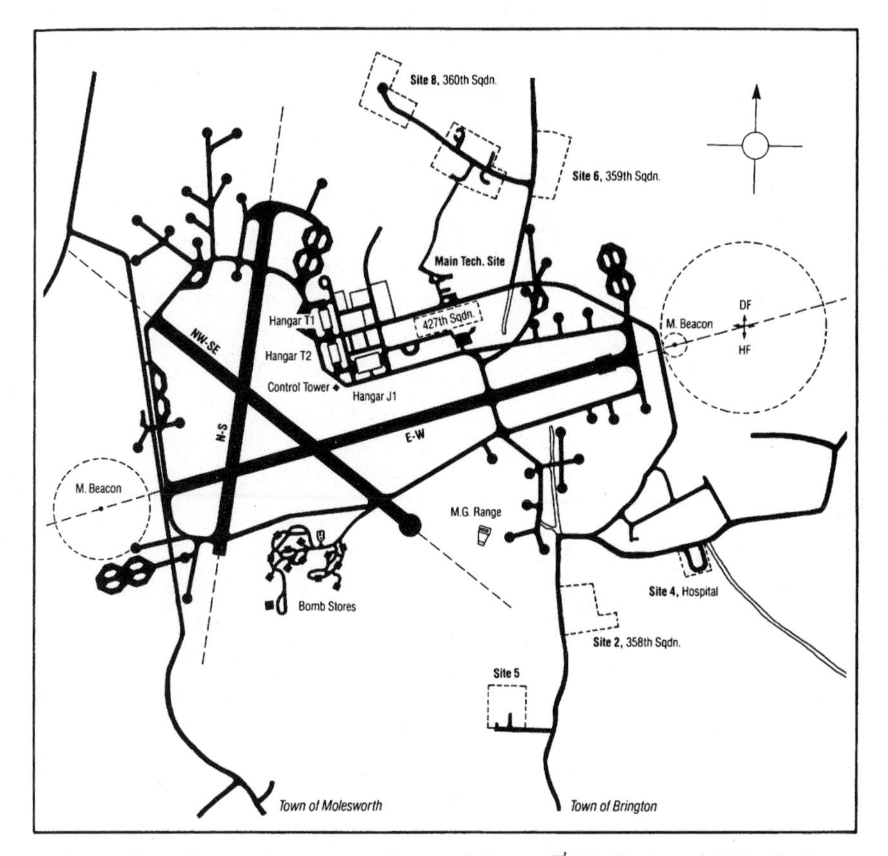

Molesworth Airdrome, Station 107, home of the 303rd "Hell's Angels" Bomb Group.
(Map courtesy Waters Design Associates, Inc.)

wooden buildings with crude doors that had antique hinges and hardware. Inside we were assigned beds which had RAF 'biscuit' mattresses that came in three separate pieces. You needed a blanket under you as well as over you to keep the cold air from coming in between the 'biscuits.' For heat we had two pot-bellied stoves with a four-day ration of coal per week. Out the back door of our barracks stood the latrine in a separate building, and to the left was the 'bomb shelter,' a dugout with a mound of sod-covered dirt rising to about six feet."

Merlin Miller recalls that there was no privacy in these accommodations. "The enlisted men's barracks were just one big room with a small room at the end for the barracks chief. We shared our barracks with the enlisted men from another crew."

The nearby officers' barracks were built just like the enlisted men's, but they were divided into four-man rooms. Elmer Brown remembers that "the officers of our crew shared one of these rooms with a single pot-bellied stove. There was another stove in our building, but we had the same coal ration as the enlisted men." Everyone soon resorted to "midnight" coal requisitions. Brown recalls that this job was handled for the officers by Mac McCormick, while the Hullar crew's enlisted men relied on two stalwarts from the other crew that shared their space: Sgt. Bill Watts and Sgt. Charlie Baggs, who flew with Lt. Edward M. "Woodie" Woddrop.

Though creature comforts were important to the crew's well-being, far more vital was the ability of the group they were joining. The day they arrived, an exuberant Elmer Brown wrote: "The men were swell here, and have an excellent combat record. Have only lost 33 Forts in about 57 missions." His instincts were right on the mark, for the crew's new unit was one of the four "new" B-17 groups sent to General Eaker when his original bomb groups were reassigned to North Africa in late 1942. By this time the 303rd was one of the most able and experienced bomb groups in the entire Eighth Air Force.

Elmer Brown's impression of the group was shared by almost every new bomber crewman who joined the organization. They included Lt. Paul W. Scoggins, navigator of Lt. Jacob C. "Jake" James's crew, who had joined the 427th Squadron the previous May. Scoggins was from Tioga,

Lt. Bud Klint poses outside the Hullar crew's officers' quarters. The 427th Squadron's enlisted barracks were virtually identical on the exterior. (Photo courtesy Wilbur Klint.)

An early photo of Hell's Angels, *B-17F 41-24577, the famous Flying Fortress from which the 303rd drew its name. A part of the 358th Squadron, she carried the Squadron code VK and the aircraft letter D, VK✪D. Note the early .30-caliber machine gun ball sockets in the Plexiglas nose, and the absence of the "angel on roller skates" nose art seen in many later photos of this aircraft. The 358th Squadron's CO was Major Kirk R. Mitchell (seated in jeep). Mitchell was one of the 303rd's most accomplished formation leaders.* (Photo courtesy John W. Hendry, Jr.)

Texas—"a very small place 65 miles north of the Ft. Worth–Dallas area," and his pilot, "Jake" James of Valliant, Oklahoma, was "a good old country boy, heavy on the country." Scoggins kept a diary, and when his crew reported in, he felt that "Everyone is so nice to us—it seems like they'd do most anything for our good."

When Lt. David Shelhamer arrived in the second week of June 1943, he likewise recorded that, "Frankly, from all indications as far as the loss of personnel, I think I'm in the best group, in the best damn squadron, in the whole ETO."

The CO of the 427th Squadron was Major Edgar E. Snyder, Jr., of Van Wert, Ohio. A prewar officer who was with the 303rd from its inception, he was very "mission-oriented" and believes "the fact that I never had a crew of my own may have made a difference in the way I looked at things over there." He describes the 427th as "a real 'can do' outfit with a tremendously aggressive and positive attitude. We had a lot of heavy depth in both the air and ground crews. The enlisted personnel in the 427th were practically all regular Army people, very experienced. The guys were awfully nice to new crews, and we tried to always indoctrinate them real well, despite the fact that we moved them along real fast to get them into things."

These 427th Squadron veterans add much to the story of Hullar's crew, but there were other combat men in the Group when they arrived whom they never met or barely knew, and their experiences are also an important part of this book. Brief introductions to them are in order now.

Most senior was Capt. Louis M. "Mel" Schulstad, a self-described "country boy" from North Dakota. Schulstad was one of the Group's original cadre and served as Assistant Group Operations Officer in August 1943. Another member of the Group's original cadre serving as part of Group Headquarters staff was Capt. Kenneth W. Davey, who was also Assistant Group Operations Officer and who served as Group Gunnery

At far right is Capt. Louis M. "Mel" Schulstad, Assistant Group Operations Officer of the 303rd, photographed after returning from one of the Group's early missions to Wilhelmshaven. Fifth from left in the top row is Major Lewis E. Lyle, CO of the 360th Squadron. (Photo courtesy Louis M. Schulstad.)

Officer, responsible for gunner training and "to see that the guns were operating at all temperatures."*

The 359th Squadron had an excellent gunner in Sgt. Howard E. "Gene" Hernan, from the Peoria suburb of Creve Couer, Illinois. He manned the top turret and was flight engineer on Lt. Claude W. Campbell's crew, ETO veterans since April 1943, with 19 missions behind them. Lt. Darrell D. Gust, from Eau Claire, Wisconsin, was the navigator of Lt. John V. Lemmon's crew in the 358th Squadron. They had likewise been in the ETO since April and had 18 raids in. Also from the 358th Squadron were Lt. John W. "Jack" Hendry, Jr., a 21-year-old "loner" from Jacksonville, Florida, cast in the role of pilot and crew commander, and his right waist gunner, Sgt. John J. Doherty, from a farm 35 miles south of St. Paul, Minnesota. Hendry's crew had been in the ETO since early July, had flown a number of missions, and had not come home without loss: on July 30th raid to Kassel, their ball turret gunner had been killed.**

The 360th Squadron included a number of replacement crews which had joined the 303rd in June 1943. Among them were Lt. Robert W. Cogswell's crew, whose radio operator was Sgt. Eddie Deerfield, from Omaha, Nebraska; and the crew of Lt. Carl J. Fyler, another Midwesterner who hailed from Topeka, Kansas.

There was a final 358th Squadron crew whose presence will pervade these pages because of the large number of raids they flew with Hullar's crew, and the number of crewmembers who kept diaries. Their pilot was 26-year-old Lt. Donald Gamble from Chickasha, near the heart of Oklahoma's Indian Territory. Don Gamble kept a first-class diary distinguished by its use of the present tense. His navigator was Lt. William D. McSween, Jr., a 23-year-old who was born and raised on a northeast Louisiana farm; he also maintained a revealing mission notebook. Finally, there was the crew's bombardier, Lt. Ralph F. Coburn, from Springfield, Massachusetts, who had a steady English girlfriend in the

*As Davey explains (and as the reader will discover in these pages) there was no technical solution to "the machine gun problem which arose from trying to fire the guns at altitudes where the temperature was 60 degrees below zero. The oil we were using on our guns, to lubricate them, was not viscous proof and the guns froze up and [often] refused to fire." More often than not, only gunners who knew how to baby the B-17's .50 caliber machine guns could keep them firing in these very difficult conditions.

**The dead gunner was Sgt. Olwin C. Humphries, pictured on p. 160. Humphries is one of a number of 303rd aircrew buried at the Cambridge American Cemetery in England.

At bottom left is Sgt. Howard E. "Gene" Hernan, flight engineer of Lt. Claude W. Campbell's crew in the 359th Squadron. Standing behind him is Lt. Campbell. Sitting at far right is Sgt. Kurt Backert, the right waist gunner, and standing at far right is Lt. Boutille, the bombardier. The ship is The Old Squaw, B-17F 42-3002 BN✪Z (Photo courtesy National Archives (USAF Photo)).

WAAF named Beryl, and the habit of noting the "significant" events which occurred on his missions.

Gamble's crew had joined the 303rd in early June, and had five raids behind them when Hullar's men reported in. Bill McSween closes this chapter with some comments about the 303rd in mid-1943.

"I will say without hesitation or qualification that we equaled or surpassed any other bomb group in the Eighth Air Force when it came to morale and leadership. Our Group CO, Colonel Kermit Stevens, was something else. I called him 'Cussin Kermit.' After each general mission briefing he took over as a one-man pep rally, and he let it all hang out! My Squadron CO, Major Kirk Mitchell, led 15 straight missions as group or wing lead without losing an airplane! Ed Snyder and Lewis Lyle were also outstanding air leaders.

"We trained hard and we had experienced crewmembers, particularly pilots, to do the training. My pilot and copilot each flew five combat

Lt. Don Gamble's crew in the 358th Squadron with their experienced "Instructor Pilot," Lt. Dave Rogan. Pictured L-R, Top Row are: Lt. Walter Kyse, copilot; Lt. Dave Rogan, Instructor Pilot (Note the RAF uniform); Lt. William D. McSween, navigator; Lt. Don Gamble, pilot; Lt. Ralph Coburn, bombardier. L-R, Bottom Row, Sgt. Vaughn Norville, left waist gunner; Sgt. Richard Scharch, ball turret gunner; Sgt. William Gilbert, tail gunner; Sgt. Charles Schmeltzer, right waist gunner; Sgt. Hugh Bland, radio operator; and Sgt. Clyde Wagner, engineer and top turret gunner. Photo taken 20 July 1943. (Photo Courtesy Dave L. Rogan).

missions with a *seasoned* 'Instructor Pilot' before we flew together as a crew.* You need to compare this to the story of a new crew that had joined our squadron around the time of Blitz Week in July of 1943. As usual, everybody who heard the news 'fell out' to greet the new troops.

"When they got off the 6×6 truck we could see they were in a bad state, just as if they had survived a rough combat mission. They explained that they had been misposted to another group that had a high

*The crew's Instructor Pilot was Dave Rogan, of Middleboro, Kentucky, and he was indeed "seasoned." At 29, Rogan was older than most combat aircrew, and he had already seen much combat flying when he was tasked with teaching the ropes to Gamble's crew. Rogan had joined the RCAF in April 1941, and had been in England since November 1941 flying twin engine Bristol "Blenheim" light bombers before transferring to the Eighth Air Force in September 1942. He had 18 missions in as a B-17 first pilot when he began flying with Gamble's crew, and after completing his 25 mission tour on August 12, 1943, he went on to fly an additional 33 missions as a B-29 aircraft commander in the Pacific.

casualty rate, and that the CO had greeted them saying he wanted to be sure to meet them and to shake their hands, because some crews had come through that he had never had a chance to know. We calmed and reassured them that they had found the right outfit, telling them that 'We flew 25 and went home.'"

Hullar's crew would not have the luxury of an Instructor Pilot and not all 303rd crews "flew 25 and went home." But the experience and *esprit de corps* of these Group veterans were critical to the Hullar crew's prospects.

As they settled into their new home, it's clear Hullar's men had a fighting chance in the 303rd.

4

Introduction to Combat

Amiens, August 15, 1943 and Le Bourget, August 16, 1943

THE DAY AFTER the Hullar crew's arrival at Molesworth, the Eighth resumed its attacks against major industrial targets in Germany. On August 12th the 303rd was part of a force of 330 B-17s sent to bomb Bonn and other targets in the Ruhr. The Group's mission was to hit a synthetic oil plant at Gelsenkirchen, and Lt. Bill McSween felt the raid was rough:

"I was a mite uncertain about the outcome of this one. The target was in the center of a valley and there was no way around anything. We flew straight through up to the Ruhr defenses, where all hell broke loose. Talk about 'intense and accurate flak'! They made us know it. Flak rattled off the plane's nose like hail. The target area was obscured and our bombs went wild. Focke-Wulfs attacked outside the Ruhr defenses and hit the high and lead groups. They didn't bother us thanks to good formation flying."

Ralph Coburn also noted that it was "Rough! Flak and fighters," and Lt. Don Gamble felt the fighters were "very eager today...We see small white bursts behind lead squadron—can see no fighters—they must be lobbing 20mm over from behind or above." He added: "25 Forts were lost today—one of our group." The missing 303rd Fortress was from the 359th Squadron: *Old Ironsides*, B-17F 42-29640, flown by Lt. A.H. Pentz and crew.

Hullar's crew missed this mission, and with it any risk of being lost before they unpacked. The first exposure any of them got to combat operations came a few days later, when Elmer Brown was sent with another crew on a mission against a Luftwaffe fighter field. Elmer Brown described it this way:

"August 15, 1943 (Sunday)—My first combat mission. Flew with Olsen—his first mission as first pilot. Went to Amiens, France, our secondary target. Missed Poix, our primary, due to evasive action.

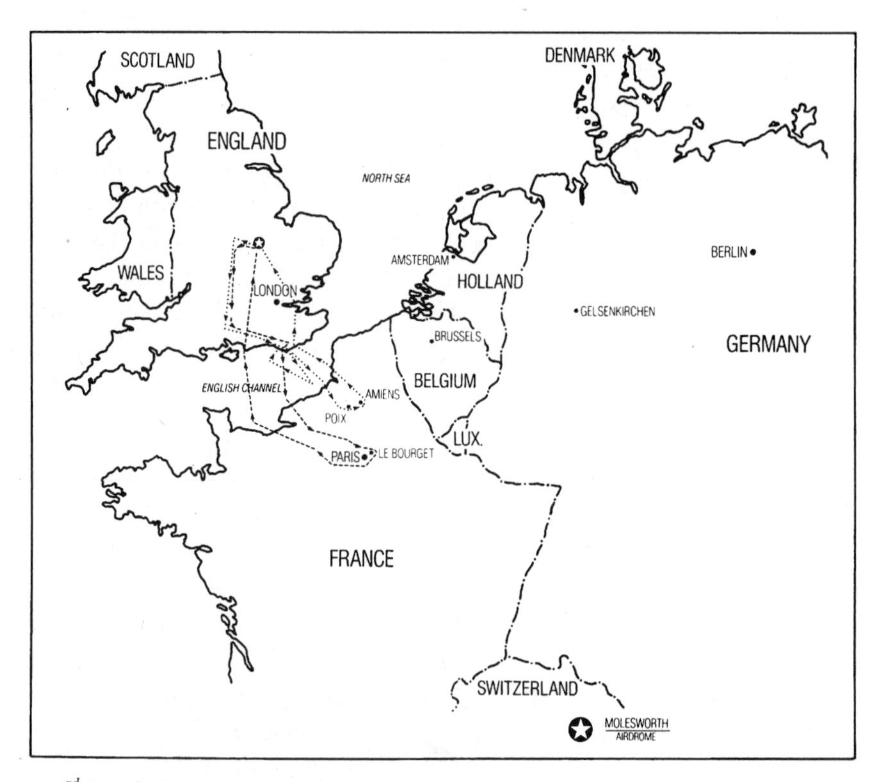

303rd Bomb Group Mission Route(s): Amiens, August 15, 1943, and Le Bourget, August 16, 1943. (Map courtesy Waters Design Associates, Inc.)

Excellent bombing. We were high group, high squadron. Flak moderate but very accurate. Bursting right around us—between our planes. Dropped 24-100 lb. fragmentation bombs—target an airfield. Had Spitfire escorts all the way. Only over France about 23 minutes. Didn't even see enemy aircraft."

Lt. Gamble's crew also went on the raid, and Lt. McSween considered it a joyride: "Everything was lovely. No flak or fighters. We had plenty of Spitfire cover—they're the stuff. Very unexciting mission—but it counts, too." Thus, though Elmer Brown had one mission under his belt, the real introduction to combat for Hullar's crew would come next day, on a trip to Le Bourget airport northeast of Paris. After the many months of working together, this was it!

George Hoyt tells how the crew's initiation began: "On August 16th we scrambled to the briefing room for our first mission. The room was set up something like a theater, and we were briefed separately from the officers. We sat down and watched the briefing officer uncover the target map."

"There it was, Le Bourget airfield, outside of enemy-occupied Paris. Intelligence had word of a large number of Me-109 fighters assembled on the ground awaiting distribution to coastal defense fields, and it was a rush job."

"We were assigned to a B-17 named *Flak Wolf*, and we were in the lead squadron. Our group was leading the wing. I began my mission preparations. I tuned my Morse code transmitter to frequency and unlocked the two spring-loaded lock pins to my .50-caliber machine gun mount. I then slid it on its track to the center of the radio room hatch, an opening about three feet by four feet, from which I had already removed the Plexiglas hatch. The hatch was left on the floor throughout the mission."

The rest of the crew made ready, and all were anxious for the early morning takeoff. Norman Sampson was "eager to go on my first mission. After all, that's what we came for."

The time arrived at last and the real beginning of the raid commenced in a sequence of sounds and sights that always came as a great thrill to George

The 427th Squadron's Flak Wolf, *B-17F 42-3131, GN✪U.* Flak Wolf *was the B-17 that Hullar's crew flew on their first mission. Note the twin .50-cal machine guns barely visible in the nose. The individuals in the photo have been partially identified as follows. L-R, Top Row: Lts. Jack Rolfson, LeFrevre, Abbott Smith, Charles Herman, and Flt. Surgeon Maj. Laird. L-R, Bottom Row: Sgts. Robert Sink, Delyn Smith, unk., unk., William Fleming, Joseph Serpa, Emery Knotts, and Joseph Gray.* (Photo Courtesy National Archives [USAF Photo].)

Hoyt: "The familiar voice of Bob Hullar called out from the cockpit, 'Clear right, clear left,' and then I heard the high-pitched whine of the Bendix starter on No. 1 engine, followed by a loud cough or two from the engine as she cranked up. There was an emission of smoke as the prop turned over and the engine caught. Then came No. 2, No. 3, and No. 4, and the whole plane vibrated with the power of those four 1200-horsepower Wright Cyclones. I could feel them right in my guts. We were on our way!

"As we taxied out to become part of a long procession of B-17s waddling along the taxi strip, I stood up on an ammo box to let my head get above the radio room roof. I saw a long, ambling line of Forts proceeding like huge, drab prehistoric birds that made screeching cries as the brakes were constantly applied to keep them on the taxi strip. It was an otherworldly scene in the dim light just at sunrise."

Bud Klint picks up the narrative from the copilot seat: "It was 0705 hours and the first streaks of daylight were just beginning to filter through the layer of broken clouds which hung over England's Midlands as the Forts began to roll down the runway. We were the second ship to take off and as we slid into place on the right wing of the lead ship, the Command Pilot began to circle the home field in order to wait for the other 19 planes to get off the ground and into their positions in the formation. It was inspiring to watch the 20 planes from our group take up their positions and to see the other two groups that completed our 60-ship combat wing join the formation. It was an added thrill as the tight formation climbed to altitude and headed out across the English Channel toward the French Coast. The sight of the other combat wings assembling and falling in trail gave me a sense of security.

"As we crossed the enemy coast, our escorting Spitfires turned back for their English bases and the Germans introduced us, very informally, to their heavy antiaircraft defenses. The billowing black puffs of smoke, which was all we saw of the defenses, looked perfectly harmless. As they boiled up throughout the formation it was hard to realize they had centers of steel and were scattering fragments of sudden death in all directions.

"As we went deeper into enemy territory, the interphone gurgled with excitement. Everyone on the ship was eager to get a crack at the German fighter planes, which the briefing officer had promised we would meet. Sgt. Rice in the top turret saw them first: 'Six fighters at three o'clock' came over the interphone. 'They are moving around toward the nose,' someone else added, and by then, I'm sure, every eye in our ship was watching them as they climbed a little above our level and continued around toward the front of the formation.

"'Bandit at one o'clock!' the bombardier blurted and, simultaneously, the .50-caliber machine guns in the nose, the top, and the ball turrets

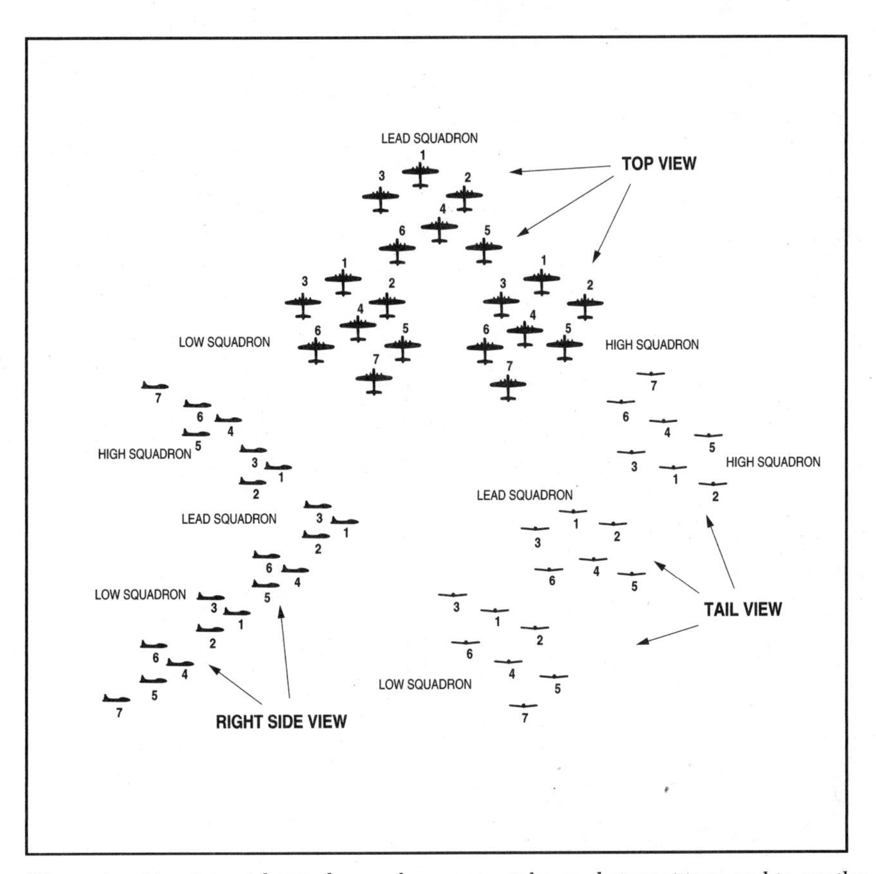

"It was inspiring to watch 20 planes of our group take up their positions and to see the other two groups that completed our 60-ship combat wing join the formation." For protection against fighter attack, Eighth Air Force B-17 groups flew this standard bomb group combat box formation of 20 aircraft in August 1943 as part of a larger 60-ship combat wing. The group "combat box" was 380 yards wide, 210 yards deep, and 300 yards high. The formation positions of the vulnerable No. 7 "diamond" or "tail-end Charlie" ships varied widely. The three-group "combat wing" was 950 yards wide, 425 yards deep, and 900 yards in height, and if the tail view of the group shown in the diagram is considered the lead group, the high group would trail the lead above and to the left, and the low group would trail below and to the right. See the photograph on p. 75 for a good view of a high group in flight and its relationship to the aircraft of a lead group. The low squadron in the low group of a combat wing was called "Purple Heart Corner" because of its susceptibility to fighter attack. (Diagram courtesy of Waters Design Associates, Inc.)

began to chatter and tracers began to snake their way toward the attacking fighters from every ship in the formation."

What combat really meant now struck Hullar's crew in radically different ways. Bud Klint saw the fighters come "in trail, with an interval of three or four hundred yards between them. Even when the leading edges of those FW-190s began to light up like neon signs, I was far more interested than frightened. But when a string of little white smoke puffs appeared across the right wing of our plane, about ten feet from the cockpit, and I recognized them as bursting 20mm shells, the Focke-Wulfs lost most of their fascination for me, as I began to realize that the air around us was filled with flying lead."

The truth came home to George Hoyt in much the same way: "From my field of vision out of the radio room hatch, I could see from three o'clock around past six o'clock to nine o'clock level, on up to an unlimited high. Flying behind and just slightly above us was our group's high squadron at about five-thirty o'clock. No sooner had the warning about the fighters hit my ears than I saw bright strike flashes on the leading edge of the wing of the No. 3 ship, to the high squadron leader's left. It was from attacking fighters coming through the front of our group formation! Then I saw an FW-190 flash by, diving for the deck at four o'clock. My adrenaline began pumping and I thought, 'This is real stuff. They're shooting to kill!' The escorting P-47s which I saw crisscrossing high above earlier were nowhere around now. I got real alert at my .50-caliber."

As the battle progressed, Elmer Brown observed "several Forts go down, especially in the low group." But he never lost his sense of detachment from the scene. "I know it sounds strange to say, but I gave no thought to danger. Peggyann had promised to pray for me every day, and I never imagined that anything would happen to me or my crew. The whole business remained something of an adventure."

In the tail, Merlin Miller had a similar reaction: "Even when I saw a B-17 get hit and go down it was a little unreal, like watching a movie."

For Norman Sampson in the ball turret, it wasn't that way at all: "After seeing a couple of Forts go down, it came real hard to me, at that very moment, that they wanted to destroy us. It scared the lard out of me!"

The fighters continued their attacks until the Group entered the flak defenses at the target. Brown felt that bombing accuracy was "excellent," and as the Group turned away from the target Bud Klint observed "a large column of smoke rising from the airfield."

Enemy fighters pursued the bomber formation after the target, but Hoyt considered the encounter with them brief: "We did not have any chances for good shots at them, and Spitfires met us coming out through occupied France. They crisscrossed over us rather close, but they were careful not to point their noses our way. As they passed by, they threw up that distinctive elliptical Spitfire wing. I waved at several of the pilots in their cockpits as they flew over the radio room hatch. It was a comforting feeling to see them giving us cover."

A view from the radio room. This scene of 303rd Fortresses in formation illustrates very well what George Hoyt observed out of Flak Wolf's *radio room hatch just before German fighters attacked during the crew's first mission to Le Bourget airport near Paris on August 16, 1943. The photograph can be dated to the summer of 1943 by the red-bordered national insignia visible on the lower right wing of the nearest B-17; this style of national marking was only approved from June through August 1943. In June 1943 VIII Bomber Command also adopted a series of geometric shapes and letters to identify the different bomb groups from the air. A "triangle C" was the 303rd's identifier; this symbol on the tail of the nearest B-17 unmistakably shows her to be one of the Hell's Angels.* (Photo courtesy Mrs. Lorraine Shelhamer.)

Other views of the mission are offered by Lts. Bill McSween, Don Gamble, and David Shelhamer. All three were rather blase about the raid. McSween felt it was "a nice early one." Gamble was equally laid back: "Another easy target... Lots of fighters attack low planes. The boys get some good shots in." Shelhamer believed "it wasn't too bad. Moderate flak. A few fighters got a little eager, but all ships returned okay, and we had good bombing."

It all depended on one's position, however. Eddie Deerfield's crew had been flying one of those "low planes," *Lady Luck*, B-17F 42-5434, in the No. 5 slot of the low squadron, and while they made it back safely, Deerfield's damage notes speak for themselves: "20 mm exploded in bomb bay damaging several bombs. Vacuum system and four oxygen bottles smashed by another 20 mm shell. Flak holes in right wing."

So passed the Hullar crew's first mission. They had been "blooded" in combat, and had stood the test well. That evening, Bob Hullar proudly wrote in his notebook, "Crew swell. Shot up a storm!"

But there was also some disquiet in the barracks. Merlin Miller thought about what he had witnessed and "realized that B-17 I saw go down could have been us. And then the seriousness of the thing really soaked in."

There would be more than disquiet next day when the crew learned about the mission then being ordered at "Pinetree," VIII Bomber Command Headquarters at High Wycombe, near London. The targets were a massive Me-109 factory located outside the ancient Bavarian city of Regensburg and the ball-bearing plants situated in the center of the small Bavarian town called Schweinfurt.

5

"We're Veterans After That One!"

Schweinfurt, August 17, 1943

THE REGENSBURG-SCHWEINFURT MISSION of August 17th had been long in coming. In mid-1943, General Eaker was still under enormous pressure to use or lose the heavy bomber force that now lay at his disposal. Schweinfurt was viewed as the ideal strategic objective, since it produced an estimated 50 percent of the ball bearings needed by the German war machine. The Regensburg Me-109 factory was equally high on the target list because of the obvious threat German fighters posed to the daylight bombing campaign. The Eighth's bomber crews had been briefed for these targets a number of times before, only to have the missions scrubbed due to bad weather.

The plan that emerged for August 17th was an intricate double strike. Regensburg would be hit by 147 B-17s from the seven bomb groups of the Fourth Bomb Wing under the command of Colonel Curtis E. LeMay. His force had late-model long-range B-17Fs with outer wing "Tokyo tanks" that made them the natural choice for the deeper penetration to Regensburg, after which they would fly to North Africa in a surprise "shuttle mission."

Schweinfurt would be struck by the Eighth's other bombers, 231 B-17s from the 1st Bomb Wing under Brig. General Robert B. Williams. His command was split into two air task forces, each of which had two combat wing formations of three groups. Since Williams had only nine groups, assembling the larger number of group formations required a "maximum effort" in the truest sense of the word. It was standard operating procedure for three of a bomb group's four squadrons to fly a raid, with the last standing down for a day's rest. For this raid, all squadrons would fly. Each bomb group would dispatch three squadrons in a regular

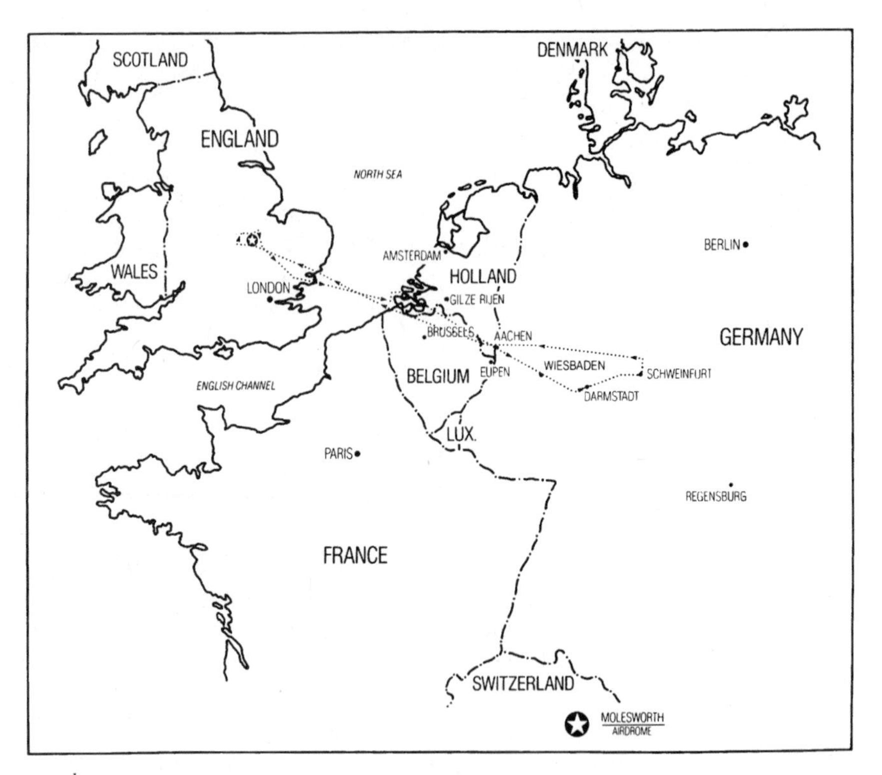

303rd Bomb Group Mission Route(s): Schweinfurt, August 17, 1943. (Map courtesy Waters Design Associates, Inc.)

group formation, while its fourth squadron assembled with those of other groups in a "composite group." Three composite groups would be formed in this way.

In the plan, the 303rd was at the very end of the parade. The 358th, 359th, and 427th Squadrons would constitute the low group of 18 ships under Lt. Colonel Stevens and Major Kirk Mitchell flying in the last combat wing. The 303rd would also contribute 11 ships from the 360th and 358th Squadrons as the lead and low squadrons of a composite group under Major Lewis Lyle. The 379th Group would add a high squadron to the two 303rd squadrons and together they would fly high group in the same combat wing as the main 303rd formation. The last three squadrons of the 379th Group would complete this wing in the lead group position.

Success of the mission hinged on close coordination between the Regensburg and Schweinfurt strike forces. Both were to take off at dawn, and the lead wing of the Regensburg force was to cross the Dutch coast at 0830. The Schweinfurt force would follow 15 minutes later and the bombers were to drop on the two targets at virtually the same time: 1013 at Regensburg and 1012 at Schweinfurt. Numerous medium bomber and fighter bomber diversions were ordered, and 183 P-47s and 97 RAF Spitfires were to escort the B-17s. But they would be on their own at maximum escort range, and everyone knew the raid was going to be difficult even if everything went exactly according to plan.

The 303rd's crews were awakened at 0300 and got the word at separate 0430 briefings for the officers and enlisted men. Bud Klint felt that "every heart in the briefing room hit rock bottom when they pulled the cover off the mission map and revealed that black tape running direct from England to Schweinfurt," and he well remembers the sobering pep talk that Colonel Stevens delivered.

"We were further impressed with the importance of this mission when the 'Old Man' told us during briefing, 'If today's raid is successful, it will shorten the war by six months—If every pilot in this group were to dive his ship, fully manned and loaded with bombs, into the center of the target area, the mission would still be considered a success.' This was our deepest penetration into Germany to date; we were to be over enemy territory for almost four hours; we were to cross the heart of Germany's fighter defense zone; and, knowing the importance of the target, we knew that it would be zealously defended."

Lt. Bill McSween also took notice, writing in his notebook that "This is the 'shaky-do' we sweated out so long."

The enlisted men in Hullar's crew learned about the target in an especially heart-stopping way. They were the only ones not present for the earlier Schweinfurt briefings, and the officers at the enlisted briefing simply announced the objective again, to what Gene Hernan remembers was "the usual ooohs and aaahs, swearing, etc." Hullar's men had no idea what the shouting was all about, so Dale Rice and the others had to find out on their own.

Years later, Rice recalled that "They gave the six of us our own briefing. It didn't take long. The officer just opened the curtain and said, 'This is it. This is the big one,' and gave a few more details. By the time it was finished and we realized how far we were going, I think we were all in a state of shock."

Norman Sampson also remembers how he felt when the curtain was raised: "That was a real shock. I didn't say too much about it, but I done a lot of thinking." George Hoyt was "apprehensive about such a long part of the mission over Germany without any fighter escort," and Merlin Miller felt for sure that "there was going to be some empty beds in the old bunkhouse that night."

Activity is frequently the best tonic for fear, but this day there were hours for Hullar's crew to consider what lay ahead. Elmer Brown noted that "fog delayed takeoff from 0700 to 1200," and this weather ruined the mission's timetable and immeasurably increased the risks the Schweinfurt force faced.

LeMay's Fourth Wing had received intensive blind takeoff practice, and he was able to obtain General Fred Anderson's approval to take off after a 90-minute delay. His B-17 force crossed the Dutch coast around 1000 and got a hot reception from the Luftwaffe—14 of the 24 B-17s he lost went down en route to Regensburg.

But the truly critical decision for the Schweinfurt-bound B-17s came when General Anderson chose to delay rather than scrub their takeoffs, for now the Fourth Wing would no longer divert attention from them. Many German fighters that had attacked the Regensburg bombers would have the chance to land, refuel, and attack the Schweinfurt force coming *and* going, and in the intervening hours more units would be able to stage south from north Germany and Holland.

The Schweinfurt force would face the largest number of enemy interceptors ever assembled: between 260 and 300 single-engine fighters and about 60 twin-engine night fighters.

For the veteran crews, the wait seemed fraught with peril. On one of the 360th Squadron's hardstands Lt. Carl Fyler's men were scheduled to fly "an old beat-up B-17F named *Red Ass,*" B-17F 42-5483, in the composite group's low squadron, and "Everyone had a case of the nerves. Frequent 'piss calls' were needed behind the planes. Others put on their oxygen masks and tried to clear their brains. Some could not get their cigarettes to their mouths because of nerves. We just waited and waited for the signal to start engines."

The full implications of the delay were not apparent to Hullar's crew at the time, and as the morning dragged on it appeared increasingly doubtful they would fly at all. Orders delaying the takeoff to 1200 hit Molesworth around 0900, but as late as 0930 it seemed to Hoyt to be "as dark as predawn." There was another briefing at 1000, and as the hours whiled away the crew concentrated on rechecking their bomber, but the only benefit they received was a lengthy introduction to *their* first B-17, a Queen who would take them home from seven missions against the Reich.

She was *Luscious Lady,* B-17F 42-5081, an early Boeing-built Fortress modified with a single .50-caliber machine gun pointing directly out of her Plexiglas nose. She got her name from a petite but curvaceous blonde painted on both sides of the nose who balanced herself adroitly on an enormous black bomb. For today's mission, the *Lady* carried 16 250-pound

Luscious Lady, B-17F 42-5081, GN✪V. The Lady *was the Hullar crew's first regularly assigned B-17. She took the crew safely to Schweinfurt and back on both of the famous 1943 ball-bearing missions and on five other raids against the Reich. She was still flying when Hullar's crew finished their tour, and returned to the USA after completing 50 missions in April 1944. Pictured are members of the ground crew who kept the* Lady *in the air. Top Row, L-R, Sgt. Charles Twesten and Sgt. Winkleman. Bottom Row, L-R, Sgts. Isaccson, Klein, and Hewitt. After the* Lady *returned home, Tweston served on the ground crew of the 427th Squadron's* Sweet Rose O'Grady, *B-17G 42-39885, a veteran of 143 missions.* (Photo courtesy *Hell's Angels Newsletter.*)

English incendiary bombs intended for the giant Kugelfischer factory in the heart of Schweinfurt. The bombing plan called for the lead wings to drop delayed-fuse high explosives to open up the factory roofs so that the incendiaries could ignite their wooden floors.

At 1200, the 303rd finally put its part of the plan into action. At 30-second intervals, 29 Fortresses roared down the main runway, throttles advanced to full power and propellers whirling in flat pitch as their blades took thin bites out of the humid air and pulled the cargoes of men and munitions into the sky. At 1216 it was time for the No. 2 aircraft of the Group's low squadron to go, and Bob Hullar began *Luscious Lady*'s run by laying the palm of his hand on her central throttle bar to "walk" the throttles up to their takeoff setting.

As the *Lady* slowly accelerated, reached flying speed, and climbed into the air, George Hoyt "searched out my radio room window for the church steeple in one of the villages near the Base while I prayed to God to protect me and all of my buddies aboard. When I caught sight of that steeple as it whipped by in the mist, I felt great exultation and relief. I knew then that God was with us and would bring us through this perilous day."

Two minutes later, Lt. David Shelhamer took off in *Son*, B-17F 42-5221, to fly as second element lead in the low squadron. Both he and Hullar were to occupy the dangerous airspace known as "Purple Heart Corner."

Assembly of Schweinfurt's two air task forces took considerable time. The first, led by Colonel William M. Gross, crossed the English coast at 1313. Slightly to the south the second, led by Colonel Howard M. Turner, met the English Channel at 1321. When the two forces rendezvoused at mid-Channel, they made a bomber stream of 220 ships that stretched over 50 miles, for there had been only 11 aborts. The B-17s began their climb to 25,000 feet, the normal combat altitude for the lead group of a 60-ship combat wing.

Colonel Gross's wings never got that high. Shortly after the Dutch coast he observed a heavy cloud bank at altitude, and rather than risk scattering his force in the clouds, he ordered a descent that brought his low groups to 17,000 feet. His force missed their rendezvous with the P-47 escorts of the 4th Fighter Group, and his wings were soon under heavy attack from up to 240 Me-109s and FW-190s. A slaughter ensued as the Fortresses flew over Belgium, crossed into Germany near Eupen, and proceeded to Schweinfurt by way of Wiesbaden and Darmstadt. The first wing lost 17 bombers and the second four more before they reached the target area.

If losses are any gauge, the road to Schweinfurt was smoother for Colonel Turner's wings. But while they made their rendezvous with the P-47s of the 78th Fighter Group, they too ran a terrible gantlet of enemy fighters en route to the target, encountering from 50 to 200 German aircraft as soon as their escort left.

Elmer Brown put it this way in his diary: "P-47s escorted us about 80 miles into France. We met an awful lot of fighter opposition. Head-on attacks by FW-190s and Me-109Es on the way to the target. Saw many a fighter burst in flames. Many Forts went down. When FW-190s and Me-109s shoot at you, the whole leading edge of their wings is aflame. The boys call them 'headlights.'"

To Bud Klint, it seemed that "from the time our fighter escort left us, we were engaged in a running battle with hordes of German planes."

In the ball turret, Norman Sampson felt the same: "The thing I remember was the many different types of planes the Germans used. Anywhere you looked, there was enemy planes coming at us. That was one day I unplugged my electric suit—too much action for that kind of heat! I saw many enemy planes go down and a lot of B-17s. It looked like an invasion, with so many airmen floating down."

The same image occurred to George Hoyt in the radio room. "We got many tail attacks from Me-110s and single-engine fighters coming in between five and seven o'clock. All the while Me-109s and FW-190s were racing through our formation head-on with guns blazing. So many guys had bailed out from both sides that it looked like a parachute invasion, and at one point I looked out my side window at three o'clock low and saw an Me-109 strafe a parachutist from one of our Forts. The Germans came at us incessantly."

To Merlin Miller the mission was "a kaleidoscope of many things. First there was the dark, clear blue sky, kind of a crazy quilt ground as it always looks from way up in the air, and lots of fighters—*many* fighters. It seemed like every place you looked, there was a fighter coming at either your group or you. There were a hell of a lot of nose attacks. I was seeing them come over and under and between and through the formation. At the same time, we had them coming in from the tail. Half a dozen fighters, maybe more, would get behind us and string out, and come in one right after the other at our group. It got so that we would just pick out the fighters to shoot at that looked like they were coming directly at us.

"There were parachutes too, many parachutes floating through the air, sometimes through the formation, white ones that the Americans had, dirty colored ones that the Germans had. You could see, oh, 40 to 60 parachutes in the air at once sometimes. And sometimes there'd be pieces of planes just floating through the formation from blown-up bombers, blown-up fighters, long columns of smoke from the ground, going down to the ground and coming up from the ground. You could see where we'd been, actually follow our track over the mission just by looking back and following the columns of smoke coming up from the ground. It was incredible!"

Lt. David Shelhamer felt that "this was the belle of the ball all right. Leading the second element in 'Purple Heart Corner' was no fun. If there's one fighter in the sky and he's going to attack a group, 99 times out of 100 he's going to hit that low squadron in the low group. We were under attack for about a half an hour before reaching the target, and in that period of time they took out four B-17s."

Lt. Don Gamble's crew was flying well above "Purple Heart Corner" in the low squadron of the composite group, and from this vantage point Lt. Bill McSween saw one of these B-17s go down. He wrote that "FW-190s and Me-109s struck as soon as our fighter support left. Those babies made us know it coming in high from the nose. They got B-17 No. 6 in the high squadron. I saw seven chutes." The other B-17s lost on the way in also came from this unlucky squadron.*

Lt. Cogswell's crew was flying *Iza Vailable*, B-17F 42-2973, in the second element lead of the composite group's low squadron, just to the right of Don Gamble's aircraft. From the radio room, 19 year old Eddie Deerfield saw other ships go down: "The 'box' formations to our far left and right seemed to be drawing the brunt of the attack. Fortresses were falling everywhere. As they dropped out of the protective formations, enemy fighters roared in for the kills. Parachutes began peppering the sky as American airmen jumped from burning B-17's. At least they stood a chance of surviving in German POW camps. What sickened me to the point of tears were the Fortresses that were exploding in midair with no hope of their crews' escape."

The Fortresses fought back, however, and Lt. Don Gamble witnessed a clean kill by his top turret gunner, Sgt. Clyde Wagner. He wrote, "Wag knocks the devil out of an FW-190—it flies all to pieces from a head-on high attack." This took place at 1456, and the Group's combat form states that Wagner "fired at FW-190 at 600 yards. Guns ran away and gunner held on FW-190. Cowling and parts fell off. Prop fell off. Plane dived down under our ship and blew up about 700 feet below."

It was like this all during the half-hour that the 303rd fought its way from Eupen to the IP—the "Initial Point" from which the bomb run began—west of Darmstadt. In *Satan's Workshop*, B-17F 42-29931, Capt. Davey was riding as Major Lyle's tail observer in the composite group lead, "for they insisted that a pilot sit in the tail gunner seat to report to the lead pilot all matters related to the formation which was following." "So many guns were firing at once...that you could feel [our airplane] shake like it had a fever from all those 50 Cal. recoils. I could see guns in all the ships behind were blazing away as well. Our pilot, Lewis Lyle, was doing a good job of taking what little evasive action he could on those fighters coming in from 12 o'clock and that was a preferred attack pattern. It was max closing speed and the Huns fired head on & rolled over, passed by and dove straight down to gain separation. The fight went on with new flights of fighters gathering at 12 o'clock high, out of 50 Cal range, to peel off and make their head-on attacks."

Bud Klint recorded that "the *Lady* was hit by a 20mm shell which tore a one-foot hole in the leading edge of the wing, just to the right of the cockpit."

---

*The 525th Squadron of the 379th Bomb Group.

On the morning of First Schweinfurt, August 17, 1943, nineteen year old Sgt. Eddie Deerfield strikes a pose under the nose of a worn-looking Iza Valiable, *B-17F 42-2973, PU✪K. (Note the equally worn-looking Quonset hut in the background.) Sgt. Deerfield flew this mission in* Iza Valiable *as radioman on Lt. Robert W. Cogswell's crew.* (Photo courtesy Eddie Deerfield.)

George Hoyt remembers an Me-110 that "put a dozen 7.9mm slugs into our port side, 'walking' them along the side of our plane from the lower vertical tail and coming right up to the side of the radio room."

Merlin Miller remembers that fighter, too: "He was a very stubborn one and he did stitch some holes in us. I was shooting at him and I think I set his right engine on fire. He was burning as he left."

Just before the target, Lt. Carl Fyler recorded another close encounter with a German fighter: "On my right I could see a row of German FW-190 fighters lining up. Then they flew out ahead of us, and turned around to attack us head on... One of them came right at me. He rolled right side up and came over my right wing, still firing. S/Sgt. Bill Addison, my top turret gunner, swung his two guns to the right and fired practically 'point blank.' He got him! I could see the pilot's face as he went past us and went down."

All in all, 22 of the Group's B-17s were hit and two had to abort, but no 303rd ships were lost as they arrived over the target and the German fighters left, leaving the bombers to Schweinfurt's flak. It was here that the Group's luck changed. Lt. McSween observed that the "wings ahead had blasted hell out of it [the target], and we couldn't see for smoke."

Worse still, Lt. Gamble wrote that the Group got "some accurate flak over the target," and while it wasn't enough to destroy any Group ships, it completely spoiled the bomb run of the main 303rd formation.

It happened as the low group was making a slight S-turn to the left past the IP to correct its course and establish proper bombing interval behind the other two groups in the wing. Lt. Lawrence McCord, Group Bombardier in the lead plane with Colonel Stevens and Major Mitchell, had just slightly more than a minute to set up a bombing solution on his Norden bombsight. Before he could do so, McCord was hit in the stomach with flak, and the formation flew past the aiming point before the lead navigator, Lt. Richard McElwain, could do anything about it. He dropped McCord's bombs, but most fell on the city instead of the target. Many in the Group didn't know it at the time (Elmer Brown thought "our bombing was excellent"), but the trip had been largely for naught. Afterwards, Lt. McElwain commented ruefully, "It was okay, I guess. Poor Mac was in pretty bad shape and after I threw the bombs out, I went back to take care of him."

Above them, the composite group had already dropped "to light a fire so the RAF could see to wipe the town off the Earth tonight," as Lt. McSween and the others had been told they would at the briefing, but the RAF had other business that evening, bombing the V Weapons Research Center at Peenemünde. Elmer Brown had also understood that the "RAF

[was] following up tonight" but he later wrote "they didn't." It rankled many B-17 crewmen.

The 303rd now faced the problem of getting home. Ahead of them lay a long,190-mile flight to their rendezvous with their P47 escort near Eupen. Lt. David Shelhamer described it this way: "After the target, all hell broke loose, and the fighters jumped us for about an hour and a half solid. They never let up for a minute. They had everything up there, FW-190s, Me-109s, Me-110s, and Ju-88s. I was sure our squadron was going to lose some airplanes. But with good hard work, expending a lot of ammunition, and smart evasive action we came through it in pretty fair shape. No ships were lost, but my engineer Barlow has a badly shot-up arm, and Barron the tail gunner, my replacement, has a bad flesh wound in his right thigh. It sure was rough."

Elmer Brown recorded "mostly tail attacks by Me-110s and a few Ju-88s."

To Merlin Miller, "it was all one long shootout. If there was a lull, it occurred when you could see 20 fighters in the air heading in your direction instead of 50. I recall Me-110s, Me-109s, FW-190s, Ju-88s, and a bomber-type aircraft, bigger than an Me-110 and painted coal black, which sat back and lobbed rockets at the formation." The Group's records confirmed that this last aircraft was a Do-217 night fighter; it apparently came from a multiaircraft type night fighter training unit based near Stuttgart.

Lt. McSween's notebook shows attacks by a variety of German fighters: "Left target OK. Me-110s and Ju-88s and FW-190s and Me-109s begin attacking."

Lt. Don Gamble wrote: "Me-110s come in. Wag gets one at five o'clock and Bill Gilbert gets one from tail."

These passages are paralleled by a line in Elmer Brown's diary: "Rice our top turret gunner, and Miller, tail gunner, each get Me-110s." This combat took place near the town of Geissen, when six Me-110 night fighters made rear attacks on the wing and three of them went down.

Sgt. Gene Hernan also saw one of these 110s fall from Lt. Claude Campbell's bomber in the second element lead of the high squadron in the main 303rd formation: "I was being very cautious, and was patrolling to the rear and the sides in my top turret. I noticed a speck, which turned out to be a 110. It flew around out there for a while, sizing things up, and finally decided to attack. Why it came in from this angle I'll never know (most of these 110s were night fighters and probably never attacked a B-17 before) but it came in to about 100 or 150 yards, still four o'clock high, and then pulled up and to the right. Of course all the top turrets, right waist guns, and any other gun that could get on it was firing. It was the perfect target and there was no way it could get through all those .50s. My right waist gunner, Kurt Backert, saw it crash in the woods with field glasses."

No one can say for sure to whom these victories should go; 31 claims were submitted for this loss and the other two Me-110s alone.

Meanwhile, the action went on and on. Lts. Gamble and McSween both noted Me-109 kills by their crew. Gamble wrote: "Dick Scharch [the ball turret gunner] hits an Me-109 coming straight up under our plane. He sees it crash into the ground and burst into flames."

McSween recorded that "Scharch got a 109. Gilbert got one."

George Hoyt remembers the tail attacks towards the end of the return trip very well: "We got really heavy attacks from Me-109s and FW-190s, usually in groups of four, six, and even eight abreast. You reacted automatically, firing your gun at one target after another like a robot. I saw my tracers ricochet off the cowlings of the FW-190s many times; when they broke off their attacks, they would roll over and dive, and the tracers fired by our waist gunners and ball turret just bounced off their armored underbellies. Some of them pulled right up into our formation and I was sure their pilots were dead at the controls. Bits and pieces of B-17s were flying past us from airplanes ahead that were damaged, and by this time our formation was loose and ragged. We had shot up just about all of the 12,000 .50-caliber rounds the *Lady* carried."

P-47s from the famous 56th Fighter Group finally came to the rescue 15 miles east of Eupen, farther east than they had ever flown before, thanks to recent improvements in their belly drop tanks. The 56th, called "Zemke's Wolfpack" after its CO, Colonel Hubert "Hub" Zemke, surprised the German fighters as they were attacking Colonel Turner's now badly-bedraggled force. One 56th squadron positioned itself above his first combat wing while the other two took their places above the second, and the fur soon began to fly. Lt. Gamble wrote: "See P-47 knock a Me-110 out—bursts into flames and pilot comes out in parachute on fire." He remembers to this day that "his chute burned up and he fell nearly six miles with no chute on."

Lt. McSween "saw one fighter crash and one explode that P-47s got."

Merlin Miller remembers "a heck of a dogfight with a bunch of P-47s and several different types of German planes. When I saw them they were almost directly above us, and a couple even dove through our formation. The last I saw of them, they were moving back behind us."

All in all, the 56th Group got 11 German fighters, fully justifying the comment by Major Lewis Lyle that "Those P-47s that came way in to meet us certainly made a lot of difference."

Getting home was problematical even without the enemy. Elmer Brown noted that it was "about a 1200-mile round trip, and several planes landed at fields on the English coast out of gas. We just had enough to make our field."

The *Lady* landed at 1808, after five hours and 52 minutes in the air.

With the end of their second raid, Hullar's crew knew they had lived through something special. The Group had been in a terrific fight, ultimately being credited with 20 enemy fighters destroyed, seven probables, and nine damaged. The interior of *Luscious Lady* was so covered with shell casings that it was difficult to move about in her. George Hoyt found "the plywood floor of my radio room covered with over 1000 spent shells. My boots rolled on them when I walked."

The *Lady* herself was, as Klint later wrote, "scarred from nose to tail, and destined to be in the hanger for repairs nearly a week." In addition to other damage, he observed "three large flak holes in the nose and a splattering of .30-caliber bullet holes throughout the fuselage."

The crew knew that 36 B-17s had been lost from the First Bomb Wing and that the Group itself had suffered some casualties. "A waist gunner in our squadron was killed and two other men injured" is how Elmer Brown accounted for them, and there were two other wounded in the rest of the Group.*

But despite this fact and "a fervent prayer of thanks to God that evening" as Hoyt "looked up at some stars that shown in the sky with the firm feeling of my feet on the ground," the crew no longer felt shocked by their experience. It *was* rough, but both Miller and Sampson now looked on the mission as "a job we had to do," and "the job" was one this new crew felt they were doing well.

Bob Hullar put it best in his notebook that very evening, when he summed the raid up in a sentence that said it all: "We're veterans after that one!"

Hullar's assessment was one his Squadron CO shared. As Ed Snyder recalls, "I had formed an opinion of this crew very shortly after they arrived. There wasn't anybody in it, in my opinion, that wasn't top-notch. After you saw a lot of crews you could almost put your finger on one that was going to survive and one that wasn't. If a crew was really a crew and knew what it was doing, you had high hopes for them. Hullar's crew had been trained well, had a good background, and a real cohesiveness. They worked together like a clock. They were one of the better crews that we ever had."

---

*The dead waist gunner was Sgt. Leonard A. Keske, from Illinois, who was flying as part of Lt. C.M. Olsen's crew in *Flak Wolf*. He is buried in the Cambridge American Cemetery.

Though the crew had good reason to be satisfied with their performance, the same cannot be said for the Eighth's leaders. Lt. McSween was in the habit of annotating his notebook with "lessons learned," and he put his finger right on the problem when he wrote: "The lead wings got hell shot out of them. We were lucky being last in a way."

The first Schweinfurt mission demonstrated unequivocally that the Eighth could not prevent prohibitive losses to its lead wings without long-range fighter escort. The operation also underscored how critical it was that tactical advantages not be lost by gross deviations from mission plans. Total losses on the two raids were a staggering 60 B-17s, with 47 more so badly damaged it was unlikely they would fly again.

Nor were the bombing results sufficient to offset these figures. The Regensburg Me-109 factory suffered severe damage, but the Schweinfurt force as a whole fared little better than the 303rd. With or without fighter escort, the Eighth would have to rebuild and hit Schweinfurt again.

6

"This One Was Going to Be a 'Milk Run'"

Gilze Rijen, August 19, 1943

TWO DAYS AFTER First Schweinfurt, Hullar's crew was on the mission roster again for an attack against an airdrome in Holland. The briefing began at 1330, and this time Bud Klint believed "everyone was happy when the S-2 officer uncovered the map and disclosed our target for the day."

Lt. McSween described it as "just a short haul before supper," and Elmer Brown identified the target as "Gilze Rijen, Holland—24 100-pound demolition bombs for an airfield."

As Klint recalls, "From all indications, this one was going to be a 'milk run.' We were to be over enemy territory only 30 minutes and we were to be covered by friendly fighters all the time."

After the raid, Klint had a different opinion: "It turned out to be not quite so easy—mainly because things didn't go off according to plan."

Lt. McSween called it "a 'milk run' that turned into one of the shakiest of all 'shaky dos.'"

The 359th Squadron led the Group under Major William R. Calhoun, Jr., the Squadron CO. He was flying with Captain Glynn F. Shumake's crew. The 358th was the low squadron, led by Lt. J.S. Nix, a 24-mission veteran who was with Lt. D.A. Shebeck's new crew in B-17F 42-3192.

The 360th Squadron was high, and there were only two 427th crews assigned to fill in extra formation positions, Hullar's and Lt. L.H. Quillen's. The Hullar crew's mount for the raid was a 359th Squadron ship called *Wallaroo*, B-17F 42-3029; Lt. Quillen's aircraft was *Stric Nine*, B-17F 42-5392, from the 427th's own stable of ships.

The single green Very pistol flare that signaled "Start Engines" went at 1550, and 10 minutes later Bob Hullar began to taxi *Wallaroo* to her

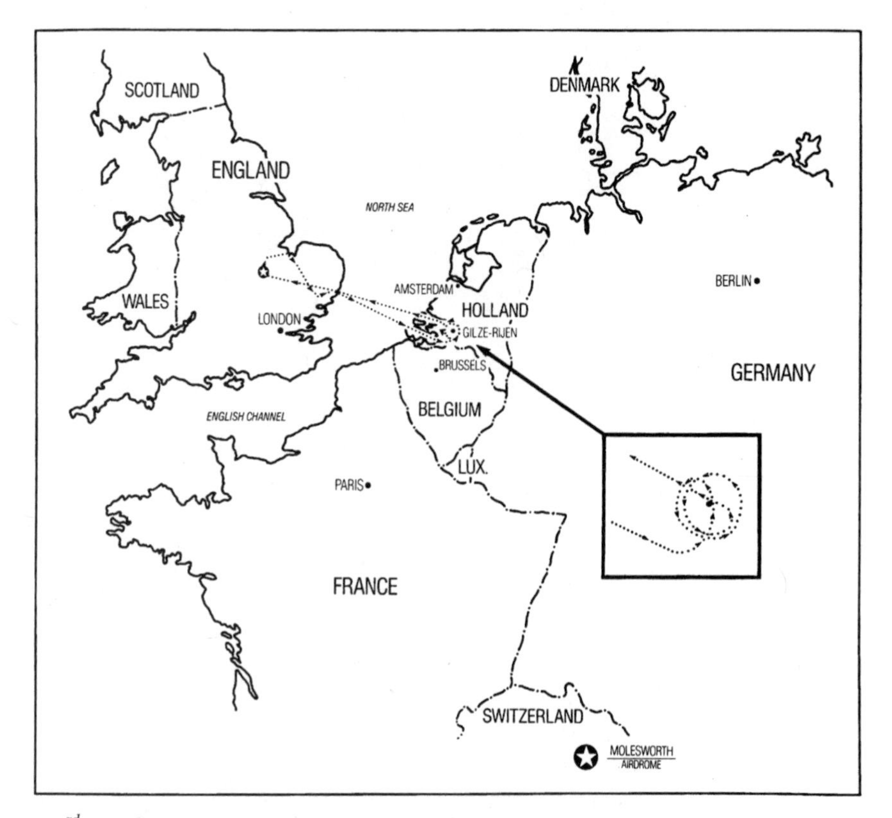

303rd Bomb Group Mission Route(s): Gilze-Rijen, August 19, 1943. (Map courtesy Waters Design Associates, Inc.)

place in the takeoff line. "We we're in the low group, high squadron, No. 7 position," wrote Brown, which meant *Wallaroo* was the 13th of 20 Group B-17s to take off. Four ships behind as second element lead of the low squadron was Lt. Gamble's crew in their favorite ship, *Sky Wolf*, B-17F 41-24562. Following them in the No. 5 position was the 358th Sqdn. crew of Lt. Louis M. Benepe, whose radioman was a baby-faced 19 year old from the Pittsburgh area, Sgt. Richard C. Grimm. Benepe's crew was on their sixth raid in *Yankee Doodle Dandy*, B-17F 42-5264. Last in line as "tail-end Charlie" in the low squadron was Lt. Jack Hendry's crew in *Black Diamond Express*, B-17F 41-24416.

The Group took off on time at 1610 but from there on out nothing went right on this "easy" raid. The combat wing did make its rendezvous with the P-47 escort, but once over enemy territory Elmer Brown observed that "it was very hazy and there was some undercast—the bombs were not dropped the first run."

Bud Klint believed this was because "the lead bombardier couldn't synchronize on the target on the first bomb run and decided to make another run," but the Group's records offer another explanation: "moderate and accurate flak coming from the center of the target."

Lt. Bill McSween wrote that "everything was OK to the IP—Haze pretty bad—and I had to look close for checkpoints. But we had a good run to target. I could see aiming point plainly. The lead navigator was hit and the lead bombardier did not drop the bombs." He added that "Major Calhoun made a circle and came over a second time."

This time Lt. Don Gamble noted that the Group was headed into the sun and the "lead plane doesn't bomb. Ships drop bombs erratically—none hit target." According to Klint, "This put us over enemy territory for an hour, and by the time we began our second run, the other groups were already out of sight on their return to England, and the fighter escort had

Wallaroo *and her loyal ground crew. These unsung, and unnamed, heroes show obvious pride in* their *B-17.* Wallaroo, *B-17F 42-3029, BN✪N belonged to the* 359th *Squadron. Hullar's crew flew her on their third mission, the August 19, 1943 attack on the Gilze-Rijen airdrome in Holland.* (Photo courtesy The Mighty Eighth Air Force Heritage Museum.)

gone with them." And that, Elmer Brown wrote, "made us a good target for the enemy fighter. We had a fight all the way back until we got halfway across the Channel."

Lt. Bill McSween described the attack graphically: "Out of the mist and haze FW-190s struck like devils from hell. From eleven o'clock to one o'clock diving at 45 degrees they really made us know it...G-192 went down and blew up. Don saved us by a violent pull-up. No. 4 was hit, caught fire and went out when Don slipped the plane—$\frac{1}{3}$ of the rudder was shot away. A 20mm burst right in the nose and knocked Coburn into my lap." That night Coburn wrote: "Almost got it. 20mm cannon shell knocked me ass over teakettle when it hit the nose and exploded in my face. By the Grace of God."

Lt. Gamble described the action this way: "Then fighters come in—mostly head-on attacks. G-192, leading our squadron, gets hit and fire starts in No. 3 engine, spreading to No. 4. We fall back and he goes down to the right. Wally [Walter Kyse, the copilot] sees the plane explode and wing come off. We try to take over squadron, but other ships are low so we fly just behind and to the left of lead ships. The enemy fighters are eager today and we see 20mms explode all around. No. 4 engine hit and catches fire but goes out. We feather the prop. A 20mm explodes outside nose and knocks Ralph [Coburn] out. A 20mm knocks a piece out of the rudder."

To Gamble's right Lt. Benepe's crew was having similar, but far more serious problems. Richard Grimm describes the damage to *Yankee Doodle Dandy* and the injuries to her crew in an action that won the Silver Star for the crew's tail gunner, Sgt. George Buske.

"Everybody has a different recollection, but this is what I know happened and also what I saw. We had six wounded in our crew. A 20mm shell went through the Plexiglas nose and our bombardier, Richard Sager, got a chestful of Plexiglas shards. He never flew again. The same shell severed the oxygen mask hose of the navigator, Ray Cassidy, just as he was leaning back to fire one of the cheek guns. The shell went on to hit just below the pilots' instrument panel. It exploded and injured the pilot and copilot wounding them in the legs and knees. Our pilot, Lt. Benepe, was hospitalized for this. Our top turret gunner, Stan Backiel, wasn't hurt and I don't believe the ball turret gunner, Frank Matthews, was either, but 20mm shells came through the waist and injured Ed Cassidy, our left waist gunner, and Francis Stender, our right waist gunner. Both were punctured by multiple pieces of shrapnel, and Stender had a large piece of shrapnel sticking out the toe of his boot. The waist of the ship had so many holes it looked like a salt shaker.

"Our tail gunner, George Buske, got hit. He was firing at a 190 that was coming in, and a 20mm from it hit the left machine gun at the handle and rendered it inoperative. Buske got a lot of shrapnel wounds, but the armor plate in front of the guns took a lot of force from the explosion. Buske got the 190 that wounded him."

The kill was confirmed by Sgt. Matthews, the ball turret gunner, who reported that an "FW-190 came in level at 5 o'clock—T.G. opened up at 1000 yards and fired about 300 rounds. E/A came into about 200 yards, caught fire, flipped over, and went down. Pilot bailed out. Same E/A that hit this A/C in tail."

Grimm continues: "I believe Buske was unconscious for a short time from his wounds because I didn't see any firing from the tail for a while, and then the one good tail gun started firing again. You won't find what I'm about to say in the records—we had six wounded and I didn't feel like reporting it, and who would believe a radioman anyway—but I got a 190 during this time. He was coming in at 5 o'clock high, and he saw that there was no firing from the tail. I know because he dropped his flaps and took a long slow pursuit curve at us. He was taking his time and was going to come right in and get us. I gave him two short bursts at practically zero deflection and hit him with both of them. I could see things fly off, but he kept coming. I kept the trigger down, and he blew all to hell, like dust in the air. I wondered, did I hit his fuel and 20mm shells?

"But as I say, I didn't report it. I was more concerned about our crew. We had six wounded. That was the end of our crew as I knew it."*

Lt. Jack Hendry's crew was just as hotly engaged in the "tail-end Charlie" slot, and John Doherty paints a remarkably vivid and similar picture of the action in his ship: "That was really a fierce fight. Those FW-190 'yellow noses' were there and anytime you met those fellows you were dealing with Goering's elite crew. Them people lived to fight.** They came right through the middle of the formation, trying to peel us out, one

---

*Right after this mission Sgt. Grimm was ordered back to the Continental United States to attend an officer's training course he had previously applied for, but now vigorously opposed since his comrades in arms were hurt. He did everything he could to return to combat, and was finally reassigned to the ETO, where he flew an additional 29 missions with the 306th Bomb Group between Feb. 19, 1945 and April 11, 1945. Forty-seven years later, in Feb. 1992, Grimm was awarded the Distinguished Flying Cross in recognition of his wartime service.

**The FW-190 "yellow noses" belonged to Jagdgeschwader JG-26, a famous Luftwaffe Fighter Wing based in Holland that had formerly been commanded by Adolf Galland. Research by Ivo M. De Jong, a modern day Dutch Army major and aviation historian, confirms the participation of FW-190s and ME-109s from JG-26 in this battle with the 303rd, as well as fighters from JG-1, the other Jagdgeschwader assigned coastal defense duties in the Netherlands and Northern France.

or two of them, and the others would be circling to pick out someone who didn't get right back into formation real quick; I was in the waist and they came in so close I could actually see the faces of German pilots, right outside the wingtip going through. You could see them with their goggles on. They had their tops back and their scarves were flying, right in your face. It was lots of shooting, lots of shooting."

It was also a day Hendry's tail gunner, Sgt. Howard Abney, will never forget: "The fighters were coming in from all directions, front side and everywhere else, but there was one FW-190 that was coming in from behind the formation straight for us in the tail-end-Charlie position. I started firing when I figured he was in range, but we were flying a replacement ship called the *Black Diamond Express* and the tail guns weren't boresighted to my specifications. The tracers were going off to the left and I had to mentally compensate for it, which made things more difficult. I got a hit in a vital spot because he was smoking profusely from the cowling. But he came on, and a 20mm shell he fired came right up by the left horizontal stabilizer and exploded right next to my compartment on my right side."

"I was knocked out for a few seconds. I remember raising myself up from the left side of the compartment and I heard a hissing sound and I saw where the shell had cut the oxygen line in the airplane. I was going to notify the rest of the crew that the oxygen on that side of the plane was out, so I reached for my mike, which was mounted on the bulkhead near my right shoulder, and my arm didn't respond. So I started pulling my right arm up with my left, and I saw the German fighter start coming around to make another pass. I knew I was out of action, that I couldn't handle the twin fifties in my condition, so I broke my oxygen hose connection to the plane, and started to get up.

"But I couldn't get up, so I knew then that I was also hit in the hip. I pulled with my right hand and my left leg to get out of the tail, and I got Doherty's attention to go back and take over the guns."

John Doherty picks up the action from here.

"Our tail gunner, Howard Abney, got shot up. When he came out of the tail he was bleeding profusely, and had blood all over him. He really looked terrible. He came to me holding his hands out and I pointed him to go to Jim Brown, our radioman, for Brown to take care of him. It wasn't that I didn't want to help him. But I immediately knew there was nobody in the tail and I knew that I had to get back there right away to get to the tail guns. The tail on a B-17 was a blind spot. That blind spot had to be covered or they'd get in there right up close and ram it right up your kazoo."

John Doherty has something to add about firing from the waist guns: "The fighters had much more speed than we did, and you had to try to follow the plane, and between the speed and the evasive action and the rolling of the airplane and everything it was awfully hard to get what you'd call a decent shot."

Even so, it was in the midst of this fierce fight that Chuck Marson scored from *Wallaroo*'s right waist window. It happened at 1820, when a fighter described as an Me-109F in the combat form "came in under right stabilizer from about 0430. Right waist gunner fired about 25 rounds into enemy aircraft. The E/A [enemy aircraft] exploded, wings coming off and entire ship went down in flames. The E/A was about 350 yards when it exploded."

Merlin Miller remembers what happened next: "As the 109 went down I heard Chuck mutter something on the intercom about 'ashes to ashes and dust to dust.' He could have a strange sense of humor at times."

The fight went on to the Dutch coast, where Lt. Gamble recorded the loss of another B-17: "We see a plane go down to the left and several chutes come out. An E/A attacks the ship once." The loss hit Hullar's crew hard, for this B-17 was *Stric Nine*, flown by Lt. Quillen's crew.

Elmer Brown wrote that "it was the first raid for Quillen's crew and they went down in Holland just before we hit the coast. Ten parachutes came out. His No. 3 engine and right wing were on fire."

Bud Klint adds that "Quillen's crew was trained in Walla Walla and came over to England in the same group with us. The fighters sent Quillen's ship down in flames just as we crossed the coast."

At this point, it seemed to George Hoyt that the action was over. "The Germans evidently ran low on fuel because they turned back to Holland. The group formation descended over the English Channel as we headed home, and our gunners came out of their positions to ride up in the radio room. Merlin was out of the tail gun position, Norman came up out of the ball turret, Pete and Chuck had stowed their waist guns, and I had racked up my radio room gun. Then someone up forward called on the intercom, 'Here come some Spitfires to escort us home, 'and a split second later Mac shouted 'Look out! Bandits twelve o'clock! Those are Me-109s!' A German flipped by overhead and there was a mad rush with everyone trying to unstow their guns and commence firing. They had caught us flat-footed. Instead of going home, they skirted wide out ahead of us to come head-on like friendly Spitfires from England. After the head-on pass, they swung around to make one final tail attack and then disappeared."

The rest of the mission passed without incident. Lt. Hendry landed at Framhingham, where there was an RAF station, to get medical care for his wounded tail gunner. Sgt. Abney would not be returning to the crew, and from here on out Sgt. John Doherty would be the man guarding Hendry's blind spot in the tail.* Sgt. Buske, in contrast, was able to return to duty and will be encountered elsewhere in these pages, on another crew.

Hullar's crew landed *Wallaroo* back at Molesworth at 1956, and Lt. Gamble's ship was the last one down at 2002. The Group lost 21 men from the two missing ships, plus six more wounded. Fourteen bombers were damaged in exchange for claims of 12 fighters destroyed, ten probables, and six damaged.

More significant were the lessons learned. To Lt. Bill McSween, the mission's moral was "one run over the target is plenty, especially if the second is by your lonesome." For Hullar's crew there was an even more vital lesson—they had made a cardinal error by stowing their guns halfway home, and were saved by the rest of the formation's firepower. It paid to fly with a well-seasoned group.

Another mission was scheduled for August 20th but was scrubbed because of heavy overcast and strong winds. No missions were slated for the following two days so, after four raids in five days for Elmer Brown and three for the others, the men of Hullar's crew finally had some time to rest and unwind.

On Sunday, August 22nd, Brown wrote that "Klint, Hullar, and I went to London. We had fun drinking beer (mild and bitters) and singing. Saw a musical, 'Strike a New Note.' There was a good song in it, 'When the lights go on again.' Hired a taxi that took us on a sightseeing trip. Saw all the famous churches, bridges, and buildings, the Tower of London (an old castle and fort) and many heavily bombed sections of London."

The enlisted men also visited London to, in George Hoyt's words, "take a Cook's tour of the City and keep the British brewers working 24 hours a day."

The one member of the crew who spent most of his free time on the base was Mac McCormick. Even at this early point in the crew's tour he was embarked on a course of study and practice that would eventually make him first Squadron and then Group Bombardier. Elmer Brown remembers that "Mac spent a great deal of time at a building on the base

---

*It was touch and go for Abney from the moment he reached the radioroom, "where Jim Brown cut my clothes back, gave me morphine, and put sulfa drugs on my wounds." Shrapnel from the 20mm had torn all through his right arm and hip, and while much of it was removed by the doctors at Framhingham, blood poisoning had set in, and it was only by arguing that Abney persuaded the surgeon not to amputate his arm at the elbow. Abney never regained the full use of his arm, and still carries enough shrapnel in his body to set off security alarms at airports.

where there were large photo maps of targets with a movable bombsight rigged overhead, so that you could practice bomb runs. He spent hours there on weekends. He was tremendously dedicated to his job."

Hullar's crew stood down on the Group's next mission—an air-sea rescue sortie on August 23rd, which turned up an empty liferaft and a German Do-24 flying boat that was caught on the water and set afire by the Group's gunners. They also missed an uneventful diversionary mission on August 24th, and sat out the next two days with the rest of the Group in the drizzle and rain.

On August 27th, however, they were slated to take part on a mission against an ultra-secret target in France.

7

"They Didn't Have a Prayer!"

Watten, August 27, 1943

THE BRIEFING FOR the August 27th attack against the "special installations" at Watten, France, was well calculated to arouse both speculation and concern among the 303rd crews assigned to take part in the operation. Bud Klint wrote that "the utmost secrecy shrouded the exact nature of the target we were to bomb that day. The CO claimed to know nothing about it, and apparently even the High Command hadn't been able to tell much about these new installations, which were beginning to appear along the Channel coast."

Elmer Brown noted that the target was "unknown" and "thought to be a new munitions dump," while Bob Hullar described it as "heavy construction work. Secret."

George Hoyt remembers it as "a group of concrete reinforced buildings hidden in a woods."

Whatever these structures were, the Eighth was going after them with a vengeance. All 16 of its B-17 groups were laid on to hit this target, and to discourage diversions from the main effort, secondary or last resort targets weren't even assigned. Each bomber was to carry two 2000-pound high-explosive general-purpose bombs on external bomb shackles, and to ensure maximum accuracy, the Fortresses were to drop from 18,000 feet rather than the usual 25,000 feet.

Since fighter escorts for this shallow penetration were supposed to be ample—RAF Spitfires and U.S. P-47s—it was the low bombing altitude that caused the greatest concern among the bomber crews. During the enlisted briefing, George Hoyt remembers "a groan going up when the altitude was announced. Everyone realized the German 88s were deadly accurate at this height."

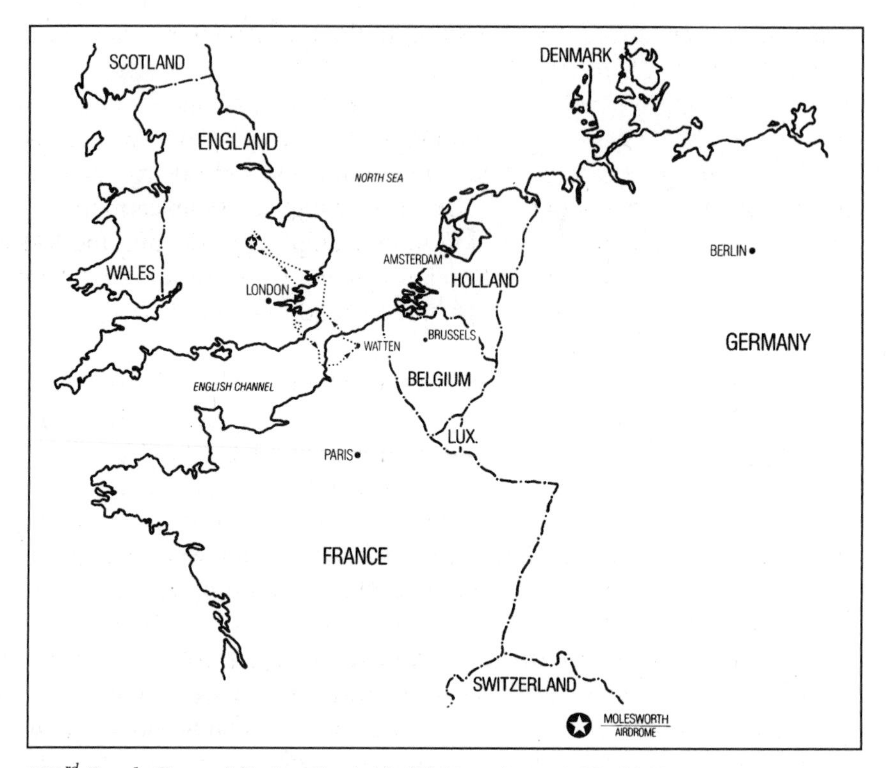

303rd Bomb Group Mission Route(s): Watten, August 27, 1943. (Map courtesy Waters Design Associates, Inc.)

Back in *Luscious Lady*, Hullar's crew was one of 20 the 303rd launched between 1625 and 1636. They filled the No. 5 high squadron slot off the right wing of *Winning Run*, B-17F 42-29944, flown by Lt. Jake James's crew. Leading the 427th in *Vicious Virgin*, B-17F 42-5341, was one of Major Ed Snyder's best crews: Lt. A.C. Strickland's. Snyder had good reason to have his top talent there, since the *Virgin* was also carrying a prime piece of top brass: General Williams, Commander of the First Bomb Wing. He was present because the 303rd was heading the entire Eighth Air Force on this strike, and the Eighth's generals were the kind who led by example, taking the same risks their men did.

The 360th Squadron was leading the 303rd, with Major Calhoun in command again, even though he had just led the Group on the August 19th mission to Gilze-Rijen. He was flying in *Satan's Workshop*, B-17F 42-29931, with a Colonel Lacey from First Bomb Wing Headquarters. In the No. 3 slot off to Major Calhoun's left was Lt. G.W. Crockett and crew in *Shangrila Lil*, B-17F 42-29754; it was the sixth mission for most of the men. Flying once again in the "tail-end Charlie" position of the low squadron was Lt. Jack Hendry's crew in *Charlie Horse*, B-17F 42-29571.

The 303rd approached the target area through hazy and cloudy skies that allowed 20 to 30 Me-109s and FW-190s—"some yellow noses, some black"—to make a sweeping tail attack at 1840, about five minutes before the formation got to the IP. The *Lady* caught a 20mm cannon shell over her No. 3 gas tank, which fortunately didn't explode, and the RAF Spitfires quickly dispersed the German fighters. Elmer Brown wrote that "we had wonderful fighter escort—three squadrons of Spit IXs, one on each side and one above."

In *Winning Run*, Lt. Paul Scoggins also felt there was "no enemy fighter opposition to speak of."

It was on the bomb run that the trouble really hit. The objective lay near the town of St. Omer and was protected by a six-gun battery of 88s (88mm antiaircraft guns) one mile to the south and a four-gun battery in the woods about a quarter mile to the north. The Group was coming in at close to 16,000 feet and, in accordance with standard procedure, wasn't deviating at all during the bomb run.

As Major Calhoun described it, "We had about a three-and-one-half-minute bomb run about as straight as an arrow. There was haze and cloud cover for about half the bomb run, and with about one minute to go, my bombardier, Lt. Fawcett [Jack B. Fawcett, Brush, Colorado], got the target in his sights and let the bombs go. I only saw about four or five fighters myself and didn't worry much about them, but that Flak was about the most accurate I have ever seen."

Just about every crew complained about making the run at 16,000 feet—with good reason. The Group's records describe the flak as being "moderate" but "very accurate, especially on bombing run," and only one of the bombers escaped damage. *Luscious Lady* was hit, returning with two shrapnel holes in her fuselage, and when Lt. Cogswell later landed *Iza Vailable* at the English fighter field at Manston with three badly damaged engines, Sgt. Eddie Deerfield counted "no less than 200 flak holes."

However, it was the ship that didn't come home that seared the souls of those who saw her end. *Shangrila Lil*, flown by Lt. Crockett and his crew, went down three miles north of St. Omer at 1848.

There is no way that statistics can convey the horror that war holds; it must be seen and felt firsthand to be understood, and in the loss of this B-17, Bud Klint gained a terrible insight into the events that he was part of.

"I saw something that day that even yet sends chills down my spine. The ground gunners scored a direct hit on one of our ships. The exploding shell

"They didn't have a prayer!" Lt. G.W. Crockett and crew from the 360th Squadron pose in front of their ship, Shangrila Lil, *B-17F 42-29754, PU✪B. This photo was taken on August 8, 1943 just days before plane and crew were shot out of the sky on the raid against the "special installations" at Watten, France.* [Photo courtesy National Archives (USAF Photo).]

blasted the cabin of the ship wide open. I could look right down into the cockpit and see the pilot and the copilot sitting there—they didn't have a prayer!"

Others observing the loss of *Shangrila Lil* reported it with the same vehemence. Lt. J.R. Johnston's crew, flying in the No. 6 slot of the low squadron, said: "Blown all to hell. Shell hit between No. 3 engine and cockpit or in bomb bay and broke ship up...Plane was hit by flak in vicinity of bomb bay, the burst practically disintegrating ship in midair. Four to six parachutes were seen to open."

John Doherty, flying next to them in *Charley Horse*, was also a witness. "I saw one time—this I don't think I'll ever forget—where a burst of flak hit a B-17, and apparently hit it someplace in the front of the plane. All that was left from the bomb bay forward was a puff of smoke, and you could see the bombs hanging in the bomb bay. The B-17 went flying right along for about 15 seconds or so, and then it peeled off on a wing and went down."

Of all those in the formation, no one had a more terrible view of what happened than Lt. Carl Fyler, who was flying *Thumper Again*, B-17F 42-5393, in the second element lead of the low squadron: "Lt. Crockett's ship directly in front of us took a direct hit in the cockpit. Chunks of flesh came back at us, across our windshield. My copilot became ill. Crockett's ship seemed to come to a complete stop. Since I was directly behind him, with wingmen on both sides, I could not turn away to avoid a collision. I cut the throttles, and 'fish-tailed' the bird, praying I'd miss the stricken ship. Somehow, it seemed to float over my right wing and was gone."*

As Lt. Scoggins put it in his diary, "The flak was very accurate for our altitude, which was 16,000." The Group's B-17s continued on the bomb run, but the bombing was only "good" to "poor;" the bombardiers were stymied by ground haze and the fact that the bomb run went into the sun. Group records state that the "Bombardiers were almost unanimous in saying it was virtually impossible to see the target until it was too late."

All in all, however, bombing results for the day were considered good. VIII Bomber Command reported that "The bombers dropped 368 tons on their pinpoint target and completely blanketed it." Four B-17's were lost.

And the target itself? Months later it proved to be a launching site for Hitler's V-1 rocket "buzz bombs," a light installation the Eighth would frequently hit with 300-pound bombs from a variety of altitudes. The road to "daylight precision bombing" was never easy.

---

*Incredibly, of those in the forward part of *Shangrila Lil*, Lt. Crockett and his navigator survived the explosion to become POWs for the rest of the war. But his copilot, engineer, and bombardier were all killed. The rest of the crew also became POWs.

8

"By This Time Things Were Really Screwed Up"

Amiens, August 31, 1943

THE TWO DAYS following the raid on Watten were quiet ones for the 303rd. One day was spent on "normal routine duties," which included lectures on aircraft recognition and security. A mission was scheduled on the other, but all aircraft were recalled due to bad weather. Then, on August 31st, another raid was laid on. It would be the first in which a mission leader would use VHF radio to improvise an attack on a target of opportunity, but it would not be one of the better efforts of the 41st Combat Wing, to which the 303rd belonged.

The plan called for the nine groups of the First Bomb Wing to attack the Luftwaffe air depot at Romilly-sur-Seine, east of Paris. Leading the effort was Colonel William H. Gross, who had commanded the first air task force attacking Schweinfurt on August 17th. It was an afternoon mission, with Colonel Gross and the lead wing flying to Romilly only to find it covered by clouds.

On the way, however, Colonel Gross observed the Luftwaffe airdrome at Amiens, and rather than waste the mission he radioed the combat wings behind to attack this target if visibility allowed. He then radioed detailed instructions to his own wing, setting up the attack.

Though Colonel Gross showed considerable initiative by taking this step, the attacks by the following wings did not go well. Nowhere was the confusion greater than in the 41st Combat Wing, where the 303rd was high group, the 379th was low, and the 384th was lead.

The 303rd put up 20 B-17s between 1520 and 1533 under Major Ed Snyder, who was in *Winning Run* with Captain Strickland and Lt. Paul Scoggins. The latter found himself "in a very responsible position—was lead navigator for our group," and while Scoggins felt that he "got along

303rd Bomb Group Mission Route(s): Amiens, August 31, 1943. (Map courtesy Waters Design Associates, Inc.)

OK," five of the Group's bombers aborted, and a 25 percent abort rate was poor in anyone's book.

Hullar's crew flew to the right of *Winning Run* in *Bad Check*, B-17F 41-24587. Elmer Brown and Bud Klint describe what took place en route to the target and afterwards. Klint begins: "There was a low overcast over England which became broken by the time we reached the English Channel, but was solid again before we got to the target area. Unable to bomb, we turned around and started to return to our base. Fighter opposition had been light, and over the Channel our Group leader decided to try to reach the airdrome at Amiens, France, where the clouds appeared to be broken."

From here, Elmer Brown wrote that "Our Group, which was originally high group, and the low group decided to bomb another target so we left the lead group and went back into France again. By this time things were really screwed up. We were directly below the other group, which had been the low group, when they opened their bomb bay doors to bomb Amiens. We got out of the way and they dropped their bombs but we were not in any position to drop ours."

Soon after the raid the Group issued a report stating that "due to a hasty run-up on the target, it was impossible for the bombardier in the lead squadron to set his sights and drop his bombs. However, the second and third squadrons had an opportunity to make a perfect bomb run and dropped their bombs directly on the airdrome." But a later 303rd report pointed out that "nine A/C bombed the target of opportunity and six A/C failed to bomb," and this mediocre result was capped by the loss of one of the Group's bombers.

The lost ship was *Augerhead*, B-17F 42-29635, from the 358th Squadron, flown by Lt. W.J. Monahan's crew in the No. 6 low squadron position. She was in formation in no apparent distress when the Group's low and high squadrons bombed, but the 303rd ran into moderate to

Bad Check, *B-17F 41-24587, GN✪P, the 427th Squadron ship flown by Hullar's crew on the August 31, 1943 mission to Amiens. This picture was taken in early September 1943, within a week of the crew's mission. Note the dark blotches of Medium Green camouflage over the top of the aircraft's Olive Drab upper surfaces, a common scheme on many early 303rd aircraft. Note also the contrast between the light grey star on the national insignia on the top left wing and the recently added white bars. The national insignia on the ship was surrounded by a red border at the same time.* (Photo courtesy National Archives (USAF Photo).

heavy flak over the target, and on the way home *Augerhead* was having trouble staying with the formation. Then the B-17 was observed pulling out and heading towards the coast, "apparently attempting to gain cloud cover as protection from enemy fighters."

Lt. Monahan's crew didn't make it. The B-17 was cut off by FW-190s and was last reported going down 12 miles east of Abbeville "under control, with two fighters attacking." Two or three chutes were seen to emerge.*

The rest of the Group landed back at Molesworth between 1929 and 1940.

The loss of Monahan's crew set Bud Klint to brooding, for even though he realized "you didn't dare dwell on such things," he knew also that "the crewmembers of the ship were all veterans in the ETO."

It was losses like these, coming in ones and twos on the "easy" raids, that contributed to the long odds against a crew completing a 25-mission tour as much as the toll on deep penetrations like Schweinfurt. For at this time in the Eighth's air war, B-17 crewmen had, on the average, a one in three chance of completing their missions.

Hullar's crew now had 20 more trips to go.

*Three of the crew escaped and evaded, six became POWs, and the tail gunner, Sgt. D. Miller, "came down dead in his parachute with his legs blown off," according to the post-war interrogation of a German officer. There is a hint that Miller's wounds may have been caused by German fighters attacking while he was in his chute. Crewmen reported that they were attacked by nine German fighters, and that some had fired at the descending parachutes. Miller was buried in Abbeville, France and later reburied in the Normandy American Cemetery.

9

"And I for One Said a Little Prayer"

Stuttgart, September 6, 1943

AS AUGUST TURNED to September and the first days of the new month passed, some of Hullar's crew began to feel they would never get their next mission behind them. On September 2nd Elmer Brown wrote: "We started out to bomb Vannes-Meucon airdrome in southern France. We crossed about 80 miles of water between cloud layers most of the time. When we were within about 20 miles of France they decided the clouds and condensation trails were too bad, so we came back."

To Bud Klint the effort was "three-and-a-half hours of flying for nothing" and George Hoyt considered it "very disappointing."

September 3rd was no better. The 303rd sent 19 Forts to attack the Luftwaffe airdrome at Romilly, but Hullar's bomber was one of four that aborted. Bud Klint noted that "we had to leave the formation about 30 minutes after takeoff because of a runaway supercharger. We returned to the base, got another ship, and tried to rejoin our Group."

Elmer Brown wrote: "We tried to catch the formation at Dungeness. We were three minutes early, but the Group left the coast about 20 miles south, so we missed them."

Klint logged "four more hours in the air, but no mission!" and felt "our whole crew was beginning to get the jitters. It was our second attempt to make our sixth raid."

The crew's third attempt was equally unsuccessful. A mission laid on for September 4th was scrubbed in the early morning hours because of bad weather over the target area.

By this time, Klint was beginning to think "we were never going to get that raid in," but the day was far from a total loss, for Hullar's crew was to meet a very famous man. Elmer Brown noted the occasion in his diary:

"It was a great inspiration knowing that 'The King' of the movies was in the air taking his chances like we were." Clark Gable photographed returning from one of the missions he flew as a gunner with the 303rd. (Photo courtesy *Hell's Angels Newsletter.)*

"Captain Clark Gable was here. He is stationed a few miles away and apparently has something to do with photography. Hullar, Hoyt, Rice, and I flew some of his cameramen up to the 'Wash' and some P-47s made some passes at us. The cameramen took some pictures of them and also pictures of a Fortress formation. A P-47 flew formation with us doing as little as 160 I.A.S. [Indicated Airspeed]."

The day's events also made an impression on George Hoyt: "I knew that Gable had flown two missions with the 303rd in the 359th Squadron's *Eight Ball* before we arrived. That ship was still flying when we arrived and Gable went on five missions altogether. For me, the morale-building effect was tremendous. It was a great inspiration knowing that 'The King' of the movies was in the air taking his chances like we were."

The next day was anything but inspiring. It rained all through September 5th and the daylight hours were spent on "normal routine duties" with lectures on target recognition, security, and aircraft identification. But the mood changed abruptly with the coming of evening. At Group Operations, the teletypes alerted the Group that "Pinetree" was organizing a raid for the morrow. Then the field order came down for another deep penetration into Germany.

The September 6th attack on Stuttgart had political as well as military objectives. General Arnold was in England for a firsthand look at the Eighth. Though he recognized the need for long-range fighter escort, it was important for General Eaker to show that VII Bomber Command could still strike hard blows at Germany's industrial capacity despite the losses of August 17th. The targets VIII Bomber Command picked were classic strategic bottlenecks: the primary was the SKF instrument-bearing plant in Stuttgart, and the 303rd's Wing was to hit the Robert Bosch A.G. works in nearby Feuerbach, a factory held responsible for 90 percent of Germany's diesel engine injection nozzles and magnetos.

The raid was a maximum effort, with over 400 heavies in the air, plus numerous diversionary strikes. The Fourth and First Bomb Wings were to send 338 B-17s against Stuttgart, with the longer range Forts of the Fourth Wing in the lead. The Second Bomb Wing—69 B-24s in four bomb groups newly returned to England from extended duty in North Africa—was to fly a diversion over the North Sea. The Third Bomb Wing—288 twin engine B-26s in eight bomb groups—was to attack four enemy airfields in France and Holland. And VIII Fighter Command was to send 176 P-47s from its four fighter groups to provide the B-17s with maximum escort protection.

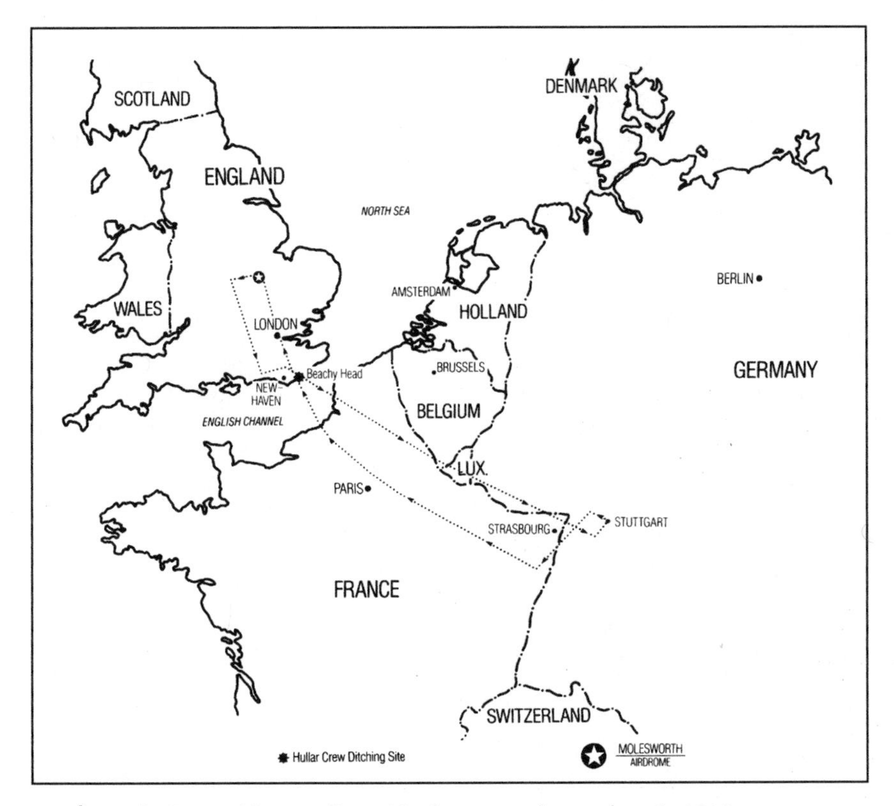

303rd Bomb Group Mission Route(s): Stuttgart, September 6, 1943. (Map courtesy Waters Design Associates, Inc.)

Weather, and the Germans, made a mockery of the carefully laid plan. The B-26 diversions went off without incident and the B-24s flew unmolested on a track that fooled no one, but things were far different for the B-17s. The short-legged P-47s made only light contact with the Luftwaffe, and as the Fortresses ploughed deeper into Germany through ever-thickening clouds, the Combat wing formations got disorganized and the whole mission began to come apart.

By the time the lead wings arrived over Stuttgart, the target was impossible to see, and German fighters, massed in great numbers, subjected the bombers to head-on attacks of a by-now all-too-familiar type. The Fourth Bomb Wing paid heavily; 18 of its bombers were lost, 11 from the 388th Group alone.

Meanwhile, Hullar's crew pressed on in one of 181 Fortresses from the First Bomb Wing that were groping their way towards Stuttgart. But the full story of their experience begins much earlier—back at Molesworth, during the 0330 briefing.

As soon as the curtain was lifted, it was apparent to Bud Klint that the target would be "our deepest penetration into enemy territory." Elmer Brown sized it up as "a very long hop into southern Germany just about 75 miles north of good old Switzerland," the neutral haven known for its hospitality to interned American airmen.

An overall view of The Old Squaw, *B-17F 42-3002, BN✪Z. Hullar's crew flew this 359th Squadron "spare" on the September 6, 1943 Stuttgart mission, and they were forced to ditch her in the English Channel on the way back. Note the twin .50-caliber machine guns in her nose.* (Photo courtesy Joseph G. Worthington, her old ground crew chief.)

The Old Squaw *on the wing. This photo shows* The Old Squaw *flown by Hullar's crew en route to Stuttgart on the September 6, 1943, attack. In the background is a fine view of the 379th Group, which was flying high group position in the combat wing on the mission. Note the straggler at the extreme center left.* (Courtesy Frank Scherschel, *LIFE* Magazine © 1943 Time Inc.)

Everyone knew fuel would be at a premium, and the significance of this fact was heightened by some bad luck on the hardstand. The crew was assigned to *Luscious Lady* again, but she didn't want to make the trip; try as he might, Bob Hullar could not get her No. 2 engine started. Thus, the crew was forced to make a last-minute switch to a 359th Squadron spare. She was *The Old Squaw*, B-17F 42-3002, built by Douglas Aircraft. She was a complete stranger to Hullar's crew, so they had no way of knowing how fuel efficient she might be, but she had been flown by Sgt. Gene Hernan and he knew her well: "The aircraft was not one of my favorite planes. She could only hold about 1675 to 1680 gallons of fuel compared to the 1725 or so you could get in others. My crew flew her occasionally on missions, but not if we could get out of it. No doubt this was one of the reasons that Elmer Brown and his crew had to take her."

The Old Squaw was one of 19 Group aircraft that got off that morning between 0601 and 0614, and soon Bob Hullar had her tucked into the No. 6 position of the Group's high squadron. This seven-ship formation was a composite of crews and planes from the 358th and 427th Squadrons, led by Lt. Don Gamble with Lt. Bill McSween doing the navigating. They were aboard *Jersey Bounce Jr.*, B-17F 42-29664.

The squadron's second element was led by Lt. Jake James, with Lt. Scoggins as navigator; they were in *Winning Run* just off *The Old Squaw*'s right wing, and they had a special visitor aboard: Mr. Frank Scherschel, a photographer from *Life* magazine. To their right was Lt. Jack Hendry's crew in *Hell's Angels*, B-17F 41-24577, and in the No. 7 "diamond" position was Lt. David Shelhamer's crew, flying B-17F 41-24619, *S for Sugar*.

The 303rd's lead squadron headed up not only the 41st Combat Wing, but also the entire First Bomb Wing. This force was led by Brig. General Robert F. Travis, flying with Major Lewis Lyle aboard *Satan's Workshop*. Travis had just taken over command of the 41st Combat Wing, and this was his first opportunity to lead a deep penetration mission. Behind him in the lead squadron's No. 4 slot flew Sgt. Gene Hernan with the rest of Lt. Claude Campbell's crew in *Knockout Dropper*, B-17F 41-24605, one of the 359th Squadron's best B-17s.

The journey into the target is well set out in Lt. Don Gamble's diary and in Lt. Bill McSween's notebook. Gamble noted: "Takeoff and assemble squadron over field at 2000 feet and catch group; 379th and 384th tag on between Molesworth and Eyebrook Reservoir. Climb to 6000 feet and head southeast over coast."

McSween wrote: "We hauled straight out of England over France with no difficulty. No trouble from fighters over France, but there was a big front lying on the Franco-German border and middle clouds extended to the target. We flew two-thirds of the way in at 17,000 feet and then climbed to 25,000 feet. We flew in off course most of the way; went over Strasbourg to the IP and must have fooled 'em. The target was covered by clouds. Groups were flying everywhere looking for something to bomb and dodging flak. We took good evasive action, but it was hard on the wingmen."

It was indeed "hard on the wingmen," no more so than in *The Old Squaw*. As Bud Klint explains: "We knew from the time we took off that gas would be a big problem. The fact that we were flying the No. 6 position in the high squadron of our Group didn't make that problem any easier. We kept a very complete record of fuel consumption, and by the time we were over the target, we knew we were going to have trouble trying to make it home. To top things off, we took a flak hit in our No. 3 feeder tank, and that made it a certainty that we would not be able to reach our home base."

Events over the target soon made even reaching the English coast a long shot. Elmer Brown wrote: "We had just about a 10/10 undercast which was thin enough in spots to see the ground from our altitude of 25,000 feet. We circled the target area several times, which proved disastrous to our fuel consumption. We didn't have any reserve gas as it was."

The rest of crew knew things weren't going well, but they were powerless to do anything except sweat it out. George Hoyt recalls "my anxiety mounting as we made these several passes looking for the target. I was aware of our limited fuel reserve after this long haul. The flak didn't help either. Several times I heard a shower of metallic hits on our plane near the radio room. It left a sinking feeling in the pit of my stomach."

Norman Sampson was also feeling the strain: "I sure thought we was over the target a *long, long* time. It was a bad place to be. I was getting nervous. We finally dropped the bombs and headed for home."

The bombs did not all go at once. Elmer Brown recorded that they "had a difficult time locating one of the targets and bombs were dropped at several different times," and Lt. David Shelhamer knew for sure that this was so. His memories bring out all the frustration and peril that the bomber crews felt as they circled over the hostile city:

"General Travis got a lot of flak from this mission—and rightfully so. Most mission leaders never flew in the left seat when leading a mission. But on this occasion, whether he was in the left or right seat, it had to be Travis flying that plane. I can still see that lead ship going into a steep nose-down position, making extreme turns to the left and to the right, pulling up sharply, and so on. It was absolutely horrendous. At one point we were making a rather sharp turn to the right. In order to stay with the formation, I instructed my bombardier to salvo the bombs, and out they went. My ship became quite a few thousand pounds lighter, but my airspeed indicator was still going from hell and gone and back and our formation was scattered all over the sky. It's fortunate that there weren't a hundred Me-109s or the like at that point or we certainly would have lost one hell of a lot of aircraft."

There were still some fighters left over Stuttgart from the force that had mauled the Fourth Bomb Wing. Lt. Gamble wrote that "the fighters hit the low group—there were about 30 enemy aircraft—some twin-engine Me 210s."

According to Elmer Brown, there were others that decided to pick on *The Old Squaw*. "The fighters included FW-190s, Me-109Fs, and Ju-88s. We were taking violent evasive action to avoid fighters on our tail after leaving the target."

George Hoyt felt that "These boys must have all hailed from Stuttgart, since they pressed their tail attacks with extreme ferocity. Merlin and I were calling to Hullar to 'kick it around' over and over, but Merlin did a good job of fending them off."

There was one Ju-88 that was not deterred. Merlin Miller remembers that "he came in from about five o'clock high in a long diving pass and kept right on going. My firing angle was bad when I saw him, and Marson would have had the best shot at him. The German got off a good burst. Out of the corner of my eye I got a glimpse of tracers from his machine guns or cannon, and I believe he hit us in the right wing. I'm sure he was the cause of the fuel leak we had in No. 3 engine. When I turned around to look over my shoulder, I could see a faint mist coming from the trailing edge of the right wing which had to be gasoline."

Hoyt also remembers that "at this point we began to have trouble with one of our engines. We were still at 25,000 feet and started to lag a little behind our formation slot, placing ourselves in an even more vulnerable position. Fortunately, the Germans must have run low on fuel themselves because they left us and headed for the deck."

Hullar's crew now faced a momentous decision: which way to turn? As Elmer Brown recalls now, "good old Switzerland" seemed the only alternative: "We knew over Germany that the best we could hope for was to ditch in the Channel and possibly not get out of France. All the way back I was figuring a course to Switzerland and the point of no return to Switzerland. About two weeks before this mission I had read news items about some of our airmen who had ended up there either by escaping from the Germans or otherwise.

"Under international law, Switzerland was required to intern these airmen for the duration of the war. We had thought they would have been put in prison, but they were not. It was more like spending a vacation in a resort. Our men had nice quarters, good food, and our government sent them money to live on. They had a real great time skiing and having a lot of fun. I thought we were perfectly justified in going to Switzerland, if we wanted to, rather than risk the possibility of crashing in France and being taken as prisoners of war, or else crashing in the English Channel, where we might not have been rescued."

Bud Klint also calculated that "Switzerland was only a couple of hundred miles to the South" and he recalls that "Brownie was heartily in favor of this alternative, and just about this time we saw one ship from our wing leave and set his course for the comparative safety of the Swiss Alps." Norman Sampson likewise remembers "Brown telling Hullar we won't make it to the coast of England, but we could make it to Switzerland. But Hullar said we were going to try."

At this point Elmer Brown protested. "I felt that the decision whether to head for Switzerland or to head for England should be decided by an election, which I conducted from the intercom. Each man was polled and asked to vote for the destination of his choice. By the time it was my turn to vote, the other nine had all voted to go to England.

"These guys were very patriotic, and that's probably why they voted the way they did. However, I am pretty sure they did not really understand what Switzerland had to offer. If they had had a better understanding of this, and the opportunity to vote again, I think we would have had some very fine skiers and yodelers by the time the war was over. As it was, there was only one thing for me to do. I voted with them, and so made it unanimous."

Thus, as Klint puts it, "We elected to stay with the Group and gamble on making the English coast."

Elmer Brown picks up the narrative again from here: "With the Switzerland matter settled, I concentrated all my navigational knowledge and experience in determining where we would probably land, after we tried to get as near to England as possible by consuming the last drop of fuel. We still had many miles of France to cover, and the English Channel. We had over 20,000 feet of altitude to lose. Meteorological changes were expected and had to be forecasted, as they would affect the temperature and the wind velocity. This in turn would affect our airspeed and headings.

"I cranked these factors into the E6B calculator, and plotted the different winds, headings, and airspeeds on my map. Klint furnished me with excellent information on fuel consumption, telling me within minutes about how long the balance of our fuel would last. This information was vital to the calculations I had to make.

"About halfway across France I recommended jettisoning excess equipment to reduce the weight and aid us on gas consumption. Soon afterwards we started throwing things out. We kept just enough ammunition to ward off fighter attacks, which, thank goodness, we didn't get."

At this point, Klint felt that "we must have created quite a scene as we jettisoned every bit of loose equipment to lighten the ship. We were still throwing clothing, extra ammunition, and spare radio equipment overboard when No. 3 engine exhausted its supply of fuel. For our own protection, we had to stay with the rest of the formation as long as we could. That became no longer possible when we lost our second engine, still over France. Luckily, there were no enemy fighters in the vicinity, but even so, an awfully empty feeling crept over me as we began to drop behind and below the other ships."

Happily, this low point did not last long, for Klint soon saw help on the wing: "Friendly fighters who had come to cover our withdrawal picked us up a short time later and two RAF Spitfires took it upon themselves to mother us home. They were a beautiful sight as they circled, protectively close above us. Since defense was no longer a problem, we jettisoned the remainder of our ammunition, and as soon as we hit the Channel coast, we got rid of our machine guns."

Elmer Brown's diary adds that "We left the French coast at 14,000 feet. By that time we had started throwing our gun barrels and everything out. I estimate we threw over 7500 rounds of ammunition out of the plane and I threw about 1500 of that out of the nose."

When not throwing excess material overboard, Elmer Brown worked nonstop on his navigation, continuously refining his calculations. "Ultimately I arrived at the spot where we would probably ditch. It was just short of England, about six miles SSE of Beachy Head. I determined the coordinates of that location, and passed them on to the radio operator with a request that he transmit the information to British Air Sea Rescue, so they would have a ship waiting for us at that point."

Armed with this information, the pressure was now on George Hoyt, who found himself "fighting a jammed frequency as scores of B-17s, low on fuel, called for assistance and to report their positions. I resorted to turning our IFF radio to Channel No. 2, which would transmit a distress signal that RAF coastal direction finding stations could monitor and take a position on. They finally picked this up and zeroed in on our course in order to plot a crash position.

Hoyt didn't stop here, however. All during the B-17's slow descent, Bud Klint noted that "the radio operator was glued to his key, sending an SOS and giving our latest position to British Air Sea Rescue Service."

Busy as Brown and Hoyt were, neither immediately noticed that their overzealous crewmates had taken the order to jettison "everything" quite literally. As Bud Klint recalls, "Mac, our bombardier, became so enthused he even tossed the navigator's chute out the escape hatch."

And back aft, George Hoyt made a similar discovery: "Bob Hullar had ordered us to throw out as much excess weight as possible to lighten our load and stretch the distance we could remain airborne. As I banged out SOS signals on the radio, I could see the boys scurrying about throwing out things. Then Dale came back to the radio room and said that Bob wanted us to snap on our chest chutes in case we had to resort to a sudden bailout instead of ditching in the Channel. I reached behind my chair for my chest pack, and it was gone.

"Just then, Marson came up from the rear into the radio room to get some more .50-caliber ammo, and I asked him if he had seen my chute.

In that Maine brogue of his he said, 'You know, Hoyt, I think we threw it out with that frequency meter sitting next to it behind your chair.' There I was with the possibility of a bailout and no chute!"

The concern George Hoyt felt for his chute evaporated as the need to confirm the ditching position with British Air Sea Rescue became increasingly apparent. Elmer Brown wrote that "soon after leaving France, the alarm was sounded and everyone except the pilot and copilot assembled in the radio room to prepare for ditching. The radio operator had been busy sending an SOS and was still doing so."

With Hoyt thus occupied, Elmer Brown and others now took up their positions. Brown lay down on the floor near the port bulkhead, face up and feet propped against the bottom of the forward bulkhead. Merlin Miller took the same position next to him, his knees slightly bent, hands locked tightly behind his head. On the other side of the room, Mac McCormick lay down next to the bulkhead, with Chuck Marson on the floor beside him. Dale Rice, Pete Fullem, and Norman Sampson sat in the middle of the room between Miller and Marson, with their backs against the forward bulkhead, their hands also locked behind their heads to prevent any whiplash. Hoyt strapped himself into his seat, using another chute to cushion the impending blow.

From the cockpit, Bud Klint could now see the English coast, "but with our remaining two engines nearly out of fuel, it was still very, very far away. We were about 200 feet off the water, trying desperately to maintain our altitude, when our third engine sputtered and quit. The crew was now in the radio room, and I was keeping them informed over the interphone. We were losing altitude and the remaining engine would turn over for the last time very soon. Suddenly, and almost simultaneously, Bob and I spotted a surface vessel on our course. The last engine quit, and we put the ship into a glide, trying to get as near to the boat as we could. Bob and I were both on the controls."

Behind them, the rest of the crew waited. Each man was lost in his own thoughts, but together they created a truly eternal moment. Elmer Brown described it this way: "Just before hitting, the pilot feathered the No. 4 engine, and we all sat tight, and I for one said a little prayer."

George Hoyt glanced at Brown and it seemed to him that "Brownie's faith shone as bright as the very sun itself. I remembered him giving me a smile as big as all outdoors. Out my window I saw the choppy and rough sea coming up towards us, closer and closer, and I prayed fervently as I continued to beat out those SOS signals."

Norman Sampson remembers: "As we were waiting to hit the water, a complete calm came over us as if 'all is well.'"

Merlin Miller recalls "a feeling of absolute calm, just like I was laying under a tree back on the farm with a shoot of hay sticking out of my mouth watching the soft fluffy clouds float by."

And George Hoyt felt "a great calm take possession of me which language can't describe. I experienced a peace that surpasses all imagination as I watched the whitecaps now flicking by under the wing. I felt Bob easing the nose up for a landing stall and I called to Brownie and the boys, 'This is it!'"

Up in the cockpit, Bob Hullar and Bud Klint could see what was coming. Hullar hit the ditching alarm buzzer and then, Klint believes, "between us, we made a pretty smooth landing. The tail hit the top of one wave, the next one must have hit about the ball turret, and then the ship settled down into the water. The nose went down into a trough, and the next wave stopped us with a terrific jolt. It shattered the Plexiglas nose, tore the leading edge of the wing, and rolled the propellers up."

From the radio room, Elmer Brown also felt that the pilots had performed well. "We made a nice landing into the wind with an indicated airspeed of about 80 mph. The tail hit first, and it didn't jar us too badly. Then the nose and whole fuselage hit and we stopped dead with a violent jolt."

That last, final impact is what made the big impression. The next thing Merlin Miller knew, "I was up on my feet. The force of the crash with my feet against the bulkhead must have stood me up without any effort on my part."

With his back to the bulkhead, Norman Sampson felt the experience "was like hitting a brick wall." For Hoyt, who was the only one sitting in a chair, it came as "a tremendous, shattering impact that was loud, abrupt and violent. We came to a stop at a finger snap, and I was badly stunned. The chest pack in front of me had buckled the plywood radio table, water was gushing up in the camera well in the floor very rapidly, and I was sitting in water up to my knees in only two or three seconds. My head began to clear, and as I unbuckled my seat belt and moved the chest pack out of the way, Brownie and the others were starting to stand up. No one appeared to be injured."

The fact of injury was, however, something each man had to discover for himself. "I thought I had hurt my insides, and spit in my hand thinking it was blood, but it was only seawater," recalls Norman Sampson.

Merlin Miller found he was unharmed, but Chuck Marson wasn't as lucky. As Miller explains, "When we hit, the water coming in the radio room splintered the door on the camera well and Marson got hit on the side of the head." Marson would soon realize that he had ruptured an eardrum and had wrenched one of his knees.

The rest of the crew were in relatively good shape, so that Elmer Brown was able to write: "We scrambled to our feet and anxiously awaited our turn to climb out of the hatch as the plane seemed to be sinking fast. That was the only time we got excited, and even then everyone was calm enough to do their job and climb out in a hurry."

Norman Sampson echoes those sentiments: "Boy, this was one time our ditching training really paid off. We knew just what to do."

Dale Rice reached up and popped the two handles to the life raft storage doors, and one by one the men in the radio room exited through the overhead hatch. Elmer Brown wrote that "we were all out of the plane within 10 or 15 seconds after landing, and I think the pilot and copilot beat us all out." Merlin Miller is certain about this: "I started to climb out of the plane and Dale Rice gave me a boost from underneath. I sailed up out of there like I was jet-propelled and landed on my hands and knees on top of the fuselage just in front of the hatch. Much to my surprise, Bob Hullar and Bud Klint were already there. One of them gave me kind of a funny grin and said, 'Where you been?' I had to say something, so I accused them of setting the plane on automatic pilot and crawling out before we hit."

In the meantime, Elmer Brown helped George Hoyt out of the hatch and everyone finally exited. Klint felt "the whole crew was out of the ship and onto the wings and fuselage within 45 seconds after we hit the water, but even then, when the bombardier left the radio room, the water was already waist-deep."

It was clearly time for the men to get into the two rubber dinghies stored in side panels over the left and right wings just behind and below the bomber's top turret. This went relatively smoothly on *The Old Squaw*'s port wing, where Hullar, Brown, Rice, Hoyt, and Miller were to deploy and man the left dinghy. George Hoyt describes part of what went on:

"After I inflated my Mae West and cleared the top hatch, I went out on the left wing in knee-deep water to help Dale Rice deploy the rubber dinghy. Four-to-five-foot swells were rolling over us, but we quickly inflated the dinghy by using its CO_2 cylinder." Hoyt, Rice, and Miller carried it to the wing's trailing edge, and it was at this point that Merlin Miller suffered a mishap:

"We had picked the dinghy up to toss it out into the water; you didn't want to slide it off the wing in case you had jagged metal from a hit or something, which would punch a hole in it. As I was helping to heave it, I stepped back a little too far and dropped right into the ocean. I went under and swallowed some seawater. When I popped up, the rest of them were in the life raft and when I reached up to grab the edge of it, two or three guys reached down for me. Before I knew what was going on, I was lying face-down on the bottom of the raft with someone's foot on the back of my head. I got that taken care of real fast, and a couple of the boys grabbed the oars and started paddling.

"The only problem was that they were both paddling on the same side and this spun the raft around a little bit. Hullar got aggravated and made a couple of remarks to them about this, and then we got the paddle straightened out."

Bob Hullar's concerns weren't merely a matter of form. As George Hoyt recalls, "We had drifted back on the large swells and had come close to *The Old Squaw*'s left horizontal stabilizer. We were worried about being fouled under the tail, and Bob called for us to shove away from the plane to avoid being dragged under by it. We then got back behind the tail, where we had a view of the second rubber raft. We saw four men in the boat with a fifth, Marson, swimming some 25 feet away in the water."

Things had gone much worse for the men assigned to *The Old Squaw*'s starboard raft. Perhaps because of flak hits in the wing or strikes from the Ju-88 that attacked, the dinghy failed to inflate fully. This forced Klint, Marson, and Fullem to jump into the water and reach for one side while Sampson and McCormick climbed aboard the dinghy's other, inflated half.

Worse still, as the raft drifted aft of the wing, the men saw that a heavy cord still secured it to the sinking bomber's storage compartment. The discovery prompted a desperate search, for, as Bud Klint recalls, "We all feared that *The Old Squaw* would go down immediately and drag the dinghy with it. We began to look for something with which to sever the cord."

Chuck Marson did more. Powered by a force one can only imagine, he performed a feat that left Bud Klint amazed: "Marson simply took the cord in his two hands and broke it. Charlie was, of course, a 'tough outdoorsman' type with a heavy Maine brogue, and I don't know what the test strength of that cord was. But it was much too heavy to be broken like that by anyone not having superhuman strength in a moment of terror. Even today I sometimes doubt the truth of this tale, but I still believe it happened this way."

Ironically, Marson's act bettered his crewmates' plight while worsening his own. Merlin Miller recalls that the heavy five-foot swells "had a tendency to crest right under the raft, and then it would tilt and slide down the side of the wave like a roller coaster," and in these conditions the freely drifting dinghy rapidly distanced itself from Marson, so that he had to swim for it alone.

But things were hardly better for the other two men still in the water near the dinghy. Bud Klint and Pete Fullem both continued to struggle against those same swells, while from the relative safety of the raft, Norman Sampson saw their difficulties. What he next observed was a moment of exceptional courage: "I remember getting into the dinghy and seeing Fullem starting to drift away from it. Klint swam right after him, got him, and brought him back. I remember thinking how brave Klint was."

With Klint's and Fullem's return, all eyes now focused on Chuck Marson, who was loudly calling attention to himself with cries of "Hey, look at me! I'm a lone wolf." Merlin Miller recalls Marson yelling this "at the top of his voice, while floating in the water with one arm over the top of a 'Gibson Girl' emergency radio, and the other waving in the air."

George Hoyt remembers Marson's call as well, but there was one member of the crew who was far from amused: Hullar was responsible for everyone's safety, and he now saw the need for stern corrective action. George Hoyt remembers that "he stiffened, tried to stand erect in the rubber raft and, summoning up all the command presence of which he was capable, bellowed, 'Marson, get your ass in that dinghy right now!'"

Merlin Miller remembers that "it was kind of hilarious to see Hullar yelling at Marson in those undignified circumstances. It's kind of hard to be authoritative when you can't even stand up."

Hoyt derived even more amusement from Marson's response: "He was thrashing around in the choppy water and said, 'Yes Sirrrr . . .' just as a wave slapped him full in the face. He finally managed to make it over to the second dinghy, and then climbed aboard."

Everyone's attention now turned to the last moments of the Queen that had taken the crew so near to friendly soil. Of her end, Klint wrote, "*The Old Squaw* stayed afloat about five minutes, then reluctantly gave up the ghost. Our last sight of her was with the tail straight up in the air and then, suddenly, she was gone."

George Hoyt remembers that "we looked at her with mixed emotions. To me it felt like attending a relative's funeral."

To Miller "it was kind of discouraging to lose *The Old Squaw*, but at the same time it was a big relief to get as close to England as we did."

Close as the crew had come, there were still doubts that they would make it. The surface vessel that Bob Hullar and Bud Klint had spotted during the last moments of *The Old Squaw*'s descent was now fast approaching.

But was she friend or foe? George Hoyt couldn't be sure. "I first saw the English coast, and in the next moment looked back to the horizon and saw a ship steaming at full speed towards us with black smoke pouring from its funnel. As it approached, it appeared to be a large gunboat or a small destroyer. I saw no flags or markings whatsoever, and as it hauled alongside I saw deckhands in black turtleneck sweaters peering at us. They did not hail us, and I began to think, after all this ordeal, that we were being picked up by Germans."

Norman Sampson was just as uncertain. "I saw the sailors with their lines, and thought maybe the enemy had got us after all."

According to Elmer Brown's diary, "the boat was 110 feet long and 17 feet abeam," and one of the record books at the Public Records Office in London shows that she was HSL 183, the only high-speed launch operating in the Beachy Head area rescuing ten men that day. She was part of the RAF's 28th Marine Craft Unit based at Newhaven, but what mattered most to the men of Hullar's crew was the first word spoken by one of her sweatered sailors; it proved unmistakably that their rescuers were British.

George Hoyt recalls that "They threw a large rope ladder over the side, and finally one of them yelled in an English accent, 'Ahoy there, climb aboard, you blokes.' We were jubilant."

Merlin Miller remembers that moment too. "We were all so happy and relieved."

Elmer Brown wrote: "I was the first to board the boat, and I estimate I was on board in less than five minutes after the plane landed. The last man in the second dinghy was probably aboard in less than ten minutes time."

Miller remembers that "the launch came alongside and they had a net on a boom sticking out, which made Rice comment about them scooping us up like a bunch of fish. They stopped pretty much dead in the water, so Rice and I reached up and grabbed the net right next to the boom.

"What we didn't realize was that the launch was rocking pretty heavily in the water. When it rocked away from us, Rice and I found ourselves quite a ways up in the air above the raft. When she rocked the other way, we were back in the water again. There were some English sailors on deck trying to help us get aboard, and one of them reached out with a long boathook.

"He managed to hook it into the back of my flying coveralls, and was pulling like mad while I was trying to climb aboard the launch. The next thing I knew I was skidding across the deck. About that time Rice came banging in alongside of me, and a couple of the others, and by the time I got up and looked around, the rest of them were all aboard the launch."

Once the crew was aboard, their joy at being rescued was fully matched by their rescuers' pride. Klint wrote that "They were thrilled to have picked up a Fortress crew. The Captain told us they had been on duty in the Channel for six months. On their first day out, they picked up a Spitfire pilot, and had picked up no one else until this occasion. He said the crew was so elated when they saw that we were going to crash-land that they stood on the deck, waving their hats and cheering."

Elmer Brown recorded this amazing incident too: "The ship's crew was cheering when we went down, and they were very happy to see us. They hadn't picked up a crew in months. In 14 months service, they had just picked up two RAF crews before ours."

Not surprisingly, Brown also wrote that "aboard ship they took excellent care of us. We went into a cabin and dried off and changed clothes (gear). We went below deck and had a shot of brandy; some of the boys had Scotch or rum. They served us beef stew and tea, etc. The boat's narrow beam made it roll and toss quite a bit. It was so rough dishes would slide across the table. We would get thrown from our seats and dishes were getting broken.

"We sure had a lot of laughs. The sea was getting rougher by the minute and some of the boys got seasick, especially Miller, also Fullem and Klint. They picked us up at about 12:45 and as no one of us was hurt, we stayed at sea until about 7:00 P.M., for if we had put ashore immediately it would have left a vacant spot there."

To this day, Merlin Miller denies being seasick. "I've never had motion sickness in my life. It was what I ate. They gave us a glass of rum and a bully beef sandwich, which I didn't care for usually but I was kind of hungry then. I drank about half the glass of rum and ate about half the sandwich, and my stomach started boiling a bit with that and the seawater I had swallowed. The boat was pitching around quite a bit, and I thought I better head up on deck because I knew it wasn't going to stay down very long. Sure enough, as soon as I got to the rail, up it came."

George Hoyt had generous amounts of food and liquor, and then: "I went on up to the bridge to take a station behind the helmsman and see where we were going. We were banging through the swells at what must have been more than 35 knots, and the helmsman told me we would be a while because the captain had been informed that German E-boats were spotted in the area, and we were going after them. I went after another brandy, and by this time I was drunk enough to go in pursuit of a German battleship. I was urging the crew on with 'Sink the bastards!'

"About mid-afternoon, a steward emerged on deck with two steaming cups of tea on a tray. He went right up to the bow gun with the craft rolling and pitching in the heavy seas. The two men at the gun then proceeded to have their afternoon tea amid the spray thrown back from the bow while we were pursuing German E-boats! It was this sort of thing that made me just love the British."

With the waning of the day, HSL 183 headed for her home of Newhaven, and as she entered the harbor her captain proved himself a prouder fisher of men than any apostle. As Bud Klint describes it, "The Captain insisted all of us line the ship's rail so that the other boat crews would see that they had picked up a Fortress crew."

Elmer Brown recorded this as well: "They were proud that they had brought a crew in and wanted the other ships in port to see us."

Recalling it now, Merlin Miller evokes a wonderfully inspiring scene. "As we went into the port, ships of all sizes and kinds docked on each side of the channel were going down. Every time we'd pass a ship, one of the sailors on our boat would toot a little whistle, and the duty officer on that ship would salute, and the commander of our launch would salute back. He really was proud of picking us up."

The launch came to a place that Miller remembers was "kind of a warehouse where we docked." There was a crowd waiting to greet rescuers and rescued alike, and after saying their goodbyes to the men of HSL 183, Hullar's crew finally set foot on land again.

Inside the warehouse they shed the remnants of their wet American flying clothes and donned a combination of civilian garb, RAF uniforms, and knee-high rubber boots. They also met a large contingent of similarly situated American airmen. Brown noted that "there were 51 men from B-17s there."

After a short delay, the crew boarded a military bus to Ford, an RAF station near Dover. It was here that the full magnitude of the day's air sea rescue effort was revealed. Elmer Brown wrote that "Air Sea Rescue had picked up over 100 Fortress crewmembers that afternoon," and Merlin Miller recounts the way the enlisted men reacted to this discovery:

"We walked in there and it was almost like old home week. We saw fellows there that we hadn't seen since we'd trained together in the States.

"The English had a band there, a small combo of some kind, and the drinks were on the house, which was a real mistake on their part. And there were a lot of English WAAFs there too. That was another mistake they made. So we ended up having quite a party that night. I think some of us partied more than others.

"After having a few drinks, I spent some time talking to some of my old buddies from the States, going around comparing different experiences with different people. I think we played a few games of billiards, too. They had a big table there, and some of our crew, Hoyt and Marson, played. Marson attracted quite a bit of attention when he kicked off his boots and crawled right up on the table to make some long shots without using a bridge.

"None of them were feeling any pain, what with free liquor. After I got tired messing around, I went to my cot and was about the first to turn in. I don't know what happened to the rest of them that night."

George Hoyt knows, and provides his own version of the evening. "We had a good dinner and they invited us to the RAF Sergeants Club, where all of our beer was on the house. We played some snooker, which is much more demanding than regular billiards, and we all got drunk. Everyone else was asleep by the time we returned to our bunks, and Dale said to me that he had brought the flare gun and some shells from the rubber dinghy.

"I was feeling quite adventurous. I asked him for the gun, and he gave it to me immediately, since his mental state was no better than mine. We went outside, where I loaded the gun, pointed it up towards the black sky, and pulled the trigger. I expected a plain signal flare to shoot out and away, but to our amazement and concern a parachute flare opened up, lighting nearly the whole RAF base as it slowly drifted down. Air raid sirens began to wail, and Dale said, 'Quick, let's jump into our beds, pull up the covers, and act innocent!' Searchers soon made the rounds but they found us all 'asleep.'"

Next morning was a time for taking stock. Marson's ear and knee injuries forced him to remain at Ford for medical treatment; because of this he would miss a number of raids with the crew, and ever after be a few missions behind. Even before his serious drinking began, George Hoyt had felt "a sharp pain in my ribs and a bump on my head" but he opted to tough it out. "I was afraid if I went to sick call, they would hospitalize me and throw me behind in my missions with my crew. The thought of finishing with another crew was something I couldn't bear. So I kept quiet."

Merlin Miller "didn't realize how hard we hit the water till I woke up next morning. I was sore all over. It even hurt to comb my hair, and I'm sure the rest felt about the same."

In the afternoon, Hullar's crew made their way back to Molesworth. Elmer Brown wrote that "Shelhamer and Dr. Lamme and some of the boys came down and picked us up in a B-17 and brought us home." It was no accident that Shelhamer was at the controls of that B-17, as he explains:

"Bob Hullar didn't just fly on my wing; I felt he was a friend of mine, too. So when I heard that his crew had ditched and had been rescued, I volunteered to go pick them up. I also made a request that Dr. Lamme, the Squadron flight surgeon, go down with me just in case there might be any medical problems with Hullar or his crew. I felt an element of personal responsibility for them, that it was my duty to go and get them."

There was another surprise awaiting Hullar's crew upon their return to Molesworth, which Merlin Miller recounts: "As we got out of the plane and walked across the field carrying our wet flying equipment in our hands, an AP photographer approached and took our picture."

It is hard to imagine a better portrait of Hullar's crew. Except for Marson's absence, the picture is perfect: Nine young men are seen striding

Hullar's crew comes home. This is the AP photo taken of the crew as they returned to Molesworth on the afternoon of September 7, 1943, after their dramatic ditching and rescue in the English Channel on the way back from the Stuttgart mission. The men are wearing old RAF and civilian clothing given them by their British rescuers, and some are seen carrying their still-wet American flying gear. Pictured from left to right are Dale Rice, Elmer Brown, "Bud" Klint, "Pete" Fullem, Bob Hullar, Norman Sampson, Merlin Miller, George Hoyt, and "Mac" McCormick. Not pictured is Chuck Marson, who remained at RAF Station Ford for a few days being treated for a ruptured eardrum and a wrenched knee. (Courtesy AP/Wide World Photos.)

across the tarmac, happiness on their faces, confidence in their steps, the end of their adventure captured for all time. The photo appeared in many newspapers, causing great excitement among family, friends, and relatives, but the crew didn't learn of it until much later.

At the time, they were more concerned with the here and now. George Hoyt was dismayed to discover that "our bedding, equipment, and clothing were all gone, as they had written us off the day we did not return. The supply sergeant had to refit us."

Bud Klint blandly observed that "Twenty-four hours after we had hit the Channel waters, we were back at our home station. After a 48-hour pass, we were back on combat status."

That's the way it was in the ETO. Individual crews had close calls, the mission went either well or awry, and within a day or so another operation was laid on. But the events of this day cannot be dismissed so lightly. The

September 6th mission had too much significance for Hullar's crew and for the Eighth as a whole. A deeper summary is required.

For the Eighth's leaders, the mission was a bitter disappointment. General Arnold considered it a total failure and General Eaker was back at square one, needing more replacements before the Eighth could strike into Germany again. Forty-five B-17s had been lost: In addition to the 18 shot down over Stuttgart, four landed in Switzerland, a fifth crashed in a Swiss lake, 12 ditched (though every crew was rescued, 118 men with three men missing). Another eight crash-landed in England, and two were unaccounted for. There were 10 more Category Es, damaged beyond effective repair.

In return, only the 303rd's combat wing succeeded in striking Stuttgart, and despite all General Travis's circling, the bombs were laid on the main city instead of the Feuerbach factory. The 303rd records noted that "Results of the bombing were unobserved due to heavy cloud cover."

For the 303rd's other crews, the trip home was almost as much of a close-run thing as it was for Hullar's men. Two of the Group's B-17s aborted early, but of the 17 that pressed on to Stuttgart, only five were able to make it all the way back to Molesworth. General Travis's ship was one of these and another was Lt. Don Gamble's. When it was over, Gamble wrote: "Flew over London and landed with two other ships, the remainder of the wing . . . Flight lasted seven hours 30 minutes. My ass is very tired, as is the rest of my body."

For Lt. Bill McSween, this mission marked number 13 and he recorded its end by writing: "This put me over the hump, and everyone in our crew made it back OK."

Ten of the Group's aircraft had to land at other bases due to shortage of fuel, including *Hell's Angels*. John Doherty remembers that "it was my 25th birthday, and it was about as hot a birthday as I'll ever have. We were getting very low on fuel, just crawling across the Channel, and were descending the whole way, trying to save as much fuel as possible."

Lt. Hendry landed at Kenley to refuel and then returned to Molesworth. Gene Hernan also recalls the long flight back in *Knockout Dropper*: "A short time after we left the target, I took a check on our fuel and figured we would be fortunate to get back. I remember transferring fuel from one tank to another as short as five seconds at a time. I suggested to Campbell that we cut back on the rpm or I didn't think we would make it. He agreed and we immediately broke formation and started on a gradual glide from altitude at 1450 rpm and about 20 inches of manifold pressure, practically all the way to an emergency landing on the coast of England."

Lt. Campbell's crew landed at Deanland, a small fighter field near Uckland. As they came in, Gene Hernan remembers that "there were planes landing—some deadstick—from every direction. We had one coming head-on into us that, somehow or another after touching down, bounced over us to complete his landing. There were several that went through those English hedges, ending up completely off the field. One had a *Life* photographer aboard, and it went through two or three of these hedges, and ended up on some farmer's front porch."

The ship with the *Life* photographer was, of course, Lt. Jake James's *Winning Run*. Lt. Paul Scoggins described her demise in his diary, writing that "We thought we would have to ditch in the Channel but we finally managed to get to the coast of England. Then we were forced to make a deadstick landing because of a complete shortage of gasoline in all four tanks. The landing was the best we could hope for under the circumstances. We all walked away from it. The ship is a total wreckage, however."

Looking back today, Scoggins recalls a detail that closes his account on a decidedly humorous note: "I'll never forget the little Englishman who owned the property we came down on. Because Jake had to avoid another plane coming in on another runway—from the nose I spotted him, yelled at Jake, and he did a marvelous job of missing them—we had hit way downfield on just a part of the runway. We went off the end of it, through one hedgerow fence, hit a ditch, bounced up in the air and across the road, through another hedgerow fence, and everything collapsed when we hit that time. The little Englishman came running down there to us and said, 'Gor, Blimey, you tore my fence down and will let my cows out!'"

It was, however, Hullar's crew who had the greatest excitement, and this chapter rightly should end with their reflections on it all. It is doubtful that Bob Hullar ever felt more pride in his men than on this mission. In his notebook he wrote with simple eloquence: "Crew tops. Everyone did their job to a T."

Elmer Brown's memories also center on his crewmates. He never tires of telling the story of the vote he insisted on, nor of his surprise at its outcome, for the strong spirit his crewmates showed has always been a source of inspiration to him. He also derives a lasting satisfaction from his own performance, which he sums up this way: "During the time I was in England, there were two or three missions where I really felt proud about the job of navigation I did. This was the first of them."

Merlin Miller also thinks of the might-have-beens had the vote gone the other way: "I remember the vote quite well. Any time we had a crew-check or something like that, the tail gunner answered first. So after Bob Hullar explained the situation to us, I was the first to speak up. I didn't

The Wreck of the Winning Run, *B-17F 42-29944, GN✪E, Lt. "Jake" James brought* Winning Run *down in a cow pasture adjacent to Deanland, a small British fighter field near Uckland. He had a good excuse for not getting her down in one piece: four engines dead due to lack of fuel, and the need to take instant evasive action to miss another Fortress making an emergency landing on an intersecting runway!* Winning Run *was a total loss, but fortunately, none of James' crew was seriously hurt. The ship was named by Maj. Billy Southworth, a Group member whose father was manager of the St. Louis Cardinals. Note the cardinal clutching a baseball above the ship's name.* (Courtesy Frank Scherschel, *LIFE* Magazine © 1943 Time Inc.)

know anybody in Switzerland, and like the brash young creature I was at the time, I said I was all for trying to get back to England. When each one voted, they all said the same thing. Since then I've frequently wondered what would have happened if I had said 'Let's go to Switzerland' instead of trying to get back across France with a plane that was damaged and not too good a chance of getting back to England. One never knows."

Though George Hoyt doesn't remember the vote, his overwhelming sense of duty preordained his response. "The only thought in my mind was of returning to England and our base. The idea of going to Switzerland and being interned for the war's duration did not take root at all. I had a simplicity of thought and purpose in those days which stood me well. For 45 years I have lived with a triumphant conscience because we made it back to fight again, which was the duty and purpose for which we were pledged."

Of all the crew, Pete Fullem probably came closest to being lost as he drifted away from his raft in the turbulent swells surrounding *The Old Squaw*. Yet, ironically, it was he who most made light of the day's events. This was due, perhaps, to the context in which he wrote. When the Hullar crew's picture appeared in the Jersey City newspapers, it put more than Fullem's family into a state of excitement. The entire neighborhood around the small family candy store at 4 Coles Street was in an uproar, with people crowding into the premises to congratulate the family and wish them well.

Word of these doings got back to Pete in England. Intending, one suspects, not to worry his recently widowed mother, he sent her a letter saying: "It really wasn't much. We found it impossible to go further, so the pilot set the ship down in the best spot he could find. It was almost the same as coming down on land except that it was a trifle damp. The skipper brought *Old Squaw* down near a boat, so it was only a matter of minutes before we had something solid under our feet again. Confidentially, I got seasick."

The day has deep significance for Norman Sampson, who looks back on it this way: "Before we hit, I of all people was telling everyone that we had better pray, but right after the crash I felt it was just luck that we were alive, and I was happy for that. I didn't know God then, but now I believe He had His hand on each one of us on that crew."

Last but not least is Bud Klint's summing up. He offers an explanation not only for Elmer Brown's uncanny navigation, which brought *The Old Squaw* down just where he predicted, but for everything else that happened to Hullar's crew that fateful September day: "I cannot help but feel that there was something bigger than luck riding with us that day. The fact that we were able to stay with the formation until there were no enemy fighters around, and the fact that our escort arrived shortly afterward might be explained by luck, but when that rescue boat appeared on our course, at just the right moment, I was sure that the Omnipotent was guiding us that day especially. I cannot help but feel, too, that a greater hand than mine or Bob's was on the controls when *The Old Squaw* sat down in the Channel.

"The Captain of the rescue boat told us one last thing which made us thank God for our safety. If we had landed a few miles to the south, we would have been caught in dangerous riptides. If we had landed a few miles to the north, we would have been in the minefields. We were just six miles from shore when we landed, and that particular area was the only place where we could have done so safely."

10

"A Horrible Fiasco"

Nantes, September 16, 1943

THE EIGHTH'S INABILITY to strike Germany after the Stuttgart debacle did not deflect it from targets closer to home. On September 7th the Eighth returned to the offensive, bombing Luftwaffe fields in Belgium and France. The 303rd hit Evere airdrome outside Brussels with good bombing results and no ships lost. Two days later, the Group was sent to hit the Vitry-en-Artois airdrome at Douai, France. Again bombing results were good and all ships returned safely.

A five-day hiatus in operations followed, and then a target of real strategic significance was found. The French Underground reported that the *Kertosono*, a German supply ship carrying crucial U-boat replacement parts, had snuck up the River Loire from the Bay of Biscay. She was under repair in a floating drydock at the inland port of Nantes, where she lay, according to Elmer Brown's diary, "near the fork of two rivers" in the harbor area.

She was, Lt. Darrell Gust felt, "a bombardier's dream target," and David Shelhamer believed "she must have been one hell of a ship because we sent about 140 B-17s down there." In fact, the 303rd was leading eight groups of the First Bomb Division (the new designation of the First Bomb Wing) with the express mission of destroying her.

General Travis led the mission, taking off at 1141 in *Satan's Workshop* with a lead crew that included Major Kirk Mitchell and Lt. Paul Scoggins. Scoggins wrote that "There was another navigator with me—just to check each other, but we led the entire works." Hullar's crew was back with *Luscious Lady* in the lead squadron's No. 5 slot off Lt. David Shelhamer's right wing; he was second element lead in *Mr. Five by Five*, B-17F 42-29955. The high squadron was led by Lt. Claude Campbell with Sgt. Gene Hernan

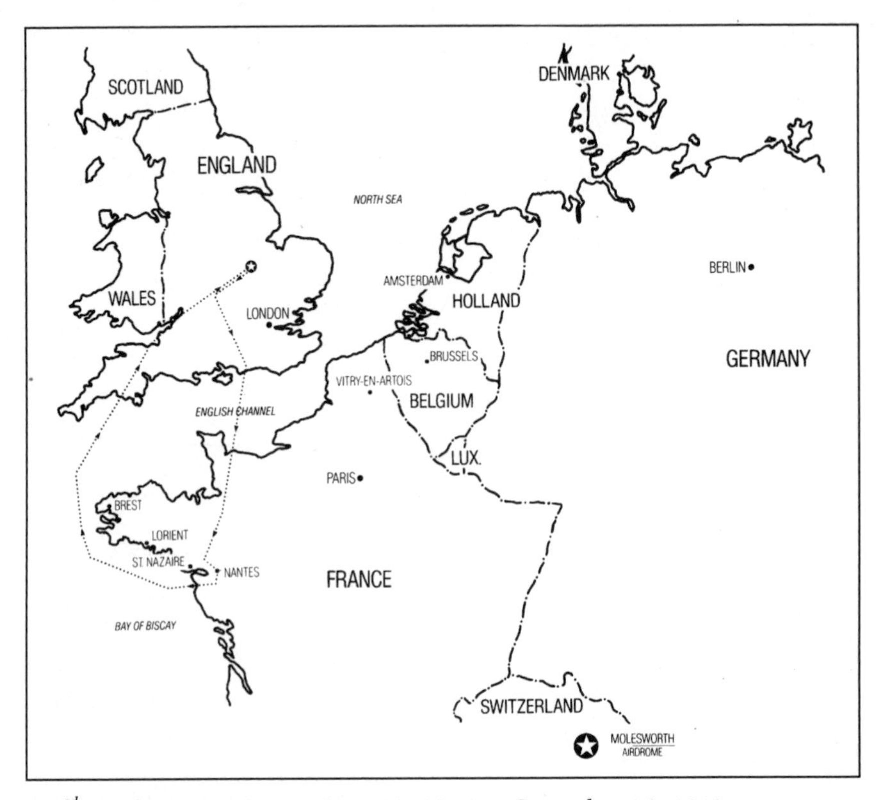

303rd Bomb Group Mission Route(s): Nantes, September 16, 1943. (Map courtesy Waters Design Associates, Inc.)

in the top turret, while down below, at the head of the low squadron, Lt. John Lemmon was with Lt. Darrell Gust in *Jersey Bounce Jr.* Behind them were Lts. Don Gamble and Bill McSween as second element lead of the low squadron in *Star Dust*, B-17F 42-3064.

Lt. Gamble wrote that the Group had a "Good assembly," and Lt. Bill McSween noted that "The haul was not too bad." But aboard the *Lady* there were problems.

"This was the mission where Hullar forgot to take his parachute pack with him," Merlin Miller recalls. "He had his parachute harness, but somehow or other he left the pack on the ground."

And while there was an opportunity to return to base after George Hoyt discovered that one of their port engines was leaving a thin trail of white smoke, when he informed Hullar, "Bob said, 'Our instruments show no trouble. We'll go on in.'"

The Group had P-47s for much of the journey over enemy territory, and Elmer Brown noted "a couple of dogfights." But the action got heavy as soon as the P-47 escort left. As Brown put it, "Enemy fighters were very aggressive. They were mostly FW-190s. They made nine head-on (nose) attacks and eleven at the tail."

Bud Klint felt the Luftwaffe gave a good account of itself, too: "Enemy fighters were comparatively scarce—we met only about 20 on the round trip—but those we did meet were unusually eager. They pressed their attacks vigorously.

"On one particular occasion two FW-190s came in abreast from eleven o'clock high, skidded their ships, and sprayed the formation with more 20mm bursts than I had ever before seen at one time. How they avoided knocking at least one Fortress out of the formation I'll never understand, but they didn't, and as they broke away some gunner sent one of them down in flames."

The Germans went after the bombers for over a half an hour, and while no 303rd ships were lost Lt. McSween noted that the fighters "knocked

Lt. John V. Lemmon's crew before Jersey Bounce Jr., B-17F 42-29664, VK✪C. *Bottom row, L-R, Sgt. C. Zeller, Sgt. V. Brown; Sgt. C. Bagwell; Sgt. W. Briggs; Sgt. A. Berzansky; Sgt. A. Beavers. Top row, L-R, Lt. Lemmon, pilot; Lt. Darrell D. Gust, navigator; Lt. W. Latshaw (not a regular crewmember); Lt. E.E. Stone, bombardier; Lt. E.E. Clark, copilot. For the fate of* Jersey Bounce Jr., see Chapter 25. (Photo courtesy of Brian S. McGuire.)

down several B-17's." Two fell from the low 384th Group, while four went down from the high 379th Group.

Then, as the 303rd's formation drew near the aiming point, the fighters were replaced by flak that Elmer Brown felt was "light and inaccurate all the way," and which Lt. McSween wrote up as "medium and inaccurate"—though it was close enough for Lt. Gamble to write: "Hear the flak go 'Woof!'" But the flak didn't cause any casualties, and since there were only small patches of clouds over the city and a heavy but ineffective smoke screen, it seemed that nothing could stop the Group from getting that U-boat supply ship.

The men didn't reckon with fate, for events were soon to make the "bombardier's dream target" a nightmare. The story Darrell Gust heard in the immediate aftermath of the mission has remained with him to this day:

"A brand-new bombardier somehow got thrust into the lead aircraft. During the bomb run he put the bombsight on 'extended vision' to see up ahead and forgot to turn it back. When the indices on the bombsight came together, it was good-bye bombs, early or not. We dropped our bomb loads right into the heart of Nantes. It was a horrible fiasco."

David Shelhamer tells the same story, and the tale is plausible since the lead bombardier was always under great pressure to steady up on course during the bomb run and lock in a bombing solution on the famous Norden bombsight. The sight had an "extended vision" knob that provided a view up to 20 miles down the bombing track so the bombardier could acquire the target and set up on it. As he got closer to the target, he was supposed to shift the eyepiece to the "normal mode," synchronize the sight's longitudinal and horizontal crosshairs on the target, and as the bombsight "locked on" and generated a solution, let it automatically drop the bombs at the calculated release point. Since the rest of the bomb group would drop their loads as soon as they saw them leave the lead ship, the entire mission depended on a single man.

But is this really what happened over Nantes that day? Paul Scoggins doesn't think so: "Although I was in that lead ship, I do not remember ever hearing the story about the lead bombardier. As I remember it, the bombardier was from another squadron, not the 427th, and I did not really know him. The General was in our ship, too."

Further mystery is added by what Gene Hernan recalls: "Our bombardier, Boutelle, told Campbell that no one was on the correct bomb run. We made a bomb run of our own and got a near miss on the target. I think it was the closest anyone came."

Hernan's account is matched by a report Lt. Boutelle made, which is in the Group's mission file: "BOMB. Held bombs 15-20 seconds after other bombs went. Ball turret said main bursts were in town and a small concentration was on the river bank with one falling across river, causing large explosion."

Perhaps the lead bombardier did drop early, or a junior bombardier "practicing" accidentally let his bombs go with the rest of the Group dropping when they saw them go. Whatever the reason, Lt. McSween noted that "The bombs hit on both sides of the river."

Bud Klint realized "Our Group missed the target completely—in fact, no one in our Group scored a direct hit."

And Lt. Don Gamble came closest to the mark by writing: "See several hits around the target when we drop the 12×500-pound bombs. We probably killed a lot of Frenchmen."

The Group assuredly did. The word David Shelhamer later got was that "We dropped 223 500-pound general-purpose bombs right smack in the far end of a public park and into an apartment complex. It was figured we eliminated about 500 Frenchmen rather than one ship." The next day German radio, quoting reports from Paris, said that more than 850 Frenchmen had been killed, more than 150 were still buried under the debris of wrecked buildings, and that the injured numbered more than 1000, 300 of whom were seriously hurt.

And no one got the *Kertosono*. When the day was done and all groups had dropped, the closest anyone came was: "A single direct hit in the SW corner of the floating dry dock" together with "four near misses which will cause some damage to the dry dock and ship." The "single direct hit" might well have come from Lt. Campbell's ship, but no one will ever know for sure.*

---

*The initial publication of this book in 1989 inspired a Frenchman, Serge Lebourg, to devote countless hours to the task of researching the ill-fated Eighth Air Force bombings of Nantes on September 16th (and September 23rd, see Chapter 11) 1943. Lebourg's family came from Nantes, and his parents survived the September 16th bombing "by a miracle." Thanks to Lebourg's efforts, many of the questions which this writer originally thought were unanswerable have now been resolved.

In brief, over 1100 Frenchmen were killed as a result of the Eighth Air Force bombings on September 16th. An additional 800 Frenchmen were seriously wounded and 1200 slightly wounded that day, whereas only 38 Germans were killed. 400 dwellings were totally destroyed, 600 were partially destroyed, and another 1200 were damaged.

According to Lebourg, there were three main reasons for the very high civilian casualty rates. First unlike many of the U-boat "sub-pen" cities on the Bay of Biscay coast such as Brest and St Nazaire, Nantes had never been bombed before. Its population had become complacent and tended to ignore air raid warnings. Second, unlike the sub-pen cities, the harbor at Nantes was not separated from the downtown area but instead ran along the banks of the River Loire right through the center of the City. Third, the bombing took place on a Thursday, which was a vacation day for French schoolchildren, and at the worst possible time, around 4 p.m. locally, a peak period for downtown activities.

Lebourg's research reveals that the 303rd's main formation concentration of bombs fell in a northern suburb of the City, short of the target area by a mile. Most of the bombs fell in a

Immediately after the 303rd dropped, Lt. Campbell's crew found themselves in serious trouble. As Gene Hernan recalls, "A piece of 88 severed the oil line to the No. 3 engine. The oil got pumped out of our supply tank before we realized it and we couldn't feather the prop. Consequently, the reduction gears broke and the prop started windmilling, due to the pistons seizing up. We went for the water, ambling out over the Bay of Biscay just above it so no fighters could get under us. The going was slow, about 120 mph indicated, with everyone else in the Group wondering what was wrong but Campbell not wanting to break radio silence to tell them. Fortunately, our Squadron stayed with us or we would probably not have made it back."

The entire Group actually got down close to the water on the return trip. Elmer Brown wrote that "We left the target on an SSW heading and came back entirely over water over approximately 400 miles. For over 250 miles we were from 500 to 1000 feet off the water due to a lower ceiling. The whole object was to stay clear of France by 40 or 50 miles so we had to go way around the Brest peninsula."

Lt. Don Gamble also wrote of a "Let-down to about 600 feet over water to get under clouds. The boys shoot at birds over the water."

Some of the Group's gunners shot at more than birds. Lt. Bill McSween recorded that "Our Group ran into several Me-110s and one or two were shot down by B-17s." One of those shooting was Sgt. Hernan, who by this time was really beginning to sweat things out:

---

residential area composed largely of small houses with gardens. Fortunately, this was not a high density population area, and part of the population was at work. Some of the Group's bombs did fall in a public park ("le Parc de Procé") where a park guard was fatally wounded. Three children were also wounded, but they survived. Other Group bombs fell in a nearby sports stadium (the Malville Stadium), where a number of players were killed. All total, Lebourg estimates that less than 100 Frenchmen were killed by the main 303rd bomb pattern. Lebourg's research also established that the Group bombardier, who shall remain nameless, was on his first mission after having spent many months recuperating from the effects of a severe attack of appendicitis. This may in part explain his poor performance on this mission.

Other casualties were caused by the separate bomb salvo of Lt. Campbell's crew, and/or by the bomb load of Lt. McClellan's crew, which "salvoed six and toggled six." The bombs from one of these aircraft (probably Campbell's) landed in the downtown area on "La rue Franklin" where they struck the "Olympia" movie theater which was showing the film "Michael Strogoff." The movie was halted due to the air raid alert, but only a few theater patrons had gone to the air raid shelter, and a number of those who remained were killed when the bombs hit.

It appears that the best bombing of the day was that of the 379th Group, and that one of its bombs was the one which damaged the floating drydock near the *Kertosono*. The worst bombing of the day, in terms of civilian casualties, was that of the 351st Group. All this Group's bombs fell in the center of the City, and some struck the main City hospital (the "Hôpital Hôtel-Dieu"). Lebourg estimates that approximately 1000 Frenchmen were killed in the 351st Group's strike area.

"When the Me-110s did come out to attack, it was the only time that I can remember my hands being warm, and when I was firing I can remember having a cigarette in my lips. We did manage to get two of them before they gave up."

His tally is confirmed by the Group's records. At 1625, one of the 110s "burst into flames" and "Was last seen spinning towards the water." At 1630 another "burned and hit water."

For Hullar's crew there was a preoccupation even more important than the Me-110s nipping at the formation's heels. Bob Hullar wrote that it was "Our first raid since ditching. The entire crew really sweated out our gas supply."

Elmer Brown noted: "We kept a very close check on the fuel consumption."

Bud Klint felt "The part of the mission our particular crew disliked the most was the long trip home over water. We were over water for nearly two hours, and after the dunking we had experienced just ten days before, that 'deep blue' looked not in the least inviting. The English coast and, an hour and one half later, our home base, were particularly welcome sights that day."

Luscious Lady landed at 1847, ending a mission that lasted over seven hours. Elmer Brown noted that "everyone in our Group got back OK." But the operation had a bitter postmortem, especially for the bombardier who had erred.

Darrell Gust believed, "They almost court-martialed him," though he is somewhat wide of the mark on this point.

As Ed Snyder recalls, "There was no real thought of a court-martial, but you can believe that bombardier had his tail between his legs when he got back and that he was given a bad time. Still, this wasn't the first time something like this happened, and it wasn't the last. In war, under pressure, you have to expect this sort of thing."

There is a postscript to this story well told by Mel Schulstad, who stayed to see the war through from beginning to end. After the Eighth dropped its last bomb in the Spring of 1945, it launched hundreds of bombers filled with food on mercy missions to the Continent. During these flights there were men in the Group who remembered what had happened at Nantes so many months before, and a decision was made to send some food-laden Fortresses to the city to make amends.

"We decided to do this," Schulstad recalls, "even though we had real questions about the reception we would get after what we had done to that place. I was in one of the B-17s that went over, and I'll never forget the sight that greeted us as we pulled into the landing pattern and set down.

"There were literally thousands of Frenchmen there to greet us, and they all raced onto the field even as we were landing, arms raised high above their heads holding bottles of champagne, cognac, and wine. They crowded around our planes and had the bottles uncorked even before our props stopped turning. You can just imagine how we all felt."

11

"Don't Let That Happen to You"

Emden, October 2, 1943

THE EIGHTH ATTACKED the U-boat supply ship at Nantes again on September 23rd, when the First Division launched the 303rd and a number of other groups against it. This time the smokescreen made it hard to see the aiming point, and the bombing was no better than on the previous raid. Afterwards, the enemy eliminated any chance of a further attack; Bud Klint wrote that "The Germans moved the ship farther up the river, according to Intelligence, and as far as I know it is still afloat."*

By this time, however, the Eighth's leaders were already looking to the next series of attacks against Germany itself. By September's end, 200 new crews had arrived in the ETO to replace the 104 that had been lost on the month's operations, and General Eaker now had 450 crews to fly 604 heavy bombers. This disparity would create hardships for the combat men, but it wouldn't prevent the offensive from going forward.

September was also a time for new tactics. The always-cloudy skies over Europe had finally convinced the daylight bombing advocates that the Norden bombsight was not the only answer. The Eighth's leaders borrowed

---

*There were actually two raids against Nantes on September 23rd, one in the morning by six B-17 groups, and a late afternoon mission by two groups in which the 303rd took part. Tragically, the only result was to raise the total number of French casualties for the missions on September 16 and 23 to 1423 civilian fatalities and over 2500 wounded. Serge Lebourg's research confirms that the *Kertosono* was not destroyed.

A further raid against the city was dispatched on September 26, 1943, but was recalled due to overcast weather. On the way back Sgt. Eddie Deerfield's crew was forced to bail out of *Lady Luck*, B-17F 42-5434, when her No. 3 engine ran away and caught fire as the aircraft approached the English coast near Southhampton. The plane crashed and burned with a full bomb load near Alresford. The only serious injury was to Capt. Robt. Cogswell, who was the last to leave the ship. He suffered back injuries which grounded him for the rest of the war. Eddie Deerfield continued flying until May 11, 1944, when he completed a tour of thirty missions.

from the British, and began to establish a "PFF" or "Pathfinder Force" built around the RAF's H2S downward-looking terrain radar, a primitive device whose cathode ray tube was able to "see" terrain along rivers and coasts, where the contrast between land and water stood out sharply.

In August 1943 the Eighth formed its own PFF unit, the 482nd Group, composed of one B-24 and two B-17 squadrons. They would employ the same tactics RAF Bomber Command used over cloud-covered cities at night. The PFF ships would locate their target cities with H2S, and the other bombers would drop on "sky marker" flares released by the PFF ships at their aiming points over the city centers.

On September 27th, VIII Bomber Command ordered the first operational use of the PFF tactic on an attack against the port area and dock facilities of Emden, Germany. The mission netted good results for the lead wing of the First Division, which dropped in unison with the pathfinder lead, but it was not a good mission for the 303rd, flying as high group in one of the follow-on wings.

Lt. Bill McSween wrote: "We were going to do PFF bombing. P-47s

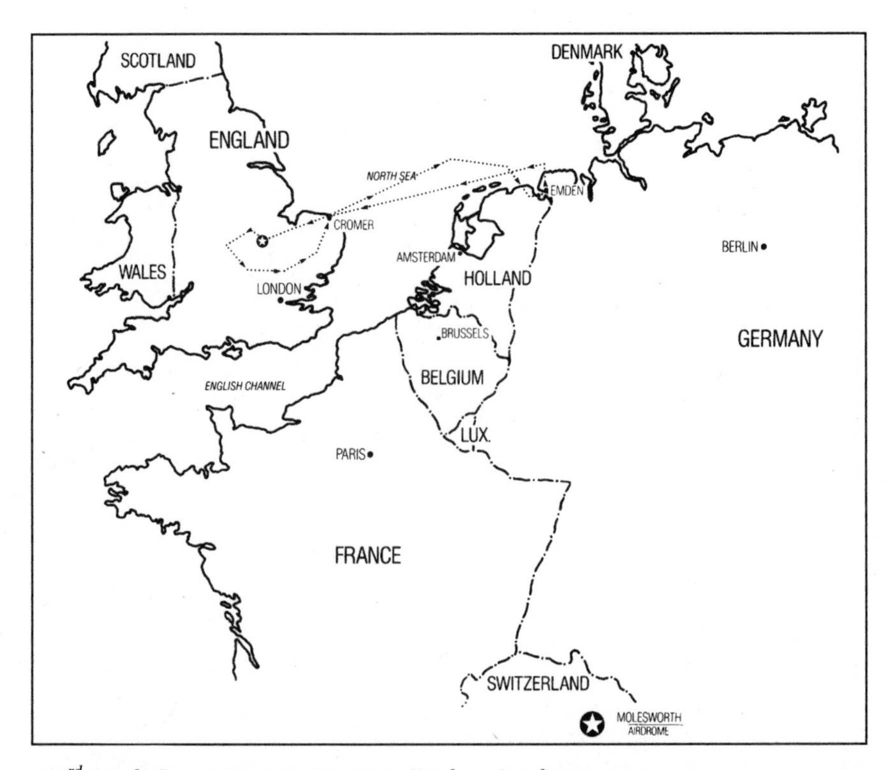

303rd Bomb Group Mission Route(s): Emden, October 2, 1943.
(Map courtesy Waters Design Associates, Inc.)

Yardbird-II, *B-17F No. 42-5260, PU✪A, lost on the October 2, 1943 mission to Emden. Note the .50-caliber machine gun in the reinforced .30-caliber ball socket in the Plexiglas nose. Note also the relocated right cheek gun.* (Photo courtesy Melvin T. McCoy.)

were all over the sky as well as B-17s. The lead outfit ran off and left us. We made a turn approximately ten miles around with the target on our left. The target had 10/10 cloud cover. No bombs were dropped."

Five days later, on October 2nd, the stage was set for a second PFF attack on Emden. Hullar's crew would go on this one, and the men would witness a sight impossible to forget.

The Eighth sent 196 B-17s from the Third Division (the old Fourth Bomb Wing) and 161 from the First Division under the command of General Travis. Each force flew with one PFF B-17 from the 482nd Group. Hullar's crew was in the First Division's lead wing, flying *Luscious Lady* in the No. 6 position of the high squadron, low group. Flying above and to their right in the No. 7 slot was Lt. P.S. Tippet's crew from the 360th Squadron in *Yardbird II*, B-17F 42-5260, a veteran Fort on its 43rd trip. Lt. Tippet had 18 raids under his belt and his bombardier had 24. The other eight crewmen were rookies on their third raid, and there was a photographer aboard on his very first combat flight.

The Group got off between 1305 and 1315 and the flight into the target was roundabout but relatively uneventful. Elmer Brown wrote that "P-47s escorted us all the time over enemy territory. It was a very long trip for them. They used extra gas tanks at first, and salvoed them. We flew around via the North Sea, then went south to the IP, then easterly, released the bombs, and then went northward out to sea again. Bombs were released on pathfinders for the first time. We had 10/10 undercast. They threw up barrage-type flak, but it was inaccurate."

When the bombs went, Bud Klint couldn't help but wonder where they would hit, but there was little mistaking what happened some six minutes after the target on the northbound, homeward leg. He described it this way:

"Our squadron had only one concentrated attack by enemy fighters. A FW-190 and an Me-109 broke across our right wing, so close that I could almost read the serial numbers on the ships. They took the No. 7 man with them as they broke away—shot his number three engine completely off and almost severed his right wing."

Elmer Brown saw very much the same thing: "An FW-190 closely followed by an Me-109 made a head-on attack on our squadron. They got our No. 7 plane (tail-end Charlie). His No. 4 engine was on fire at first and it must have gotten to his gas tank because the whole right wing was aflame. Within a few seconds the No. 4 engine dropped off and the plane started down. A few parachutes came out but we don't know the exact number."

The rearward visibility in a B-17 was poor from the nose and equally bad from the cockpit, and it was the enlisted men in *Luscious Lady*'s rear who saw what really happened as *Yardbird II* began her final dive. George Hoyt remembers that "Right after the two German fighters slipped through, the next thing I saw was a B-17 just a few feet behind us with the No. 4 engine and wing engulfed in flames. He climbed a little behind us and I saw two guys jump out the open bomb bay doors. I saw their parachutes open immediately after they cleared the open bomb bay doors. One chute was burning. I saw it, very close. I knew what it meant, but I had no analytical reaction, no emotional qualms. I thought simply, 'That's the way it goes; get very alert on your gun, and don't let that happen to you.' Then the plane plunged downward out of my field of vision toward six o'clock."

In the *Lady*'s tail, Merlin Miller saw everything to the very end: "We were flying along under fire, with another B-17 in the diamond position. All of a sudden one of his engines started flaming. I believe it was the right inboard, which is No. 3. There were flames in the engine and quite a bit of black smoke, and I could also see a white misty spray coming out from the trailing edge of his right wing. The plane pulled up and over to our left, a little above us and even ahead some till it was about 100 yards behind, close enough for me to get a real good look.

"A gunner stuck his head out the right rear door in front of the tail. He took a quick look and then jumped. He was hardly underneath the horizontal stabilizer when his chute popped open and he started to float way. He couldn't have been more than about 20 yards behind the plane when the chute disappeared in a puff of smoke. I couldn't believe it. It just vanished in a puff of smoke. Then another guy jumped out the waist and the same thing happened to him, almost exactly."

"It was startling. I'd seen guys jump before, open their chutes, have their shoes fly off and be sagging in their harnesses from the speed they were going, but I'd never seen chutes pop open and then just disappear. I guessed when these guys pulled their ripcords, they were in that white misty spray of gasoline from the wing, and that it was burning, though I couldn't see any flame. I made a mental note that if we ever got on fire and I had to jump, I would wait before opening my chute. The only other thing I could do was take a deep breath, swear to myself that fate would play that kind of a trick on a couple of fellows and then *forget it*!

"After that, the plane pulled away from us in a shallow dive to the left, with the wing burning more all the time. All of a sudden it was gone. There wasn't enough left to stick in your hip pocket. It was amazing how a plane as big as a B-17, with a 104-foot wingspan and a 75-foot fuselage, and four giant engines, could all of a sudden be a big ball of fire and smoke in the sky. Then all I could see was just a little trace of smoke going toward the ground. It looked like paper fluttering down."*

The Group continued up the northbound leg out to sea, then turned to a southwesterly course toward Cromer on the English coast and Molesworth. Nothing of consequence took place during the overwater flight, and the Group landed back at Molesworth between 1838 and 1858. Of the 18 ships that returned, six had suffered minor damage.

For the Eighth as a whole, the mission ended with mixed results. *Yardbird II* was the only B-17 lost by the First Division, and the Third Division only lost one as well, but the bombing had not been good. One of the 482nd Group's H2S ships had released too early and many bombs fell short. The aim of the other was nullified by high winds in the target area, which scattered the smoke of the sky markers.

A month would pass before the Eighth would resort to PFF tactics again, but in the interim the Flying Fortresses would press their assault against the Reich in the bloodiest aerial battles of the Second World War. Hullar's crew and the 303rd would be in the thick of them.

---

*The Missing Air Crew Report ("MACR") for Tippet's crew confirms that all of the crew were killed.

12

"Black Week" Begins

Bremen, October 8, 1943

WHILE THE EIGHTH was rebuilding in the wake of the September 6th Stuttgart raid, the Luftwaffe's leaders were not idle. Recognizing the threat daylight bombing posed to targets such as Schweinfurt and the inability of their day fighter forces to prevent future attacks, they undertook a major reshuffling of defensive strength. They stripped the Russian and Mediterranean fronts of single-engine fighters, and nearly doubled their twin-engine "destroyer" strength, adding Me-110s and newer Me-410s (known to the bomber crews as "Me-210s").* They kept their night fighter force constantly on call, adding Ju-88s, Do-217s, and other more exotic types. And, finally, they centralized their forces for defense in depth along the approaches to the Ruhr and other critical objectives.

To this day, estimates of total German strength in the West vary—some sources put the number at nearly 800 fighters, others at over twice that figure—but there is no doubt the bomber crews now faced greatly increased odds. At VIII Bomber Command, General Eaker and the others knew of these developments, but felt the offensive had to go forward.

They really had no choice. There was a war to be won.

The first clear weather raid of the month took place on October 4th, when the First Division sent 155 B-17s against various targets around the cities of Frankfurt am Main and Wiesbaden. The Third Division sent 168 Forts to attack in the Saar region near the French border. Thirty-eight B-24s of the Second Bomb Division flew a diversion. The 303rd sent 21 ships to attack a plant that assembled aircraft components in the Frankfurt suburb of Hiddernheim.

*The unsuccessful Me-210 and the later, improved Me-410 were very similar in appearance.

Strike photos showed "much destruction by demolition bombs" and the Group's war diary stated that a "heavy concentration of incendiary bombs fell in the target area and many fires were started." However, the Group encountered "moderate but intense flak," and the guns got B-17F 42-29846, flown by Lt. V.J. Loughnan's crew, from the 359th Squadron. Lt. Loughnan was on his 20th mission, and most of his men were one mission behind him. The Eighth as a whole lost 16 bombers.

There was a three-day pause in operations, and then the bombers were off again. Between the 8th and 14th of October 1943, the Eighth Air Force would fly four raids in which the greatest number of B-17s ever would be lost. It was the seven-day period now known to history as "Black Week."

The first mission was mounted against a target Hullar's crew would visit many times during their tour—the German port of Bremen. The Eighth dispatched 344 B-17s from the First and Third Divisions to various targets in the city, while the Second Division sent 55 B-24s against the U-boat yard in the Baltic town of Vegesack.

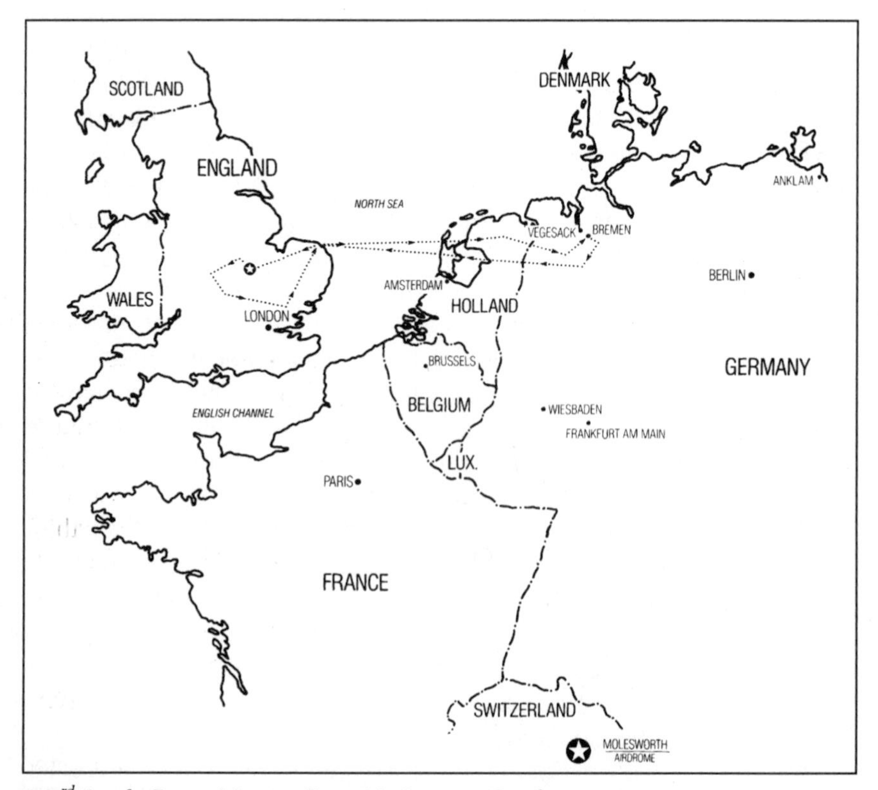

303rd Bomb Group Mission Route(s): Bremen, October 8, 1943.
(Map courtesy Waters Design Associates, Inc.)

The 303rd's mission looked like a rough one. The target was a submarine shipyard located in the heart of the port area three miles northwest of the city center. Bud Klint wrote that "The S-2 officer told us at briefing we would be in range of nearly 600 enemy fighters, and, at the target, nearly 300 flak guns."

This was the Hullar crew's ninth raid, and their sixth in *Luscious Lady*. Bob Hullar was beginning to describe her as "Old 081" in his notebook, and by this time the whole crew was happy about another mission in "their" Flying Fortress.

George Hoyt "always felt secure in her, even though she had the obsolete 'constant-flow' type of oxygen system with the rubber bladder that could freeze at altitude. She made up for it by being fast."

Merlin Miller remembers her as "one of the earlier type B-17s, but she was lighter than some of the later models, and she was a bit faster. She got good gas mileage, too. We were all quite delighted with her."

The crew's hopes for another smooth trip in the *Lady* were soon to be dashed. Elmer Brown recorded that "On this day we started out in plane 081 (*Luscious Lady*) and had to abort because of a gasoline leak. We were in the No. 6 position of the high squadron at the time. We came back to the field and got plane No. 221. We caught the formation at Splasher No. 6 and filled in at the No. 7 position in the low squadron, high group."

While it all sounds rather matter-of-fact, Merlin Miller was angry: "We didn't like changing planes. We didn't like changing anything after a mission was set up. Once you got familiar with your airplane, you were comfortable. You knew how the guns were aligned in it, how far out the bullets were set to converge, and where everything was. Each plane was a little different, and you didn't have time to check the new aircraft out or get comfortable in it. You just grabbed it and went. It made me mad."

The switch also touched a nerve in Norman Sampson: "The changing of planes and of our position in the formation disturbed me very much. When we aborted, we should not have had to go again. You never did when you aborted! Catching up put us in a bad position in the formation, and I didn't know anything about this plane or the guns in it. Everything was going wrong, and I was afraid things would continue to go wrong over enemy territory. This was 'not a good start.' It gave me an empty feeling."

The flak was rough over the target, but the crew's reactions varied. Elmer Brown noted: "We were over 27,000 feet and my windows were so frosted up I could not see out. The flak was predicted fire barrage type, and didn't bother us too badly."

Bud Klint felt more threatened. "I was certain every one of those 300 flak guns was trained on us as we made our bomb run. Those black puffs were as thick as flies before a rain and the gun crews really had the range.

We could hear the bark of the exploding shells on all sides and the fragments peppering the metal fuselage sounded like hail on a tin roof. The ground gunners carved quite a few notches on the smoking barrels of their 88s and 105s that day. One of the ships in our wing must have taken a direct hit among its control cables—it did a perfect slow roll, fell off into a spin, and then very shortly broke up completely."

George Hoyt considered it "very heavy flak," and Group records describe it as "intense and accurate." Nine Group ships were damaged, and a combination of flak, cloud cover, and industrial haze caused the Group's bombing to be off. The 303rd's war diary states that "bombs fell somewhat to the east of the Aiming Point."

Luckily, the Group was largely unmolested by the large number of fighters the enemy had about. "We had just a few fighter attacks," wrote Elmer Brown, and Bud Klint concurred:

"The fighters were out in force. They didn't bother our wing very much, but I could see them crawling all over wings that were flying lower than the 27,000 feet which was our assigned cruising level. They were taking their toll too—I counted ten chutes at scattered points within three minutes after we were out of range of the ground defenses and after the fighters had begun their attacks."

These impressions were shared by Lt. William C. Fort, Jr., a taciturn 26-year-old from Ft. Meade, Florida, whose crew had reported to the 358th Squadron in September. Fort was flying under Lt. Calder L. Wise, his "Instructor Pilot." He remembers that "Being this was my first mission, I didn't know what to expect. There seemed to be a tremendous amount of flak when we got to the target area. We weaved in and out, and it didn't look like a fly could get through, but we made it some way or other. On the way back, I saw two or three of our planes shot down right close by. Other than that, nothing out of the ordinary happened."

By the time all the bombers had gotten back to England, their ranks had been thoroughly riddled. Despite good escort coverage, 27 B-17s from the First and Third Divisions were missing and three B-24s from the Second Division didn't return. It was the beginning of a very long week.

The second raid got underway October 9th as the Eighth launched 378 heavy bombers against targets in central Germany, East Prussia, and Poland. The Third Division sent five groups against the Focke-Wulf factory at Marienburg, a city on the western border of the "Polish Corridor" in East Prussia. Seven more groups—two from the Third Division, three from the First Division, and two from the Second—were sent to attack the port area of Gdynia, Poland.

Two other groups from the Second Division were assigned to strike the U-boat yard at Danzig. Meanwhile, six groups from the First Division were sent to strike the Arado aircraft factory at Anklam, on the Baltic coast of central Germany, in what was termed a "diversion" to mask the deeper penetrations to the east.

The 303rd led this force under General Travis, and it met the roughest opposition. German fighters intercepted the B-17s en route to Anklam, but they were not strong enough to disrupt the strike. The 303rd snuck in at 13,000 feet and bombed the target "with good results." The return trip was far different; 100 to 125 enemy fighters were encountered and Lt. Bill Fort found it very rough:

"I saw more different types of German fighters on that mission—Ju-88s, Me-210s, FW-190s, Me-109s—than on any other raid I went on. I can remember one FW-190 that made a diving attack out of the sun. His tracers hit the tail of a B-17 behind us in the Group, and it fell back out of the formation. They tried to get back in the formation, and managed to rejoin it for a while, but they could only fly sort of sideways because of the damage to their tail. They fell back again and the German fighters went after them. The plane headed down and I saw four or five chutes.

"I also saw an Me-210 get another B-17 in our Wing by making an attack from under the tail. He pulled up and put a lot of shells into the Fortress and set it on fire. There was so much flame I thought the German was on fire too, but when the B-17 started down, I saw it wasn't. We wasted a lot of ammo on a Ju-88 that was hanging around, but we managed to hit an Me-210 pretty badly that came in and lobbed a few shells at us. My tail gunner also managed to shoot out the engine of a B-17 that was flying behind us, and they weren't very happy about it back at the base.

"After this mission I still didn't know too much about what to expect. But my attitude was 'Take it as it comes and hope for the best.'"

The Anklam raid was Sgt. Gene Hernan's last, and he ended his tour on a positive note. He flew with General Travis and was personally credited by the General with the destruction of a Ju-88. It was one of 12 fighters claimed by the 303rd for one of its B-17s.

However, the total score for the Anklam attack was much more disheartening. By the time the fighters were finished with the Anklam force, 18 B-17s had fallen—15 percent of those sent and the equivalent of an entire group. The other bomber forces came home minus ten ships, so that 28 heavy bombers were lost for the day.

The missing 303rd ship was flown by Lt. B.J. Clifford's crew from the 427th Squadron. Their ship "was seen to go down after persistent attacks by E/A." It was last observed beginning a long glide into the clouds near the Danish coast with its wheels down, and an engine on the left wing smoking. Three chutes were seen to emerge, two of which opened. Lt.

Clifford's crew was on their fourth mission, and they were never anything other than strangers to Hullar's men, even though they were in the same squadron.

As Merlin Miller explains: "By this time we weren't interested in getting to know new crews. The people in our own barracks, the guys on Woddrop's crew, and a few others who we associated with were our only friends. If we got to know new crews, it only meant that we would wind up losing more friends when they were shot down, and we didn't want that to happen. It was better if they stayed strangers."

As Bud Klint puts it, "Losses just came to be accepted as a way of life in the ETO."

Despite these harsh realities, Hullar's crew couldn't help but know the plane that was lost on the Anklam mission. It was *Son*, B-17F 42-5221, the same No. 221 they had flown to Bremen just the day before.

13

"We Were All So Scared"

Münster, October 10, 1943

DESPITE THE LOSS of 58 heavy bombers on October 8th and 9th, the Eighth's leaders saw no reason to reduce the tempo of attacks. Telling blows had been struck against important strategic targets—in addition to the damage at Anklam, the Focke-Wulf plant at Marienburg was all but annihilated—and General Arnold was clamoring for more.

Eaker and Anderson were ready to oblige; the planners at Pinetree were putting together another maximum effort for October 10th. The target was Münster, a "Major rail center" in the Ruhr. The Aiming Point was Münster's 1000-year-old cathedral and the reason for this choice was, quite literally, to hit the city's railway workers where they lived. It was hoped that a heavy bombing of the residential districts around the cathedral would kill many workers, disrupting the morale and efficiency of the survivors.

The Third Division was leading the mission, followed by the First, while the Second Division was to stage yet another feint over the North Sea. The Third Division's seven groups were arrayed in three combat wings, or "CBWs," making up a total force of 133 Fortresses.

Flying 15 minutes behind were the nine groups of the First Division, 141 B-17s strong in three combat wings. The 41st CBW was last, with the 384th Group as lead, the 379th Group high, and the 303rd low. At the very end, in the No. 7 slot of the low squadron of the low group in this final combat wing, flew Hullar's crew in *Luscious Lady*.

Hullar's men were well aware of their vulnerability to fighters in this position. "We were tail-end Charlie in the low squadron, low group (Purple Heart Corner)," noted Elmer Brown.

But enemy fighters were not the source of the crew's woes this day. When it was over, Bud Klint recorded that "We had fighter escort all the

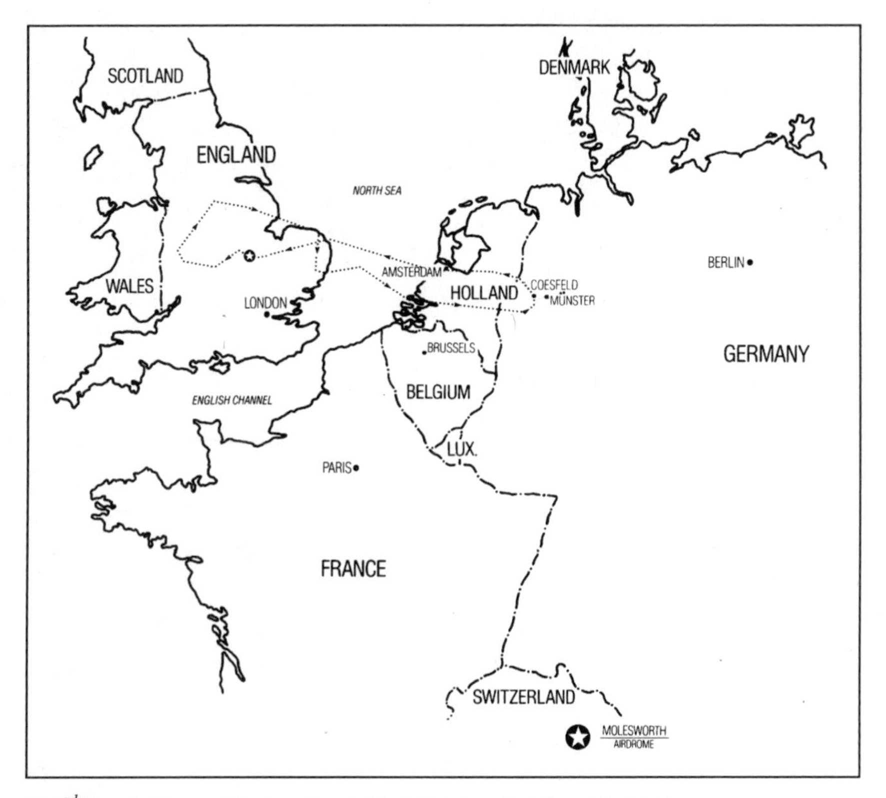

303rd Bomb Group Mission Route(s): Münster, October 10, 1943.
(Map courtesy Waters Design Associates, Inc.)

way and they did a very good job of keeping the enemy 'peashooters' away from us."

Instead, Elmer Brown wrote, with good reason, that "the most outstanding thing on this raid was the flak." For on this mission the 303rd would run a fiery gantlet of flak eclipsing anything the Group had experienced before.

The key to avoiding trouble among the flak batteries lay in good mission planning and navigation. The staff officers setting up mission routes at Division and Wing Headquarters took care to plot tracks that would avoid the worst flak areas, and the lead navigator was responsible for keeping on course both to get to the target and to minimize exposure to antiaircraft fire.

In the air as on the ground, however, the road to Hell is often paved with good intentions.

The 303rd sent 20 B-17s off between 1130 and 1142 under Major Kirk Mitchell and Captain Merle Hungerford in *Jersey Bounce Jr.* The bombers included *Hell's Angels,* flown by Lts. Calder Wise and Bill Fort; Shelhamer's crew in *Vicious Virgin,* as second element lead of the low squadron; and Lt. "Woodie" Woddrop in *Flak Wolf,* to Shelhamer's left in the No. 6 squadron slot. The launches were made into fog. Elmer Brown wrote that "We had to make an instrument takeoff and landing as the ceiling was about 500 feet and fog—visibility about 1,000 yards. The top of clouds was about 4,000 feet."

From the very beginning, the mission showed every sign of being a bad one, because the 384th Group was Wing lead and had far less experience and depth than the 303rd. Major Mitchell's frustration with the early flight comes through quite clearly in his Group Leader's Narrative:

"The Combat Wing completed its assembly on line Molesworth-Eyebrook, climbed on course cutting one turn about five minutes short for no apparent reason. This put us approximately five minutes ahead of schedule at bombing altitude. The Combat Wing S-ed for about 15 minutes, crossing the Channel to lose time.

"The 384th Group had a 'bunch' of airplanes rather than a 'Group formation.' There was a three-ship low squadron of airplanes in the 384th Group which shoved us out of formation twice. An A/C out of the high squadron of the 384th Group aborted about two-thirds of the way across the Channel and dumped his bombs over our formation, barely missing some of our A/C."

One of the ships the bombs barely missed was *Hell's Angels,* just behind Major Mitchell in the No. 4 lead squadron slot. As Bill Fort recalls, "Somebody cut loose a 1000-pound bomb above us coming through the formation, and it almost took all of us with it."

Much worse was to come. The briefed mission route had the 41st Wing entering Occupied Europe south of Rotterdam and flying a course almost due east to the town of Haltern—the IP from which it was to fly northeast on the bomb run into Münster. But just before the IP, the 384th Group turned almost due north on a track towards Coesfeld, 20 miles west and slightly north of Münster.

According to Major Mitchell, this erroneous course took the bombers "over Hüls, subjecting us to the fire of approximately 100 flak guns. The fire was extremely accurate for altitude and deflection and was tracking type. The lead Group took no evasive action." He later commented, "It's the first time in my life that I've really been scared. That flak was the roughest I've ever seen."

Elmer Brown wrote: "We went along the north edge of the Ruhr Valley ('Happy Valley') for about 20 minutes and we were in a solid cloud of flak explosions for the full time."

As the run began, Norman Sampson looked out of the large circular glass of the *Lady*'s ball turret and saw that "the flak was both barrage-type and tracking. Very close stuff. It was so thick I remember saying to myself, 'We will never get through this.'"

Merlin Miller recalls that "the black smoke from the antiaircraft fire up there was so thick the sky was hazy. I could hear the fragments from the flak shells hitting the plane like someone throwing rocks at it. You had a feeling of helplessness, frustration, and anger. You had to just sit there and take it."

George Hoyt felt "the flak just filled the sky, and it became quite frustrating not being able to answer the German fire directly."

The frustration finally got to Sampson, whose battle station left him more exposed than anyone. Miller remembers that "it got so bad Sammy pointed his ball turret guns straight down and fired a couple of bursts in the general direction the antiaircraft fire was coming from," and Sampson admits today that "it's entirely possible I did. You do some funny things to get rid of the tensions."

Up forward, the *Lady*'s officers weren't faring any better. Elmer Brown wrote that "Mac said he looked out once and I had my head down studying my maps very hard about that time. Also about that time I saw some fighters, so I called 'Enemy fighters at 10 o'clock' in a very calm voice. I knew damn well they would not attack us with all that flak around us, but we were all so scared I just wanted to see if I could still talk."

That evening Bob Hullar admitted his fears, too, confiding to his notebook that he "saw the light on flak. We hit the north end of 'Happy Valley.' The flak was thick enough to walk on. It really scared me today."

Bud Klint felt the flak was "so thick we were almost on instruments. It was, without a doubt, the worst flak I ever saw. At the time, I held little hope of ever seeing those White Cliffs of Dover again. One burst nearly saw to it as it shattered my windshield."

Off to the *Lady*'s left, in *Flak Wolf*, a shell came a lot closer to the cockpit than this. Elmer Brown recorded that "Woddrop had a 155mm shell come through the hatchway below him, then between him and the copilot, and out the top between the copilot and the top turret. They were very lucky it didn't explode."

Not surprisingly, the whole base was abuzz with talk of the incident that evening, and David Shelhamer was one of those who spoke directly with Woddrop about it. As he recalls, "At this particular period of time I was aware of bursts of continuous tracking flak that were occurring immediately in front of us. The Group lead was taking no evasive action whatsoever, so I pulled the flight out of the main formation. Woodie decided not to follow but stayed with the main Group, and that's when he got that thing through his ship.

"When we got back to the ground, Woodie came to me and mentioned what happened and showed me a hunk of something, a piece of fabric that he had behind his seat when the shell came through, and it was quite shredded. The shell came immediately behind his seat up through the top center of the fuselage, immediately in front of the top turret. And Woodie said to me, in so many words, something like 'Whither thou goest, I will go.'"

George Hoyt also heard all about it that evening from Sgt. Bill Watts, the man in the top turret: "I remember Bill telling me Woddrop called to him on the interphone, 'Watts, what is that cold draft behind me blowing on my ass?' Bill said, 'I don't know sir, but I'll check on it.' He came down out of his turret, saw a hole in the bottom of the airplane with the ground showing through, then looked up and saw a similar sized hole in the roof of the cockpit. He called to Woddrop, 'Sir there's a hole in the floor behind your seat and one in the ceiling. It looks like we got a direct hit from a flak shell but it didn't explode.' Woodie replied, 'Well, cram something in the hole. It's freezing my ass!'"

Woddrop's copilot, Grover C. Henderson, Jr., completes the story of the close call with an account still fresh after 45 years:

"We had just dropped the bombs and our leader had made a turn as soon as we did to try to get out of the flak. We were in a steep bank to the left and a large flak shell—from the hole it made we assumed it was a 155 mm—went through the bottom of the B-17 and came up through the floorboards squarely between my seat and Woodie's seat.

"Bill Watts was right behind us in the top turret and the shell centered the three of us. It passed through the only open space in the airplane and went out the top of the cockpit, taking the overhead radio equipment with it. It was amazing. We were in this 45-degree bank, and the shell must have been almost spent, going at the same angle at the end of its trajectory.

"The shell damaged Woddrop's parachute and he also had a canteen of water on the floor. It tore that up and spattered the water all over him. Woodie joked about it later on, saying he felt under his arm and it was all wet up there and he thought, 'Oh hell, they finally got me.' Then he realized it was water and not his own blood.

"When the shell came through it sounded like somebody hit the bottom of the airplane with a sledgehammer, or like an automobile crash. It scared the hell out of me. Then there was this terrific whistling sound from the air rushing up through the hole in the top of the cockpit, and Woodie told Bill Watts to cram something up there to stop it. Watts had on one of the big, heavy fleece-lined jackets, which was the winter-type jacket we used. It was a real heavy jacket. He pulled the jacket off, doubled it over, and crammed it in the hole in the ceiling, which was about 12 inches in diameter. But the wind sucked it right on through, and we had to fly all the way back with the wind whistling through that hole and coming up through the floorboards."

Miraculously, not a single 303rd ship was lost, though 14 suffered flak damage to varying degrees. And for all this, the Wing failed to get its bomb loads anywhere near the real target. Major Mitchell reported that "we took interval for the target. The 384th Group turned back to the left, which put the 303rd between them and the primary target. We turned left to get behind the 384th Group again and they bombed a little town, believed to be Coesfeld, Germany."

Bud Klint recorded that "We headed for Münster, but something else went awry. The lead bombardier of our Group synchronized on a small town to the northwest of the city—Coesfeld. He may have hit it, but most of the other ships in the Group were in a turn at the time of release, and it is hard to say where their bombs may have landed."

Elmer Brown described the fiasco best of all by writing: "The Wing got all screwed up before the target and the bombs were dropped everywhere."

The B-17s now turned for England, and as they did Bud Klint glanced out the *Lady*'s right cockpit window, where "the town of Münster was plainly in sight. It was plainly evident, too, that someone had hit the target, for there was a terrific column of smoke rising from the center of town."

Not until the 303rd got home under the protective wings of its P-47 escort, however, would the men learn how terrible a price had been paid for that "terrific column of smoke." When he found out, Elmer Brown wrote pensively: "Over 30 B-17s were lost. Only one plane came back out of one group."

Hullar fingered the unfortunate unit when he noted: "The 100th Group was wiped out by fighters in seven minutes."

The day had been a disaster for the Third Division, whose lead Wing—the 13th CBW—suffered 13 aborts en route to the target. Then its fighter escort failed to show—the 355th Fighter Group, given the critical leg over the target, never got off the ground due to fog. Finally, there was a snafu in the Second Division's North Sea diversion. The lead B-24 lost all electrical power, and with it the ability to communicate by radio. Despite frantic hand signals and flares, the rest of the Division followed the lead ship home when it aborted.

Thus, when the 13th CBW got to Münster, over 350 German fighters were waiting. They zeroed in on the low 100th Group—known as "The

Bloody 100th" forever after—shooting down 12 of 14 ships. They next took on the high 390th Group, destroying eight out of 18 B-17s. They turned to the lead 95th Group, knocking down five of 17 Forts. Four other Third Division bombers fell, plus another one from the First Division. Total losses were 30, plus three Category Es.

The price of daylight precision bombing was becoming prohibitive, but there was one more mission to go before "Black Week" would be over. The First Division would occupy the eye of the storm on this one, in what many consider the fiercest aerial battle of all time.

14

"Schweinfurt Again!"

Black Thursday, October 14, 1943

FOG COVERED ENGLAND the morning after Münster, and all over East Anglia the Eighth's bomber bases were quiet—there was no mission on. The losses the Eighth had suffered during its last three raids—88 heavy bombers, plus eight Category Es—far exceeded those of any comparable period, but Hap Arnold was still focusing on what had been achieved rather than its cost. On October 11th he sent General Eaker a telegram which confidently predicted a turn of the tide:

"The employment of larger bombing forces on successive days is encouraging proof that you are putting an increasing proportion of your bombers where they will hurt the enemy. Good work. As you turn your effort away from ship-building cities and towards crippling the sources of the still-growing German fighter forces the air war is clearly moving toward our supremacy in the air. Carry on."

This was exactly what Eaker and Anderson intended to do. The weather remained bad on October 12th, but that afternoon General Anderson and VIII Bomber Command put the finishing touches to a new plan and then briefed General Eaker on it. On the next clear day over the Continent the Eighth would return to Schweinfurt, the most heavily defended target in Hitler's Reich.

The attack appeared to be a clear military necessity. General Eaker was aware that the Germans were putting maximum pressure on neutral Sweden for its total ball bearing output, and that they were scouring the Continent for other sources of supply. He knew too, that the damage done to Schweinfurt on August 17th had not been that great, and photo-reconnaissance flights confirmed that Germany would soon have its most important ball bearing factories back up to full production.

Thus, if the Eighth was to destroy the Luftwaffe by cutting this strategic jugular, the time to do so was now, regardless of cost.

October 13th offered no opportunity. Fog continued to blanket England, and the still cloudy skies over Europe gave VIII Bomber Command only two options: Waste another day or try a PFF mission. General Anderson opted for the latter, but the effort was stillborn. Elmer Brown wrote that "we started out to bomb Emden, but it was recalled before we left England."

That afternoon, however, the picture started to change. There was no reason to think the fog over England would lift, but the 1600 weather briefing at Pinetree showed encouraging signs: It appeared the sky over Germany would be clear the next day. And so General Anderson put out the fateful order—there was a mission tomorrow and the target was Schweinfurt.

The word went through the chain of command as on any other operation, finally being passed by teletype to the bomb groups, where the plan would be completed down to the last detail. At Molesworth, Captain Mel Schulstad was one of those playing his part as Assistant Group Operations Officer.

"I was working with the squadron operations officers, telling them how many planes we needed from each squadron, and figuring out where they would be assigned. These discussions became fierce fights and got pretty personal because so much was at stake. I'd tell one guy we needed so many airplanes and crews from his squadron, he'd say it wasn't possible because of battle damage or whatever, and we'd go around and around till things were worked out."

This day Major Ed Snyder was slated to lead both the 427th Squadron and the Group, flying *Mr. Five by Five* as part of Jake James's crew—James, now a Captain, had taken over for Strickland's crew when they finished their missions and went home. Lt. Carl Hokans was the lead navigator and the lead bombardier was none other than Mac McCormick, who had traded places with James's regular bombardier, Lt. Walter Witt, for this all-important effort. Leading the Squadron's second element was David Shelhamer (he was a Captain now, too), while Hullar's crew flew the No. 6 position off Shelhamer's left wing in *Luscious Lady*.

The 360th was high squadron, but ships and crews from the 358th helped complete its formation. Among them were Lt. Roy Sanders's crew filling the No. 5 position in *Joan of Arc*, B-17F 42-29477, and Lt. Bill Fort's crew (with Lt. Calder Wise as Instructor Pilot again) flying the No. 6 slot in *Yankee Doodle Dandy*. The 359th filled six of the seven low squadron slots. The No. 5 position was taken by a crew on their fifth raid, Lt. Ambrose Grant's in the *Cat O' 9 Tails*, B-17F 42-5482, while the No. 6 position was filled by Flight Officer T.J. Quinn's crew in *Wallaroo*. Completing the formation in the tail-end Charlie position was Lt. Jack Hendry's crew in their favorite Fortress, the *War Bride*, B-17F 42-5360.

The 427th Squadron's Mr. Five by Five, *B-17F 42-29955, GN✪T, lead ship of the 303rd during Second Schweinfurt. Pictured are Capt. Jake James and some of the crew who led the Group on the mission. (Photo was taken ten days before "Black Thursday.") Bottom row, L-R, Sgt. F. Kuehl (not on mission); Sgt. F.B. Knight, ball turret; Sgt. J.J. Scheueuer, top turret; Sgt. J.E. Tripp (not on mission); Sgt. A.J. Hamilton, right waist. Top row, L-R, Lt. Paul W. Scoggins, (not on mission); Lt. R.W. Whitcomb, (not on mission); Capt. J.M. Strickland (not on mission); Capt. J.C. James; Col. Kermit D. Stevens, 303rd CO (not on mission); and Lt. B.K. Butt (not on mission).* (Photo courtesy Paul W. Scoggins.)

At the time these assignments were made, most of the men had no idea what lay ahead; only lead crews were put in the know as part of the standard mission routine. But Molesworth was alive with men in the early morning hours beginning to get the planes ready to go. It was impossible to see them going to work in the darkness and the fog, but as he toiled through the night, Captain Mel Schulstad kept an ear cocked for their first stirrings.

"When the ground crews got out to the hardstands, one of the first things they did was to start a portable gasoline engine. They used it to provide electric power to the airplane and lighting to work. You'd hear the noise as a simple 'putt-putt-putt' when the first engine was started, and then soon it would be joined by others and grow until you heard this 'hummmmm' coming in from the hardstands. To me there was always a terrible drama about it greater than the opening of Beethoven's Fifth, or the beginning of any symphony. When you heard that sound, you knew for sure that today men were going to die."

On no day of the Eighth's air war was this truer than "Black Thursday," October 14, 1943.

The 303rd's ordnance personnel bomb Vicious Virgin, *B-17F 42-5341, GN✪Q, up for a mission. One of the 427th Squadron's best B-17s, she later became the Hullar crew's second regularly assigned ship. Note the portable gasoline engine just to the right of the bomb dolly beneath the* Virgin's *nose. These were used to provide power at the hardstand for lighting and the aircrafts' electrical systems. Of their noise in the mornings before a mission, Mel Schulstad recalls: "When you heard that sound, you knew for sure that today men were going to die."* (Photo courtesy Mrs. Lorraine Shelhamer.)

The bomber crews were roused at 0630 for breakfast at 0700 and then the crews filed into the officers' and enlisted briefing rooms to get the word at the 0800 briefings. Mac McCormick already knew what the objective was, but there was no way to soften the blow which hit the other officers in Hullar's crew when the curtain was raised.

"Schweinfurt again!" is how Hullar began his notebook entry, and Elmer Brown started his diary for the day with exactly the same words.

Klint's reaction was visceral: "When they uncovered the map, and showed us going back to Schweinfurt again, it was like getting hit in the face with a baseball bat. Again, the keynote of the CO's introductory remarks was: 'Today's raid, if successful, can shorten the war by six months.'"

At the enlisted briefing, Miller remembers "sitting down with the rest of the fellows and wondering what that map was going to look like when they took the cover off. A couple of officers came in and someone yelled 'Attention.' We stood up, then sat back down, and my heart started beating a little faster as the suspense of waiting to see where that red line went up.

"Then they peeled the cover off the map, and we were a bit startled to see the line go deep into southern Germany again—to Schweinfurt. I knew then that we were in for a bad day."

George Hoyt recalls "being told at the briefing that the ball bearing factory at Schweinfurt, which we had knocked out on August 17th, was now back to 65 percent of production, and we were going back to destroy it again. This was *not* agreeable news."

Norman Sampson remembers feeling that "this mission was going to be the roughest of them all. They knew we would be back, and would be ready for us with more planes than they had the first time."

Intelligence estimated that the bombers would be facing 1100 German fighters, and since U.S. fighter escort couldn't penetrate deeper than Aachen, all the B-17s had to oppose this strength was their own numbers. The plan called for a mass assault on Schweinfurt with every bomber the Eighth's three divisions could muster: 360 B-17s and 60 B-24s.

The First Division was designated the first air task force and was to send its three combat wings—the 40th Wing leading, followed by the 1st and 41st Wings—on a more-or-less direct route to the target running just north of Aachen. The Third Division's three wings formed a second air

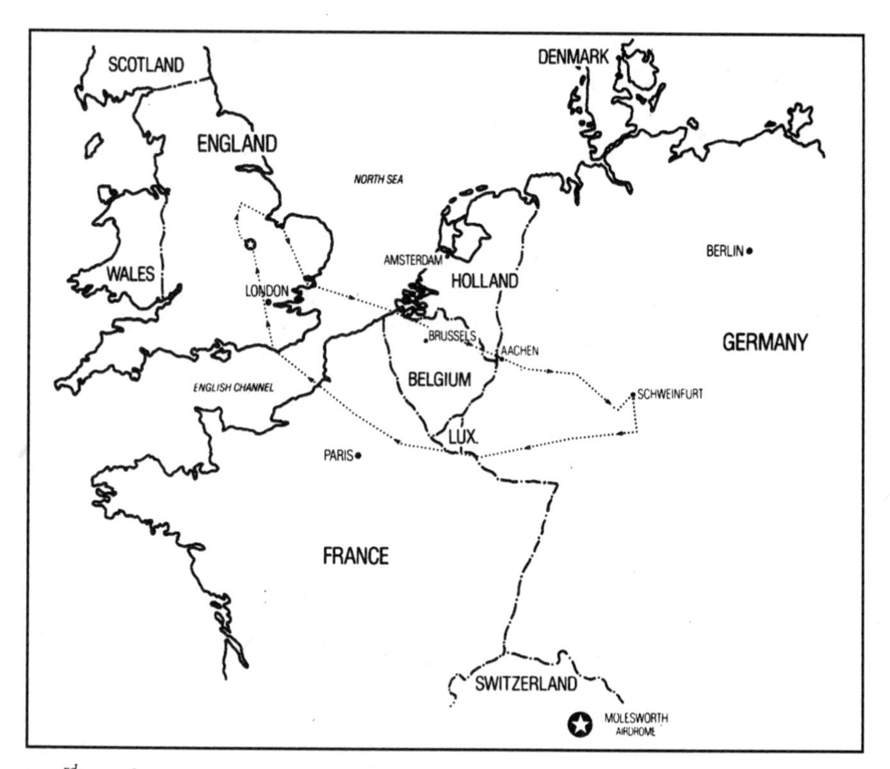

303rd Bomb Group Mission Route(s): Schweinfurt, October 14,1943.
(Map courtesy Waters Design Associates, Inc.)

task force. They would follow the First to Aachen at a 10-minute interval, taking a parallel track 30 miles to the south. During this time the two forces were to fly with division formations as near to line abreast as possible, creating a gigantic wedge going into Germany.

Beyond Aachen the First Division was to continue along its original track while the Third cut sharply south, flying along the Belgian-German border. The First Division was to draw off most enemy fighters as it flew to a point north of Frankfurt am Main, where it would take a southeasterly course to its IP west of Würtzburg and southwest of Schweinfurt. The Third Division would continue south just inside the German border halfway down Luxembourg, where it would swing on a course due west to its IP northwest of Würtzburg, following the First Division into Schweinfurt.

Meanwhile, the 60 B-24s of the Second Division, constituting a third air task force, would use their faster cruising speed and greater fuel endurance on a more extended route to the south. They were to arrive over Schweinfurt shortly after the Third Division dropped its bombs.

After the plan was outlined, John Doherty remembers that "the officer who gave us the briefing said, 'Gentlemen, look to your right. That man won't be coming back.'"

George Hoyt "did not see anybody flinching," but in the officers' briefing, David Shelhamer's keen eye took in a correlation between the length of the tape going out to the target and another, human line.

"Our Catholic Chaplain, Father Ed Skoner, would wait at the end of the briefing hut to pray over any crewmen who wanted a blessing before a mission. The longer the ribbon, the more crew were back there kneeling in front of the priest. On the day of Second Schweinfurt, the men who had been unfortunate enough to go on the first Schweinfurt mission lined right up like a British queue. It was really something to see."

The bomber crews arrived at their aircraft around 0935, an hour and ten minutes before takeoff. Once there, Hullar, Klint, and Rice conferred with the *Lady*'s ground crew and went through their preflight routines. Miller and the other gunners "checked all our equipment, our guns, and everything we could so we could feel a little bit better, a little bit safer."

As Norman Sampson explains, "Each gun on our plane meant the life of our crew."

Extra ammunition was especially important. The *Lady* had two extra ammunition boxes installed in the tail gunner's compartment, which were filled to overflowing, but more vital still was the extra ammo secreted aboard in portable boxes. On First Schweinfurt the *Lady* had about 12,000 .50-caliber rounds aboard; today she was crammed with over 14,000 rounds.

Other crews followed suit. David Shelhamer remembers that "On Second Schweinfurt I loaded about half a ton of extra ammunition aboard because, having had the pleasure of going on the first mission, I knew quite well what to expect."

John Doherty likewise explains that "We had been on the first Schweinfurt mission and most of us took new gun barrels and extra ammunition, a considerable amount more than we normally carried, which was a real lucky thing because we had used it all up by the time we got back."

Ed Snyder would have intervened, had he known. Weight loading for this mission was especially tight. Elmer Brown wrote that "This time we had a bomb bay tank on the left side. The bomb load was three 1000-lb. bombs and three 100-lb. incendiaries."

And yet, as Snyder recalls: "The gunners, everyone who fired a gun for that matter, tried to take additional ammo. It was done in an emotional, unscientific way. They didn't ask, 'Will this airplane fly with this much weight on board?' The airplanes were overloaded with fuel and bombs as it was, so we had to go out and take ammunition off because many guys would have creamed themselves otherwise. It made a lot of them real unhappy but they would have stalled out on takeoff and never gotten off the ground. The old B-17 did all kinds of things they said it couldn't, so they figured it always would. But there was a limit."

Hullar's and Shelhamer's crews escaped their Squadron Commander's notice, and there finally came a point where all the preparations, good and bad, ended in the foggy stillness. As on August 17th, the crews were left to wait for the word—to go or not to go.

In *Luscious Lady,* Bud Klint put his thoughts down on paper: "The fog still hangs on, and we're still trying to get a raid off. Today we're scheduled to go back to Schweinfurt. With 1100 fighters in range of our course, and 56 flak guns over the target alone, it promises to be rough.* If we get it off. Our first trip to that target was the roughest one I have been on to date. Maybe we'll hit a new high today. My only prayer is that we come through as well as we did then."

*There were, in fact, far more flak guns than Klint knew about. On First Schweinfurt, August 17, 1943, the Germans had sixty-six 8.8 cm flak guns in eleven, six-gun, single batteries defending the City. By October 14, 1943 the Germans had increased the defenses to the equivalent of 22 single batteries of 8.8 cm guns—broken down into four single batteries, four double batteries, and three triple *Grossbatterien*—for a total of 126 8.8 cm flak guns. In addition, there were two, four-gun batteries of the larger caliber and even more dangerous 10.5 cm flak guns. Thus, there were 134 rather than 56 flak guns defending Schweinfurt on Black Thursday. According to a German historian, at this time Schweinfurt was "the best defended town in Western Europe" in relation to its size. Friedhelm Golücke, *Schweinfurt und die strategische Luftkrieg 1943*. Schöningh, Paderborn, 1980, p.173.

Manning these batteries were approximately 250 regular *Luftwaffe* personnel and another 1500 schoolboys, fifteen to seventeen years old, euphemistically called *Luftwaffenhelfer* or *Flakhelfer* ("Flak Helpers"). Five hundred of the *Luftwaffenhelfer* served the 8.8 cm guns directly while the remaining 1000 worked their associated radar, optical tracking, and battery control switchboard equipment.

In *Yankee Doodle Dandy*, Lt. Bill Fort was similarly preoccupied: "we were scheduled to go to Schweinfurt, which had a reputation for being a real tough target. The weather was terrible, and I never thought we'd get off the ground, figuring they would scrub it most any time. But they put us in the air anyway."

This decision was made after a weather report was passed to VIII Bomber Command from a British Mosquito reconnoitering over Germany. The sky was crystal clear. General Anderson put out the fateful, final order: "Let the bombers take off."

All over East Anglia the Eighth's bomber bases now rang with the noise of radial engines as Wright Cyclones and Pratt & Whitney Twin Wasps coughed and came to life on more than 400 B-17s and B-24s. They taxied from their hardstands and waddled to their runways, turning onto them and disappearing into the foggy air one by one.

At Molesworth, Major Ed Snyder and Captain Jake James's crew was first, speeding down the main east-west runway in *Mr. Five by Five* and climbing aloft right on time at 1045. Three bombers later, Captain David Shelhamer got *Vicious Virgin* up at 1047. The timing of these takeoffs is all the more remarkable when one learns how they were done. David Shelhamer describes the procedure:

"The airdrome was fogbound, and the ceiling was zero. When we taxied to the takeoff end of the runway there was a jeep there that would get ahead of the aircraft about 100 to 150 feet. It would then start right down the center of the runway at about 15 to 20 mph, while the aircraft followed directly behind it. We established a direct course down the center of the runway, locked the tailwheel, and set the gyrocompass to 'zero it out.' The gyrocompass was extremely sensitive and would react immediately to any change in course. The jeep then pulled away from the aircraft and we went on full power."

At 1048 Bob Hullar got a signal that the runway was clear and *Luscious Lady* began her run. As the overloaded *Lady* lifted off, the crew encountered the first of the mission's many hazards. George Hoyt remembers that "we were violently buffeted about by the propwash of the plane that took off seconds before us. I always sweated this out, since it took place only 30 to 50 feet off the end of the runway, with treetops and rooftops whipping by just under our wing."

The Group takeoffs continued until 1057, when Lt. Jack Hendry got the *War Bride* in the air, the Group's 20th and last ship. Now the 303rd faced the difficult task of assembling above the cloud cover and joining up with the 379th and 384th Groups to complete the Wing. Captain James's Group Leader's Report describes this phase:

"We climbed to 6,500 feet before breaking out of the lower overcast. Then we continued our climb through a cloud-free space of about 700 feet, at which time we began to climb through another cloud layer, which we broke out of at 9,000 feet. Our assembly was made at 11,500 feet and we left the base on course for Eyebrook on time. On leaving the base we had spotted the other two Groups of the Combat Wing. We got into Combat Wing Formation slowly en route to Eyebrook. We experienced a lot of high cloud around the English Coast and gained and lost altitude intermittently to avoid this cloud. We left the English Coast about seven minutes late, probably due to turning south to go around a very heavy cloud layer at the intended point of departure from the English Coast."

Though assembly went reasonably well for the 41st Wing, the same cannot be said for the First Division's other two wings. The 40th CBW was to head the Division, but when its three groups and those of the 1st CBW sorted themselves out in mid-Channel, the latter was in the lead with the 305th Group of the 40th CBW flying an unorthodox "double high" position in its formation—the two remaining groups of the 40th Wing, the 92nd and 306th, were trailing behind. The First Division was flying into battle in disarray, its lead wings in poor condition to meet the German onslaught.

The Third Division's assembly went more smoothly; its seven groups crossed the English coast in proper order on track 10 miles to the south of the First Division. In contrast, the B-24s of the Second Division never assembled at all—of the 60 Liberators scheduled for the raid, only 29 came together above the cloud cover. This paltry force was ordered to abandon the main effort, and instead was fated to conduct an ineffectual feint towards Emden.

Aborts reduced the weight of American numbers further as the B-17s crossed the Channel. The Third Division lost 18 out of 160 aircraft. The First Division's ranks were cut from 164 to 149. In the 41st Wing, the 384th lost two of 19 Forts, the 379th pressed on with 17 ships, and the 303rd had two of 20 B-17s return.

Of the cross-Channel flight, Elmer Brown wrote: "We switched the gasoline over and dropped the [bomb bay] tank before we reached Holland. We had P-38 escorts at first, and then P-47s."

George Hoyt and Merlin Miller also saw the twin-boomed fighters, which were from a new unit in the ETO, the 55th Fighter Group. The Lightnings had been ordered to escort the Fortresses as far as Flushing, Holland, but their mission was canceled at the last moment because of problems the big fighters still had operating in the ETO's frigid air. From what Hullar's crew observed, however, some of these eager fighter pilots either "didn't get the word" or chose to ignore it. Unhappily, George Hoyt recalls that "little if any enemy fighters were spotted during the P-38 escort time."

The primary defense would remain the Fortresses' firepower, and readying this defense was the final prebattle ritual. In the *Lady*, Norman Sampson had long since gotten into the ball turret.

"When we got off the ground, the turret was locked into place with the guns pointed towards the tail. As soon as possible I entered the turret through the door, put the safety belt on my back, and locked the door from the inside with two handles. I decided to wear a chest-type parachute. It made for close quarters, but I felt more comfortable with it. If something happened I could just roll back, unfasten the safety belt, open the door, and roll out. I put the power on, and was free to move the turret all around to protect the bottom on the plane."

Dale Rice got into the top turret, perched on its "bicycle seat," turned on its electrohydraulic power drives, and was soon rotating his twin .50s to protect the top of *Luscious Lady* fore and aft. Merlin Miller also took to his bicycle seat in the tail, and prepared himself to protect her rear with his flexible twin guns:

"Crossing the Channel we had fighter escort, but I knew they couldn't stay very long, and it was going to be a long fight in and out if the last Schweinfurt mission was any indication. So I sat there, anticipating things, wondering what was going to happen, how the mission was going to go, and I got a little tense.

"Then all of a sudden I realized the battle was going to start pretty soon, and I said to myself, 'Merlin, get mad, get angry.' I started thinking about the Germans, what they were trying to do to us, what I was going to do to them, anything to stir up my blood and get anger coursing through my system, anything to take away from that feeling of fear I otherwise would have had. By this time I had figured out that it was difficult for a person to feel two emotions at the same time. In combat I thought I was better off being angry than afraid.

"Pretty soon I heard a call on the interphone. It was Hullar saying, 'Crew, check your guns.' So I hand-charged each machine gun and fired a short burst from each one, and said, 'Tail guns okay.'"

The process was repeated all the way up to the *Lady*'s nose as each man at a machine gun tested it and reported in. Hullar's crew was as ready as they would ever be for the fight which lay ahead.

At the head of the First Division, the 1st CBW crossed the Dutch coast at 1250 in the company of P-47 escorts from the 353rd Fighter Group. At 1311, the 41st CBW also crossed the enemy coast, and Elmer Brown wrote that "P-47s took us into just south of 'Happy Valley,' or rather the town of Bonn. We saw several dogfights."

The Thunderbolts stayed with the Fortresses approximately 25 minutes, but even as they did, B-17s were leaving the 41st Wing from the

384th Group. Four of their Fortresses turned back due to various problems over enemy territory. Their loss left the 384th terribly exposed. It now had only three B-17s in its high squadron, four in its lead, and six in its low.

While this was occurring, Captain David Shelhamer had a few brief moments to observe events in the wings ahead.

"The visibility was unbelievably beautiful. You could see the Swiss Alps about 200 miles to the south. I was personally aware of at least four B-17s swinging out of formations from other groups heading right straight down to the Alps. With the fighter opposition that was only moments away from us, it is very questionable whether they were able to make it to Switzerland."

A short time later, Lt. Bill Fort saw what Shelhamer was talking about: "There was a wing ahead of us about a couple of miles. They seemed to be hit real hard. We saw a number of planes explode, and others going down."

Meanwhile, George Hoyt saw "a lone Me-109 paralleling our course on our starboard side about four o'clock high. I swung my flexible .50-caliber to frame him in my ring sight. He appeared to be 800 yards or better out, and I hesitated to see if he would turn to attack. In a split second he did, and I hosed him with two good short bursts, giving a good allowance for the trajectory drop at this range. Much to my surprise, he appeared to be hit. He abruptly broke off his attack by throwing one wing up and made a long dive out of my field of vision. I breathed a sigh of relief."

This single encounter was like a raindrop before a cloudburst; after the P-47s left, the storm struck with indescribable fury, progressively engulfing each wing of the First Division as it flew to meet the enemy.

The first two combat wings were decimated. The 1st CBW and its tagalong, the 305th Group, lost 13 of 55 aircraft. Twelve of these came from the 305th, leaving it with three ships. The 40th CBW fared even worse. On the way to Schweinfurt its 92nd Group lost five of 19 and the 306th Group 10 of 18.

The story of the 41 CBW's ordeal is less well known. What happened to the Hell's Angels and to the other groups in the Wing is a tale worth telling, for on Black Thursday no part of the sky was still.

Elmer Brown's diary sets the tone for what took place around 1345, just east of Aachen: "We had fighter attacks from that time to the target and deep in France. We were fighting for over two and a half hours constantly." They came, quite literally, in the hundreds: FW-190s, Me-109s, Me-110s, Me-410s, Ju-88s, Do-217s, and "at least 3 FW-189s," a twin-engine, twin-tailed reconnaissance plane whose presence shows how completely the Germans were pulling out the stops.

Official 303rd reports describe it this way: "At least 300 E/A were seen. Generally speaking the E/A seemed to have tried just about everything...The attacks by E/A were made from all around the clock generally from the high and level positions. Most of the T/E [twin-engine] fighters attacked from the side—chiefly on the nine o'clock side. FW-190s attacked from the nose. The E/A carrying rockets made their attacks from the tail. Several T/E fighters slow-rolled through the formation to break it up. The rocket carrying E/A attacked in trail usually three at a time...The attacks are reported as particularly vicious and the pilots gave every indication of being experienced and clever. The Yellow Nose FW-190s were particularly effective in their attacks."

Up to this point in the Eighth's air war, Major Ed Snyder felt that he had pretty much seen it all. Though he missed First Schweinfurt, he had flown so many other missions that they had almost become routine. Now he realized he was in the center of something truly exceptional.

"There was a tremendous amount of aerial combat activity going on the whole time. On all the missions I had been on before, there always seemed to be goodly periods when nothing was really happening. You just anticipated the enemy being there. On this mission there was always something going on, flak or fighters all the way in, and all the way out."

The impression was the same from front and rear. In the *War Bride*'s tail, John Doherty remembers that "going in, we'd be picked up by a group of fighters. When they got to where their fuel supply was getting low, they'd go down and you could see another bunch coming up. It was kind of a comical situation in one way; the one bunch would leave you, and the other bunch would start coming up at about 10,000 feet. That's the way they kept us all the way into the target.

"You could also see these flak guns shooting, because you were looking right down into the barrels when they'd shoot and you could see the fire. And then you would sit there and say to yourself, 'Well, I wonder where that one's gonna hit? I wonder how close that one's gonna be?' You could see them all along in different places like that."

Ed Snyder continues: "You could see fires all over the place, and parachutes dropping all over the place. You can't believe the number of aircraft that were being shot down, the number attacked, and the different kinds of aircraft that were involved. The Germans came at us with just about everything in their inventory. On one occasion I saw a German fighter come through our formation that was badly hit. The pilot must have been wounded or killed because he jammed the controls. As he went past under our wing his left wing was where his nose should have been. He just skidded through the formation sideways."

Just behind Snyder in *Vicious Virgin*, David Shelhamer had a different kind of head-on encounter: "I very vividly recall this one FW-190 who obviously was attempting a head-on ram on my aircraft. He was not firing at all when he was well within his own range. Our rate of closure had to be about 400 to 450 mph, and I eased the aircraft down, realizing full well that he was attempting a ram. It required split-second timing, but my timing was good. I pulled the aircraft up and he went about 50 feet under me. I will never forget the green scarf that German pilot had around his neck."

Frightening as these experiences were, many veterans, particularly the pilots, say that they were simply "too busy to be scared." As Ed Snyder explains, "I was concerned about my own safety, but that had to take its place at the back of my mind as long as we were not on fire, or incapable of flying. I was too busy trying to make everything in the formation work properly to get too concerned. I don't mean this to sound macho or anything. That's just the way it was. There was just too damn much going on."

Every crew was now locked in the ultimate test of teamwork. Hullar's fought with hardly a moment's pause. Lt. Witt was at the *Lady*'s single .50-caliber in the nose; while he fired at fighters making head-on attacks, Elmer Brown was "crouching behind him looking to see which way they would break. I would jump to either the left or right cheek gun and try to get a few rounds off." In his diary, Brown also wrote that "Jerry would cue up at about one-thirty [o'clock'] time and again and attack our lead group."

In the cockpit, Hullar and Klint were consumed by the twin tasks of staying in formation and taking evasive action. At debriefing they complained of "too much variation in airspeed over enemy territory. Varied from 130 to 170 mph." Their job was made doubly hard by the need to take evasive action, as Bud Klint explains:

"We were tucked away in the No. 6 position of the lead squadron, and while this offered a definite advantage from the standpoint of enemy fighters, we were hemmed in on all sides by other B-17s. The Forts were packed in a tight defensive formation, and everyone was using violent evasive action, which was an added hazard. Bob and I were trying to get the most evasive action possible without ramming one of our own planes.

"Evasive action was largely a psychological thing. Probably our chances of kicking the airplane into the path of enemy shells were almost as good as kicking it out of their path. In addition, those violent gyrations certainly made an unstable platform for our gunners, and disrupted their aim. In spite of this, it was a tremendous psychological boost for everybody on the airplane, and it was virtually impossible for Bob and me to sit in the cockpit and hold the airplane straight and level while we were under attack by enemy fighters. We just felt that we had to do *something*.

"We both were on the controls almost constantly doing evasive action. The sky was literally filled with aircraft, and we were trying to hold as tight a defensive formation as possible. Bob was responsible for aircraft to our left and I was responsible for those to our right. When he would kick the plane to the right it was up to me to kick it back when we got too close to some other aircraft. There was no method or plan to the evasive action we took. We just did whatever we felt was necessary—kick the rudder, drop a wing, pull the nose down, anything that occurred to us that we could do in the limited airspace we occupied.

"The B-17 had no boosters on the controls. It was a very stable airplane and easy to fly on a straight-and-level path. But when you did strenuous evasive action it really required a lot of muscle and a lot of perspiration. Evasive action would wear one pilot out in a very short period of time."

Bob Hullar offered a more succinct explanation of these efforts in his notebook, where he wrote that "Bill and I worked like Hell on evasive action."

Amazingly, however, Hullar found time for humor during the worst of all this. In his diary, Elmer Brown noted that "They were firing a lot

"Bill and I worked like Hell on evasive action." Interior of a B-17F cockpit. (Photo courtesy Mrs. Jean J. Hullar.)

of rockets at us. We sure took plenty of evasive action. In the midst of the battle the flying got a little rough, and on the interphone Hullar said, 'Oh, my aching back,' just as innocent as he could be. Rice said, 'Is someone calling the engineer?'"

When he wasn't trading quips from the top turret, Dale Rice was firing nonstop at enemy fighters, and using his position's superior visibility to call them out to the other gunners. Merlin Miller vividly recalls Rice's help with some Me-110s making head-on attacks:

"I heard Rice call out, 'Fighter over the top!' Then a 110 came over the top of our tail upside-down. He seemed to hang there for a moment and I could see the pilot so clearly that if he hadn't been wearing an oxygen mask I could have recognized him on the street. But the plane was going too fast for me to react.

"So I called to Dale and said, 'Rice, one of those fighters comes over the top again, tell me where and when and maybe I can take a crack at him too.' Rice said okay and the next time an Me-110 came over I was ready. My guns were pointed up, and when I saw the German I held the triggers down. He flew through the bullets, and pieces flew off of his airplane."

It wasn't all give for Rice, however, as George Hoyt explains: "At one point Dale ducked down out of his Plexiglas dome in the top turret to get some reload belts of ammo on the floor. While he was down a couple of slugs went through the Plexiglas right where his head would have been." Rice fought the rest of the mission with the wind screaming through the splintered dome.

For Hoyt, the big shock was the large number of rocket attacks from the rear:

"I kept thinking, 'How can the Jerries have so many rocket-firing Me-110s?!' They would hang out beyond the effective range of our .50-calibers and fire their rockets into our formation.

"As soon as I saw the flashes when they fired, I would call to Hullar to 'Kick it around!' You could track the rockets as they came in. It always seemed to me that Hullar would dive out of the way just as the rockets zoomed over the top of the radio room.

"After rocketing us the Jerries would bear in with guns blazing, and that was when Dale, Merlin, and I really zeroed in on these boys. Through all of this it was very hard to hold one's fire to short bursts to keep from burning out the machine gun barrel."

The first encounter with rocket-firing fighters also took Merlin Miller by surprise: "I looked up and saw two Ju-88s sitting in the back wingtip-to-wingtip. I wondered 'What are those bastards up to?' because they didn't seem to be closing in. They just sat there. All of a sudden I saw big smoke puffs from under the wings of both planes, like they'd been hit. Then, to my dismay, I saw four 'gizmos' coming at us which looked like black softballs. I pushed down on my mike button and yelled to Hullar, 'Kick it, kick it!' He bounced the airplane around and the four rockets exploded above us.

"I got used to it after this, and would tell Hullar to get ready to either climb or dive when the rocket ships lined up. When I said 'Kick it!' he would abruptly maneuver up or down, depending on which way he was heading at the time. As things turned out, none of the rockets actually exploded too close to us."

While all this was going on, George Hoyt "could feel through the soles of my flying boots Norman down in the ball turret firing away at Jerry. He had a lonely and detached station down there with his Sperry computing gunsight." The *Lady*'s evasive action didn't make his job any easier, as Sampson himself explains:

"The gunsight was a frame which had two lines, with one on each end of a box. You turned handles to 'frame' an enemy fighter between the lines as it came in. When you had the fighter framed, you shot at him by pushing the firing buttons, which were on top of the handles. I searched mostly to the sides and the tail because there was a lot of blocked-out area up front from the body of the plane and the propellers. The turret had stops that wouldn't let you fire there. Also, the head-on attacks came too fast for the sight.

"I didn't have any trouble spotting the enemy fighters. You could see them way off like vultures before they came in. But shooting them down was another thing. Just about the time I got a good frame on one of them, someone would call out for evasive action."

Back in the waist, Marson and Fullem were busy at their hand-held guns, and the image of them in action is something George Hoyt will always carry with him:

"Through the open radio room door I caught glimpses of Chuck and Pete firing away at their right and left flexible guns. Chuck Marson was a bit superstitious about the door, and insisted that I keep it open despite the slipstream that blasted back through my open upper hatch. But seeing him and Pete at their guns gave me a great feeling of confidence. Our tracer ammunition had been eliminated before this mission, since the powers that be had decided that it threw our aim off in aerial combat. So all of us at the flexible guns were on our toes to use our sights to the best of our ability. You had to be good to hit anything with these guns, and both of them were.

"Through the open radio room door I caught glimpses of Chuck and Pete firing away at their right and left flexible guns." The interior of a B-17F waist looking forward toward the open radio room door. In the center is the frame and upper portion of the ball turret. (Photo courtesy Mrs. Jean J. Hullar.)

"Marson was a real pro in every way. The .50-caliber had a recoil plunger and spring device which impacted against the metal discs that were on the backplate. Marson had taken the gun's backplate off and he added some coins to the discs to make the plunger and spring go faster. This upped the rate of fire, but it also increased the danger of the gun 'blowing' or the barrel burning out unless you stayed down to very short bursts. Marson knew just how to handle it. With him at the gun I knew I had nothing to worry about.

"Pete didn't have the familiarity with guns that Chuck had, but he was a quiet, tenacious gunner who never complained or backed down in any way. Bob Hullar always showed complete confidence in him, and that was good enough for me. Pete was a great guy."

In the *Lady*'s tail, Merlin Miller was performing his one and only real duty: "My job was to protect the airplane. The fun really started after our fighter escort left. I looked back and saw maybe six to eight enemy fighters behind us, so I called to the crew, 'Fighters, six o'clock.' I watched them as they strung out one behind the other, and started in. Then I called, 'Bandits, six o'clock.'

"When they got to within 500 to 700 yards I fired a couple of short bursts to discourage them from coming in closer. But they came in hard, and fast, and that's when I started to get really angry, because the closer they got, the more dangerous they were. From here on in it was just a matter of shooting and calling the fighters out to the other gunners as they came in and went past.

"We all had to work together. I relied on Sammy to see below and behind us, and on Rice to see above and behind us. Marson and Fullem would keep me posted too, and I would call out any fighters I saw, particularly when more than one was coming in at the tail so that Bob Hullar could skid the plane sideways a bit and maybe give one of the other gunners another shot. We worked together that way all the time. It was the only way we could survive."

So it went all during the flight from east of Aachen to the IP between 1345 and 1437, each crew doing their utmost to stay alive. In the low squadron, Lt. Jack Hendry moved to the No. 6 slot from tail-end Charlie after another bomber aborted, but the Group's records show him waltzing the *War Bride* with *Wallaroo* all through the fighter-filled sky that day. A formation diagram note says: "Ship #029 and #360 flew Nos. 3 and 6 position alternately due to intense enemy opposition."

It was near the IP that enemy opposition reached its greatest intensity, for the Germans knew this was their last chance to stop the 41st Wing. Their all-out assault finally bore fruit against the low 384th Group. Three B-17s went down—at 1425, at 1430, and at 1431. Shortly thereafter, Bud Klint witnessed other losses from German rockets.

"As we began our bomb run, I saw three '17s from the Wing ahead completely disintegrate and fall earthward in flaming shreds. This served as my introduction to the 'rocket gun,' which played an ever-increasing part in the European air war."

This happened as the Wing's three groups shifted out of the combat box into a line-astern formation in order to begin individual runs on the target—the Kugelfischer ball bearing plant in the center of Schweinfurt. The 379th went first, the 384th second, and the 303rd last.

Captain James wrote in his Group Leader's Narrative that "At the IP we made a left hand turn and made our run on the target on a magnetic heading of 40 degrees at 24,000 feet using AFCE...Flak in the target area was moderate and accurate. Enemy fighter attacks began on us before we reached the IP and continued to the target." In another report, he added: "I have never seen the like of those fighters in my life."

Ed Snyder recalls these moments well: "As we got into the target area and got onto the IP, we were attacked very heavily by twin-engine

German fighters. One of them put a cannon shell or two into our radio room and blew a good portion of the radios up. But it didn't disturb the bomb run.

"Back in the tail I had a pilot, Lt. John Barker, riding as an observer to help me with the formation. He was going crazy because there were these 109s, 110s, 210s, and 410s back there shooting at us. He kept screaming, and I finally did have to cut him off. I can remember Mac very calmly saying, 'Please tell him we can't do anything now. I'm on the bomb run.' There was only one thing on Mac's mind, and that was to put those bombs on the target."

Lt. Barker later offered this assessment: "It was a hell of a day. I'll bet there were over 300 enemy planes."

The twin-engine fighter that hit *Mr. Five by Five*'s radio room may well have been an Me-410 that attacked at 1438. It came in from about four o'clock level to about 400 yards and the right waist gunner, Sgt. Daniel Harmes, fired about 100 rounds in two bursts. He told the interrogator: "He went down out of control—parachuted out about 1000 feet below, and plane was observed to crash." Sgt. Harmes got credit for a kill.

The 303rd's mission file contains so many combat reports from this phase of the fight that a chronological account of what took place would be meaningless. But the experiences of individual crews, as reflected in the records and their own words, do provide one way of sensing what Black Thursday was really like. What follows is a selected chronology, set against the larger backdrop of events in the Wing, showing *part* of what happened to six of the 303rd's crews: Bob Hullar's, Lt. Bill Fort's, Lt. Ambrose Grant's, Lt. Jack Hendry's, Captain Lake James's, and Captain David Shelhamer's.

At 1442, a Do-217 went after *Luscious Lady*. Dale Rice told the interrogator: "He came in from seven o'clock high. He was about 1200 yards away. I kept firing as he came in. At approximately 700 yards he began to smoke. He burst into flames—two men bailed out of E/A. Plane went down." Rice got a "probable."

At 1445, the 379th dropped its bombs on Schweinfurt. Just after they went, two of its Forts were lost. Moments later the 384th Group dropped its bombs. Three of its B-17s were shot down, but the action was so hectic they weren't seen leaving the formation.

Just before the 303rd dropped at 1446, calamity struck Lt. Ambrose Grant's crew in the *Cat O' 9 Tails*. The crew's engineer, Robert Jaouen, tells what happened:

"For some reason that I do not remember, I was a waist gunner that day. Just before the target, I turned to yell something to Woodrow [Woody] Greenlee, who was on the other waist gun, and there was a bright flash between us. Woody was hit in the right side of his face by an exploding 20mm shell. Ed Sexton, the Radio Operator, came back and bandaged Woody's face and eye. We didn't have much time for Woody, as we had to keep shooting in hopes the fighters couldn't tell one of our crew had been hit.

"They threw everything at us, and the fighters swarmed down in wave after wave. It was a chaotic time for the entire crew, both in trying to protect the plane and in keeping it flying. The *Cat O' 9 Tails* was like a sieve. Someone stated there were around 200 holes of various sizes. Ed Sexton, being a devout Catholic, often signed himself. I had seen him do this before, and it hadn't bothered me. However, things being as they were this day, it scared the hell out of me. I thought that maybe he knew something that I didn't, and that things were even worse than they appeared."

In the same seconds before the 303rd's bombs dropped, a Ju-88 went after *Luscious Lady*. From the left waist, Pete Fullem saw it coming in "from about seven-thirty o'clock, slightly high, but almost level."

He said later: "I started firing at him at 600 to 700 yards. Fired several bursts at him as he was coming in. At 200 yards he began smoking, burst into flames, and went down."

Fullem was credited with a damaged fighter, but the *Lady* was hit, too. George Hoyt recalls that "Pete threw four .50-caliber slugs from his left waist gun into the leading edge of our left horizontal stabilizer. The armor-piercing bullets ripped through the inner tail structure and came out through the rear, tearing away a section of the elevator some three feet in diameter, and leaving a gaping hole with shreds flapping in the slipstream. I can remember Merlin calling Bob Hullar, saying 'Hey, we got a hole in the tail right beside me big enough to crawl through.'"

Miller remembers the incident as well and says, "It didn't seem funny at the time, though we joked with Pete about it later. It was an easy thing to do in the heat of battle."

Meanwhile, it was Mac McCormick's moment at *Mr. Five by Five*'s bombsight. As Ed Snyder recalls, "Mac laid the bombs right on the target."

Bud Klint later learned that "The bomb run was perfect. Strike photos showed our bombs completely blanketed the target area."* The 303rd then made a sharp right turn off of the target, returning to its high slot in the wing formation.

---

*The 303rd's target was the same Kugelfischer ball bearing plant that the Group had bombed on August 17th. According to Georg Schäfer, whose father owned the factory, the October 14th bombing could not have come at a worse time. The night before the factory had received a precious shipment of fuel oil which completely filled a large tank on the plant premises. The bombs exploded the tank, creating a massive fire that could be seen for miles.

"The bomb run was perfect. Strike photos showed our bombs completely blanketed the target area." 303rd strike photo, October 14, 1943.
(Photo courtesy Mrs. Lorraine Shelhamer.)

"As we came out to the south," Ed Snyder continues, "you could see the fires from the aircraft that had been downed, and the smoke plumes going up. And, of course, we were still under fire."

The next minute was Merlin Miller's. At 1447 an Me-109 with a 30mm cannon slung under each wing came in from five o'clock high. These large-caliber guns were truly lethal; three hits, on average, were all it took to kill a B-17.

Not surprisingly, the 109 was, Miller remembers, "one of those fighters you saw every once in a while who was determined to add a B-17 to his score. He came in straight and level with our tail. He throttled back, and I could see his prop slow down. I could see him fishtail as he started to aim at us.

"There wasn't much doubt who he was going to shoot at, but to put it bluntly, I sneezed first, before he could pull the trigger. I hit his plane just at the base of the left wing where it joins the fuselage, and it blew the wing completely off the airplane. He immediately flipped upside down and spun away. I didn't have time to see whether or not the pilot bailed out." Miller got credit for a kill.

At 1449 the Germans claimed the first 303rd Fortress. It was *Joan of Arc*, flown by Lt. Roy Sanders's crew. To her left in *Yankee Doodle Dandy*, Lt. Bill Fort saw the events leading up to *Joan of Arc*'s demise.

"The sky was covered with quite a number of planes, and debris from planes that had exploded. It was real tough, close fighting. We were being hit from all sides. The slower German fighters were hitting us from the back with 20mm cannon shells and rockets. The Me-110s kept picking on this one plane off our right wing. They knocked all the fabric off its elevator and rudder, and eventually set it on fire, but apparently this didn't slow the plane down much, since they stayed with us another five to ten minutes.

"Then this German plane came back of us and under us, in a vertical bank that almost cut our wing off, and set the plane on fire again. It was mostly smoke, coming from the bomb bay, and then it flared up. We tried to move over to get out of the way of this plane, and shortly after I looked around and it was gone." Others in Lt. Fort's crew reported *Joan of Arc* being hit in the tail by an Me-110 rocket and going down on fire. Ten chutes were observed.

At 1450, the 379th lost another Fort. An Me-109 collided head-on with the second element leader of the Group's lead squadron, destroying the 109 and cutting 15 feet off the B-17's right wing. The Fort dropped out of formation; five chutes were seen.

While this was occurring, Lt. Grant's *Cat O' 9 Tails* was being assailed by FW-190s. Two with yellow noses came in at six o'clock level and Sgt. Francis Anderson opened up on them 600 to 700 yards away from the

Cat's tail. One FW broke away but the other kept coming in, and at 300 to 400 yards Anderson's fire took effect. The 190 nosed up and seemed to stop in midair. It went into a spin and as it spiraled down, the pilot bailed out. Sgt. Anderson got credit for a kill.

Another yellow-nosed FW-190 attacked *Luscious Lady* at 1452. From the right waist, Chuck Marson took aim as it came in from five o'clock high. He later told the interrogator:

"I opened fire at 400 to 500 yards, bullets going into engine—Parts of engine began falling off—He began to smoke—His prop stopped completely—Then he rolled on his back and fell down."

Miller, Hoyt, and Rice were all certain the FW-190's engine was knocked out and Marson got a "damaged."

In the 14 minutes between 1453 and 1507 there were no less than 15 recorded attacks on the 303rd. One occurred at 1500, when an FW-190 came in to attack the *War Bride* from "six-thirty o'clock high to about 100 yards." In the top turret, Sgt. Loran Biddle "gave him 200 rounds. E/A started to smoke—rolled over on his back and pilot bailed out." Sgt. John Doherty confirmed from the tail and Biddle got a kill.

Germans weren't the only ones hitting the silk. The 379th Group lost a ship "at 1510 hrs, with No. 2 engine out and a large hole in the wing, going down under control. Five chutes seen." A short time later another B-17 was seen "with No. 3 engine feathered, dropping out of the formation."

And, in the most telling observation of all, the 379th's Group Leader, Lt. Colonel Louis M. "Rip" Rohr, stated: "There were reports of other B-17's going down, but action at that time was so intense crews could not keep track of them. As many as 10 to 15 B-17s were seen going down at one time."

It was sometime during this phase of the fight that Merlin Miller accounted for another Me-109. Neither he nor George Hoyt reported it, but both remember the incident.

As Hoyt describes it: "An Me-109 flew up our tail so close that I was afraid his prop would chew it off. The vertical stabilizer was in my way, so I could not safely shoot at him. But Merlin let go with a burst from his twin tail guns that riveted the pilot back in his cockpit seat just as he was releasing his canopy and starting to push himself up to bail out."

Miller remembers it differently. "An Me-109 came in at six o'clock, at just about our altitude. I fired at him, and I think I hit him a couple of times. Pieces of his canopy flew off and he stopped shooting. But he kept coming in, closer and closer, nearer than any other fighter had ever come to us before. I just sat and watched, wondering what was going on. He was wobbling around, about 20 to 30 feet below us.

"Then all of a sudden I got this terrible sinking feeling. I realized all he had to do was pull back on his stick and hit his firing button, and I'd be a dead duck and the rest of my crew with me. So I put about 20 rounds into the cockpit, pounding it in on top of him. I remember the canopy flying off, but I don't remember the pilot trying to get out. My only thought was, 'He won't do it to us now!' After that he just continued on, wobbling away below us."

This fighter didn't get the *Lady*'s tail gunner, but another almost did. Miller recounts that "I was back there with my head on a swivel, looking right to left, left to right, up and down, all the time. My head was never still for a moment watching for fighters. The tail position had windows to the left and the right, and it had a thick, flat bulletproof window in the back I looked through to see my post and ring sight. You could see quite well.

"Suddenly, out of the corner of my right eye, I caught a twinkling, like a flickering neon sign. I leaned back and turned my head to see what it was, and the next thing I knew I was lying on one of the ammo boxes, half on it and half on the floor wondering what the hell happened. My face didn't feel right. I rubbed it and realized that my oxygen mask was half off, so I put the mask back on. My right shoulder tingled and I had a sore spot above my right ear.

"I was still trying to figure it all out when I saw that my side windows were gone. The twinkling had to be a fighter sliding in at nine o'clock. He blew out my windows, and almost blew my head off. The lump on my head had to be from bouncing off the ammo box. I worked my right arm a bit, checked my shoulder, and found no holes in my uniform, so I must have been hit by flying Plexiglas from the windows.

"I really got angry at this point, because I was now getting a blast of cold air. I wanted that fighter to make another run so I could take a crack at him. It was a futile wish, but I hoped someone else would set him on fire."

Other crews in the Group continued to score. One of the victors was Sgt. Howard Zeitner in *Yankee Doodle Dandy*'s ball turret. At 1515 an FW-190 passed to the rear of the high squadron 600 yards out. Zeitner started firing at six-thirty o'clock, and at five thirty o'clock the 190 went into a spin and exploded, with the tail and rear fuselage breaking off.

Sgt. Zeitner said afterwards, "He turned his back to me as he banked around and made a perfect target. I let him have a long burst and he tumbled down and blew up."

In the same minute an Me-110 made a tail attack on *Vicious Virgin*. The German came in at six o'clock level, firing rockets from 300 yards out. In the *Virgin*'s tail, Sgt. Robert Humphries opened fire, hitting the 110 at the same time its rockets exploded under the B-17. The 110 went straight down with its left wing and engine on fire. Sgt. Humphries got credit for the crew's second "damaged" fighter.

It was also at 1515 that the *Cat O' 9 Tails* got still another Me-110. Sgt. Anthony Kujawa told the interrogator:

"I was in the right waist gun position firing the guns. My regular place is top turret, but it was out of commission and I had gone to the waist guns because [the] waist gunner had been wounded. I was firing his gun.

"An Me-110 was out around four o'clock. He was just standing there shooting at us when I came to the waist gun position. I turned the gun on him and began firing away. He was very near. I kept firing. Then he began smoking and he burst into flames. He began to spin towards the ground. He was disintegrating in midair." Kujawa got credit for a kill.

So it went, interminably. To Merlin Miller the return flight was "a constant battle, back and forth, fighters from all directions, lots of head-on attacks, and lots of attacks from the tail. I checked my ammo and realized that all I had left was in my auxiliary boxes behind the main boxes. So I called Marson, and told him to bring me back some ammo.

"When he came back, I told him to put the ammo in the boxes with the bullet points out, because that's how the belts feed into both guns. He made an appropriate remark, because that was like telling someone to put sod down with the green side up."

Chuck Marson came back at least two more times with extra boxes of ammo, and George Hoyt did too. "Later during this grueling dogfight, Merlin ran low on ammunition. I grabbed a wooden box full of .50-caliber belts, clipped on a portable oxygen bottle, and headed back to the tail with it. After I crawled past the tailwheel, he greeted me with a big wave of his hand. I gave Merlin the spare ammo, and felt relieved, as I knew he was the most capable defender of our most attacked point on the plane, the tail."

Hullar's crew stuck it out, joking to ease the tension, but finding themselves slowly worn down by the sights, sounds, and emotions of endless combat. Miller's other memories evoke "the way it was" better than any work of fiction:

"I remember hearing from Rice in the top turret. He was getting a blast of cold air just like I was. It didn't help his temper any more than mine, but he stayed pretty calm, and joked with the fellows.

"During a crew check I said, 'Everything's okay except the sons of bitches are trying to freeze me back here. They shot my windows out.'

"Rice said, 'Yes, I got a hole up here. They're trying to freeze me, too.'

"Marson piped up with, 'I don't know what the hell you fellows are griping about. Me and Pete don't have any windows at all—never did.'

"It was quiet for a moment, then Rice said, 'Don't pay any attention to him, Merlin. He's just mad because we're having all the fun.' And we were all still shooting at fighters as they came in."

"It was really cold up there. I was uncomfortable. All of a sudden I wanted a drink—drink of water, drink of beer, drink of whiskey, a drink of anything, I didn't care, I was thirsty. Why did I get so thirsty when I couldn't drink? Then I thought about smoking a cigar, but I knew I couldn't do that either. I tried it once at altitude, but when Hullar found out how I lit the thing with my oxygen mask on he threatened to kill me if I ever tried it again. He could smell the smoke through the oxygen system all the way up in the nose. All I could do was shoot.

"Off to my right I saw a B-17 flying alongside us. He got hit. His outboard engine on his left wing, No. 1, caught fire, burning back through the wing. I started yelling to myself, 'Come on, you guys, get out of there, get out of that plane!' It started down in a shallow dive, burning more all the time, and I counted the chutes as they came out. I felt a big relief when I saw the tenth chute pop out, because now at least they'd have a chance. A few seconds after that, the burning wing folded over the top, the plane went into a steep dive, and blew up. But I'd seen that happen before, so it wasn't that much of a shock.

"I was getting more and more uncomfortable. My nose itched and I couldn't scratch it. My oxygen mask felt like a dull knife gouging at my cheekbones. My shoulder still ached a bit, and I felt generally bad all over. I'd been in this one position too long, and I was hoping this would get over soon.

"I was still shooting at German fighters, which were coming in with pretty fair regularity. They were shooting down a lot of bombers. They weren't hurting our Group so much for some reason, but I could see others all around going down.

"Off to my right I noticed a Ju-88 pull up under a bomber and hang there, shooting into the bottom of the bomber. The ball turret gunner was shooting back. I could see both his guns flashing. They got each other. I saw the German burning, and cannon shells exploding inside the ball turret, flashes inside, so I know the fighter got him.

"The bomber peeled off, burning badly. As it peeled away below me I could see a glow behind the pilot and copilot from flames inside, and I could also see flames coming out the top of the radio room. The plane was burning from behind the cockpit clear through the bomb bay and out through the top. I started thinking, 'Get the hell out of there,' like I did on the one a while earlier, but before I could even think it the plane cleared the formation and blew all to pieces. Outside of what looked like the four engines dropping, there was nothing left.

"I took a deep breath, was glad it wasn't me, and continued watching for fighters, shooting at them when they came in, worrying about my ammunition, and wanting to get this raid over with. It began to seem like it was lasting forever when I heard Brownie say we were about 20 minutes from the French coast."

Each bomber crew has a similar story of courage and teamwork, a tale that must be told. The *War Bride* lost her No. 3 engine and Lt. Hendry could not feather the prop. She fell out of formation and was trailing below the Group, a prime target for fighters. John Doherty offers this account of her fight:

"Of that second Schweinfurt mission, what can be said? That raid will stand out in my mind for as long as I live. By all dimensions we should have gone down, but we made it back with a tremendous effort by every man on the crew.

"You couldn't single anybody out, but the pilot that day went far beyond what most pilots would do. I remember one time the call came over the intercom, 'Prepare to bail out!'

"I called back and I said, 'Don't give up!' That, I remember. The plane was vibrating so badly, at any minute a wing could have fallen off. But it didn't. Nobody did more for any group of people than Hendry did that day. That's why we all put him in for the Silver Star when we got back.

"I started to run out of ammunition. I called for the waist gunner to bring me back more. And the action kept going but it never seemed to get there. So I called on the intercom again and I told them I was leaving the tail. I was out of ammunition.

"Jim Brown, our radioman, called back and said, 'Stay where you are, I'm coming.' So Brown got me ammunition back there. But I had one gun, the left one, that was completely out [of ammunition]. The ammunition was all used up in the belts. The other gun had eight or ten or twelve rounds left. When we put a new box of ammunition in, we used to like to string it into the old bandolier. We took one shell out and put one from the new one in and it was just like a chain, it kept going.

"But I had this left gun completely out, so I had to get down on my back and crawl down in there where the gun was to get this new ammunition in. It didn't take too long, and as I was crawling back out I thought I would kind of roll over to the other side and link up my other box while I was at it. But something told me I had better check up and see what's happening, and as I looked up there was a plane coming in on us.

"This guy had got in there in our blind spot, and nobody had seen him. He had a kill—he knew he had a kill. He was getting right up to where he was going to give it to us with 'both barrels,' and there wouldn't be a question. He'd have us dead to rights.

"He was close enough to where I could see the silhouette of his face. I grabbed both them guns and shot right into the windshield. Never aimed, just grabbed them and shot. The windshield splattered, and I must have hit the pilot right in the face. The plane tipped, went into a dive, the wing broke off or something, and I forget the rest of it."

"The rest of it" is revealed in a 1602 combat report. The fighter was an Me-109 that had closed to 50 yards. Sgt. Doherty "gave him 200 rounds" and the 109 "went into a spin and then went over and over. Pieces of plane came off and it seemed to break. John Doherty got a "probable."

By this time, the *War Bride*'s No. 3 propeller had broken off its shaft and she had fallen completely away from the formation.

"We were using the clouds as much as possible for cover from the fighters, " Doherty remembers. "And our navigator, Lt. McNamara, done a tremendous job of getting us *out of there*. We were by ourselves, and he got us out of enemy territory as best he could, directly."

The remainder of the 41st Wing crossed the French coast at 1655, south of Bologne. Now the bomber crews had to get back to England and land safely with damaged aircraft in weather conditions that rivaled the morning's. At 1720 the Wing crossed the English coast between Beachy Head and Dover flying at about 18,000 feet.

Captain James reported that "From the English coast to the base, we worked our way through the clouds and landed at the base at 1758 hours."

In his diary, Elmer Brown described this return with unintended poignancy: "This was another foggy day when even the birds were not flying, but we made another instrument takeoff and landing."

Bill Fort remembers that "Eventually we got back to England in about the same kind of weather that we left in. Luckily, we got down OK, along with a number of other planes. We were very fortunate not to get a bullet hole. Others were landing in any field they could find that was open."

There was one 303rd bomber that didn't land at all—Lt. Grant's *Cat O' 9 Tails*. On the return flight, the *Cat* had also strayed from the Group due to damaged engines, and she was little more than a flying wreck by the time she arrived over England. Robert Jaouen describes her last moments:

"We were struggling along, trying to keep the engines going. Several passes were made over the field, but we couldn't see the landing lights. Either the fog obliterated them or they weren't turned on, as there were reports the Germans might try to follow planes in.

"During these landing attempts, as we were circling, a break came in the clouds. We saw we were passing by a church steeple, and the pilot struggled to gain more altitude. Besides the crippled engines, we were running low on fuel, our control cables had been damaged, and keeping up altitude was near impossible, as was landing. Finally, after once again struggling to gain altitude, Lt. Grant ordered the crew to bail out."

"First out was Woody. Being semi-conscious, he was ejected with a static line, and I was ordered to follow him down. After leaving the chaotic conditions of the plane and the mission, I never have experienced such silence, before or since, as floating down through the clouds.

"I watched Woody disappear into the fog and was certain I knew where he was. After landing in a cow pasture, I started looking for him in the opposite direction from where he landed and I never did find him. He landed, apparently revitalized by the cold air, not knowing if he was in Germany or England. He was found by a farmer trying to read his compass. The farmer took him to a nearby air base on the back of a tractor. He was sent to an English Hospital and when well enough was sent to the States, where he underwent extensive plastic surgery.

"My landing was my first good break of the day, as I hit the mud near a gate where the cows had been standing. Thus, I couldn't have had a softer landing, except for the inconvenience of a little manure on my clothes. When I arrived at a farmhouse, Chester Petrosky [the ball turret gunner] had entered a few minutes earlier. The people were very nice, but stayed a long ways away from me. I finally deduced it was from the 'cow perfume' I'd picked up in landing. Eventually a constable picked us up and took us to a Canadian Air Force base, where we were hospitalized for the night, given sedatives, etc."

Lt. Grant's crew all landed within a four-mile radius, not far from the *Cat*'s point of impact near Riseley, a small English village 10 miles south of Molesworth. Her end made an indelible impression on a young English boy named John Gell, for the *Cat* came down in his family's backyard:

"I was a young lad nearly six years old, just old enough to be aware of what was going on. We lived in one of a gathering of cottages, my parents, myself, my two baby brothers, and my grandma. We had been watching the Forts come home that miserable evening, when we heard this particular B-17 making a lot of popping noises. I remember going to the front door as it passed over our group of cottages. A very loud crash was heard soon after it passed over, and on looking through a back window a cloud of dust was visible, with a large metal object nearly 80 yards away.

"The *Cat* came down in our back garden. She broke up on hitting an oak tree, which ripped off the starboard wing. The port wing separated from the fuselage when it chopped an elm tree off eight feet above the ground. The fuselage finished in two pieces 150 yards away in an adjoining field.

"My father was a special policeman who was involved with many plane crashes in the area. He and the wife of a neighbor went—holding

The wreck of the 359th Squadron's Cat O' 9 Tails, *B-17F 42-5482, BN✪W.*
(Photo courtesy John T. Gell.)

hands—to the wrecked fuselage, not knowing what they would find inside. To their amazement, nobody was in it. My father put out a small electrical fire, and gathered up some used dressings. Soon the MPs arrived and darkness fell.

"The wreck was taken away two days later, on October 16th. Some days after, my father noticed something lodged in the crook of the oak. It was one of the *Cat*'s propeller blades. I still live in that cottage today, and I still have that prop blade."

The *Cat* was not the only 41st Wing bomber abandoned over England. The 384th Group lost three in this way, plus six shot down over enemy territory. Three of its bombers landed away from its home base, and only the Group's lead ship returned to Grafton Underwood! The event had an enormous impact on the 384th's ground echelon that is well recalled by John F. Bell, a Lieutenant then serving as Assistant Squadron Engineering Officer in the Group's 547th Squadron:

"Our ground crews were waiting and waiting at their hardstands. And *nothing* was coming back. It was a pretty sad looking sight. These people had worked their butts off preparing these aircraft over a long series of missions that summer. They really thought of themselves as part of a team with the flight crews. And when the other half of the team didn't come back, it was a hard thing to take."

The 379th Group paid a high price, too. Six of its ships were lost and one landed away. Lt. Colonel Rohr summed up the day's events as well as anyone when he reported: "This, in my opinion, was an extremely rough mission."

At Molesworth, the mood that evening was far more upbeat. The Hell's Angels had come through the roughest mission ever with minimal losses and much to be proud about. Thanks to McCormick, the Group's bombing had been superb, and the prolonged aerial combat yielded a bumper crop of fighter claims. When all the crew interrogations were completed, the 303rd sought credit for 20 fighters destroyed, four probables, and 13 damaged.

There were other, more personal reasons for celebration as well. Lt. Carl Hokans and Lt. Walter Witt had both beaten the long odds against completing a 25-mission tour. Elmer Brown wrote that "The boys took Hokans' pants off and he was running around without any at interrogation." Two other Group crews also marked this mission as their last. There were, of course, others in the Group affected by the loss of the one 303rd crew. That evening Lt. Ralph Coburn wrote, "Sanders went down. Good boys, we'll miss them."*

For Hullar's crew, the feelings were mixed. McCormick took satisfaction in his accomplishment, saying afterwards: "I think we did the job right today."

Norman Sampson summed his opinion up in a word—"Rough!"—and Bob Hullar was equally taciturn when he wrote: "Rough do this trip."

Bud Klint thought about the losses: "I saw what seemed to be a fabulous number of planes go down, and later I found out how right I was."

Elmer Brown remembers that he was "very impressed by the number of planes going down. It was just terrible to see all those B-17s on fire and to know our people were being killed. It really hit home that the high command thought we were expendable. All we could do was fight the Germans as best we could." In his diary he noted: "Our plane shot down five enemy fighters and we shot up more ammo than any other plane in our squadron."

Merlin Miller recalls thinking at the time, "'The raid's over. We've done it again. We're all still alive.' It was something I wouldn't have bet on a few hours earlier. We shot up all our ammo. Even the spare boxes in the radio room were empty. There were less than 150 rounds left in the whole plane. I must have fired at least 1000 rounds from each tail gun, and Rice reloaded his guns a number of times too. When I came out of the tail to the waist the brass was knee-deep. We figured we shot down at least five German fighters, and though we really didn't keep count, some of the other crews tried to give us credit for a few more. We used up more ammo than any other plane in the Group."

*Lt. Sanders' right waist gunner and tail gunner were killed in action; the rest of the crew became POW's.

Pete Fullem took some ribbing about the holes he shot in the *Lady*'s tail and was rather sheepish about the incident. He felt even worse about the duel between the Ju-88 and the ball turret gunner, which he had seen just as Merlin Miller had. Miller remembers that "Pete was sitting on his bunk comparing notes with me. He talked about wanting to shoot at that Ju-88, but not being able to. I know he felt bad about it, but that was just the way it went sometimes."

Hoyt didn't want to think about any of it. "I wasn't shaky, but I was badly fatigued, thoroughly zapped. I wanted to forget it all, to wash it out completely. So when someone said we should go to the 'NAAFI,' I was all for it. NAAFI stood for the British 'Navy, Army, Air Force Institute,' but it really meant the enlisted bar on our base, the 'NAAFI Club,' which the British kept open for us. As I recollect, all the enlisted men went, and we had quite a time. They had a kind of potato beer there—you had to watch for the sediment in it—and they later broke out some Canadian ale. By the time the evening was done I had washed a lot of the day out."

The "someone" who suggested the NAAFI was Merlin Miller. "I went to eat my evening meal with the rest of the crew, and a bit later on in the barracks I said to Rice, 'Let's go over to the NAAFI and have a couple of beers.'

"He said, 'That's a damn good idea. Only I'm going to have more than a couple! He really liked the mild and bitters that they served warm. So we wandered over there and drank a few beers, and we had some cookies too. And after a while we all headed back to the barracks.

"Rice looked at me and kind of grinned, and he said: 'Merlin, you better get a good night's sleep, 'cause you know, we might have to do this all over again tomorrow.'"

The following days brought a sober reassessment at all levels of the Eighth. The losses had simply been too great. Second Schweinfurt was to prove a true turning point in the daylight bombing campaign.

15

Schweinfurt Postmortem

October 15–19, 1943

IT WAS NOT long before the Eighth's bomber crews discovered the full extent of Black Thursday's bad news.

On October 15th, Elmer Brown recorded that "Sixty B-17s were lost yesterday. We were in the first task force composed of nine groups; 45 B-17s were lost by this task force. Our Group lost one."

These numbers established that the First Division had indeed fulfilled one part of the October 14th plan. It had drawn the greatest number of enemy fighters, though there were still plenty left over for the bombers that followed.

The Third Division bombed Schweinfurt's factories five minutes after the 41st Wing left the target area; the balance of 15 B-17s lost had come from its ranks. Total losses amounted to a stunning 19 percent—of the 291 B-17s that pressed on to Schweinfurt, nearly one in five failed to return.

Nor was this all. Seven bombers were written off as Category Es and many of the planes that *did* come back were out of commission for two to seven days undergoing repairs.

A total of 139 B-17s suffered battle damage, including *Luscious Lady*. Bud Klint recorded that she "suffered a broken wing spar, a shattered top turret, and a wide assortment of holes which sent her off the field to a Unit Repair Depot for a major overhaul. The *Lady* even had to be patched up before she could be flown to the Depot."

While the ground crews and repair units tackled the huge task of getting the Eighth's aircraft ready for the next raid, the combat crews tried to unwind. The day he recorded the Eighth's losses, Elmer Brown also wrote: "We have a 48-hour pass but we officers didn't even leave the post. Just laid around and relaxed."

In contrast, the crew's enlisted men did spend their time off-base, as George Hoyt recounts: "The next day we enlisted men set out on a 'special mission' to London, a two-day pass. These passes, with their escapades, provided very important relief for the stress that built up from high-pitched missions like Second Schweinfurt."

At the same time, America's leaders were trying to manage the public relations crisis that had grown from the October 14th losses. The toll, when combined with the aircraft lost on the preceding three missions, amounted to 155 B-17s. It was a serious military setback, for no combat organization could long endure losses on such a scale, and the military problems were made to appear worse by a series of gaffes President Roosevelt committed during a press conference on October 15th.

The President inadvertently confirmed the loss of 60 bombers on the mission and then stated that the Eighth could not afford to lose that number on every operation. He confused the issue further by denying that it was losing that many aircraft an every mission, and then tried to offset the losses by mentioning the 100 German fighters the Eighth had claimed were destroyed on the raid.

The President also mentioned the severe damage inflicted on Schweinfurt's ball bearing factories, but he was unable to convey the impression that a true knockout blow had been delivered. And, in fact, this objective was not achieved, despite the excellent bombing by both the First and Third Divisions and the fears of German leaders such as Albert Speer, Nazi Minister of Armaments and Munitions.

Three days later it was General Arnold's turn to fumble during a press conference. He was greatly concerned about the impact the loss of 60 B-17s would have on the public's attitude toward the air war, but his comments were poorly conceived. He opened by stating that high losses were inevitable against an objective such as Schweinfurt, and then implied that even a loss ratio as high as 25 percent might be acceptable. His closing, exaggerated comment—"Now we have got Schweinfurt"—offered neither solace nor reassurance.

Back in England, Eaker and Anderson also sought to quell public unrest, and to this end VIII Bomber Command called upon a surprising source: Major Ed Snyder. He recalls the occasion very well:

"I was called down to Bomber Command Headquarters after Second Schweinfurt to make a propaganda-type broadcast to the folks back home. I understood that Bomber Command was interested in getting one of the group leaders who participated in the mission to give a personal account. It was done on the radio, and was sent on to the broadcasting systems in the United States.

"I think the point they were trying to get across was that this was a very important mission and, yes, we had all those people go out on it, and it was quite hairy, but even for the guys that didn't get back there was some hope. I believe I was put on mainly as an eyewitness who saw a lot of parachutes come out, who could say that just because we lost all those airplanes, it didn't mean no one was going to make it back."

"I had no reservations about making the broadcast. I realized the gravity of the situation with us losing that many airplanes, but I felt this was something that just had to be. You couldn't win the war without taking some losses, and this happened to be one of the tough ones. I always had the feeling that the losses were justifiable some way or the other. I had faith in the people who were planning these missions and I believed they wouldn't have sent us there unless they really thought it was necessary, and that they could get us back with reasonable safely. And you have to remember that we were practically the only Americans fighting the Germans at the time. The Invasion wasn't until June of 1944."

Despite the determination of men like Snyder, it was now absolutely clear that the Eighth needed new resources to make deep penetrations with acceptable losses. Until new escort aircraft—the P-38 and P-51—became available, the Eighth would be restricted by weather and limited fighter range to shallower raids into Occupied Europe and Germany.

In the week after October 14th, the Eighth was incapable of even this kind of effort. On October 18th Elmer Brown wrote that the Group "started out for Düren and was called back at the English Coast. Weather."

But this explanation merely heightened doubts among the press and the people back home and the crisis of confidence continued. It would not be cured two days later, when the Eighth set out a second time for Düren on a mission that would prove a costly fiasco for both the 41st Wing and the 303rd.

16

"I Could Even See England"

Düren, October 20, 1943

OF ALL THE 303rd's crews roused out of bed at 0730 for the mission on October 20, 1943, few relished the day more than the "short timers" of Lt. Jack Hendry's crew. For three men in the crew—Lt. Bernard McNamara, the navigator; Lt. Richard Webster, the bombardier; and Sgt. James Brown, the radioman—this was the end of the line, mission number 25. For three more—Jack Hendry himself; Sgt. Loran Biddle, his engineer; and Sgt. John Doherty—it was mission number 24.

The Army Air Force was always alert for favorable publicity—no more so than in the idle days after Second Schweinfurt, when public opinion about daylight bombing hung in the balance. Thus, when an Associated Press newspaperman arrived at Molesworth a few days after Hendry's heroic return from that mission, his crew's situation seemed a perfect success story in the making. *This* was one of the crews that was going to make it, and they would do so in a way that made great newspaper copy.

After meeting Lt. Hendry and the test of the crew, the reporter learned of Hendry's fondness for the dangerous "diamond" position in the squadron formation and put together a story entitled: "'Tail-End Charlie' Has System for 'Bringing 'Em Back Alive.'"

Sgt. Jim Brown was especially looking forward to his last raid. Twenty-seven years old, he had enlisted in the war and had declined a Stateside posting as a gunnery school instructor because he felt his proper place was in a combat billet overseas. He now had very strong reasons for getting back home to the small town of St. Mary's, Georgia. He had a fiancée whom he planned to marry as soon as his combat days were done. He had a good business, which he and his older brother Charles had built up. And he had his widowed mother, for whom he felt a great responsibility as the

Lt. Jack Hendry's crew in front of their favorite ship, the War Bride, *B-17F 42-5360, VK✪Q. Bottom row, L-R, Sgt. John Arasin, waist gunner; Sgt. John J. Doherty, waist/tail gunner; Sgt. James J. Brown, radioman; Sgt. Olwin C. Humphries, ball turret gunner (KIA 7/30/43); Sgt. Howard L. Abney, tail gunner (WIA 8/19/43); Sgt. Loran C Biddle, engineer. L-R back row, Lt. Calder L. Wise, copilot; Lt. Hendry, pilot; Lt. Kruse, Instructor Pilot; Lt. Bernard T. McNamara; navigator; and Lt. Richard E. Webster, bombardier.* (Photo courtesy John W. Hendry, Jr.)

sole unmarried son. Brown had even written to her on October 18th, saying he would "have good news soon."

Brown had worked out very well as Jack Hendry's radio operator. Hendry saw him as "one of my crew's two mainstays. Loran Biddle was the other one, but Brown was the oldest and the most mature. I considered him a good friend, and I had the very highest regard for him."

John Doherty held Brown in equally high esteem: "He was very intelligent, and he seemed to have a very sound and basic understanding of life. He was a person who would always go out of his way to help you, and he had a very positive, dynamic, magnetic-type personality. He was the one we all seemed to gather around. He was the leader."

Everyone on Hendry's crew was happy for Jim Brown when they arrived at their aircraft. The *War Bride* was still under repair, so the crew was assigned to B-17F 41-24629, but no one seemed to mind. The bombers were to have continuous Spitfire and P-47 escort to and from the target, and "it was supposed to be a milk run," Jack Hendry recalls.

"We weren't expecting any trouble," John Doherty adds, "and when we got into the plane I shook hands with Brown. This was his last trip."

The First Division was leading the Eighth on a PFF strike against Düren, a "factory-railroad centre" to the east of Aachen, Germany. The 41st Wing was Division lead, with the 379th lead group, the 384th high and the 303rd low. The 360th was the Group's lead Squadron and the Group Leader was Major Walter K. Shayler, the Squadron CO.

While *Luscious Lady* was under repair, Hullar's crew drew a 360th ship named *Dark Horse*, B-17F 42-29498, in the tail-end Charlie slot of the high 359th Squadron. Don Gamble, newly promoted to Captain and fresh from a week's R&R ("Am anxious to get going after rest home and laying around"), led the low 358th Squadron in *Sky Wolf* along with Lt. Bill McSween.

To their left in the No. 3 slot was Lt. William Hartigan's crew on their seventh raid in *Charley Horse*. The Squadron's second element was led by Lt. Calder Wise and Lt. Bill Fort in *Yankee Doodle Dandy*, with Lt. Hendry's crew to their left in the No. 6 squadron slot.

The Group got 19 B-17s off beginning at 1100 and had a reasonably good flight over the Channel, though Major Shayler had to radio the 379th Wing Leader to slow down so the 303rd could tuck tightly in formation before crossing the enemy coast.

Things continued to go well as the formation entered France; Captain Gamble noted: "Spits take us in part of the way and P-47s pick us up. Form good close top cover."

But John Doherty "did not feel comfortable as we got into formation. To me it seemed that we were boxed in tight. Most of the missions we flew were tail-end Charlie. It gave us more flexibility. We were looser. In our position on this day it seemed that we were boxed in tight when it came to maneuverability."

The trouble began when the formation ran into clouds. Bill Fort recalls that "We were having a time keeping our formation together in the clouds," and Elmer Brown recorded that they "got so bad we could not even fly formation, so we turned and started for home."

It was in these conditions that the Germans struck. Major Shayler reported that "We encountered a cloud bank which stretched upward to 28,000 feet or higher at Cambrai and we climbed to 27,500 feet without reaching the top. While climbing, we turned north and it was at this point that we were hit by enemy fighters."

There were many witnesses to what followed, but the accounts vary, as is inevitable when a group falls victim to a well-executed surprise

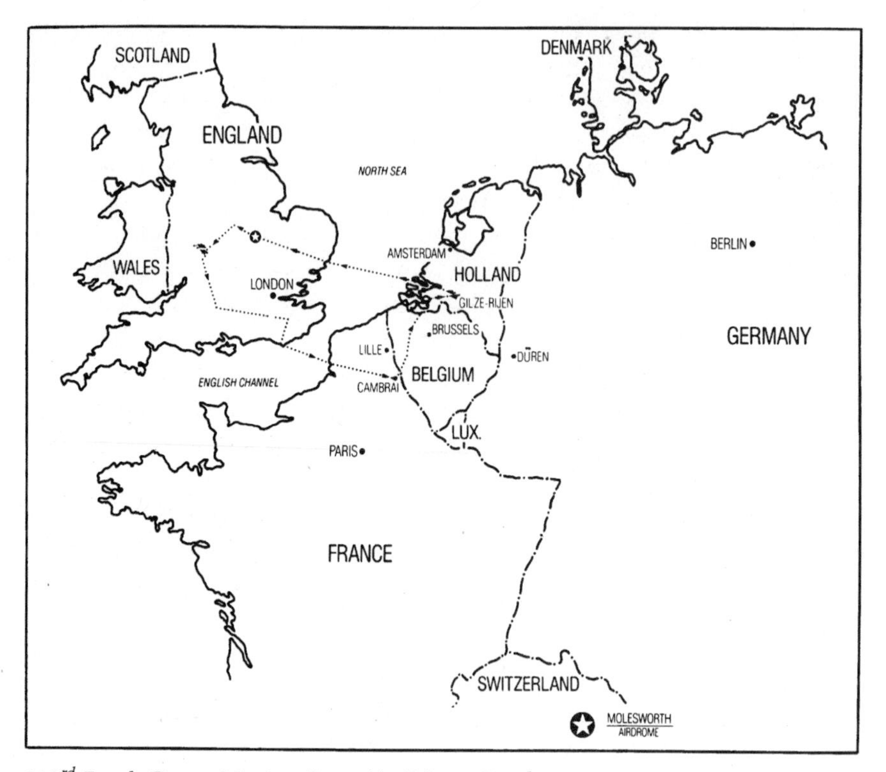

303rd Bomb Group Mission Route(s): Düren, October 20, 1943.
(Map courtesy Waters Design Associates, Inc.)

attack. Elmer Brown reported that "about 15 to 20 Jerry planes dropped out of the clouds at about the twelve-thirty high position, and took us by surprise. They made just one sweep, and knocked down two planes from our low squadron."

Major Shayler wrote: "These E/A came down through the thin layer of high cloud, and were not seen until within 500 yards of our Group. It is estimated that there were six to 12 Me-109s and they were painted gray to resemble Spitfires."

Bud Klint saw "a group of about 20 of Goering's yellow-nosed FW-190s pour off the top of that cloud bank, make one pass at our Group, and dive for home. In that one sweep, they took two Forts out of our Group."

Those with the most immediate experience of the ambush were the crews of the low squadron. Lt. McSween recorded that "Low clouds forced us to climb up. We couldn't see ahead. A low fighter sweep hopped the clouds and hit our Squadron. They knocked down Hartigan and Hendry."

Captain Gamble wrote: "Have to climb to 28,000 over a cirrostratus cloud bank. Just after we get over it about 10 enemy fighters hit us. We slide over under lead squadron and do a quick pull-up on the way to miss several 20mm shells. Hartigan, on my left wing, slides under and gets it. Ralph [Coburn] sees plane on fire—spinning down. Hendry goes down too, but we don't know what happens to him."

In the confusion not even Bill Fort knew what happened to Hendry, though he was flying just off Hendry's wing. Fort remembers:

"I was in the No. 4 position. We went through one cloud bank, and then went through the next one. And then I saw four or five Me-109s. They were right level, right on us. One of them was right on our nose. We were eyeball-to-eyeball. I never saw an enemy fighter come as close in a frontal attack. How he missed us I don't know. They all came in so fast, all we could do was yell 'Fighters!' and we didn't get but a few bursts off.

"They went around, snuck back, and got in front of us to make another attack. We were ready for them this time. I know one of them got hit because he was flopping all over the sky, real erratically. I think the pilot must have been killed. When it was over and we got to looking around and getting our wits together, so to speak, we noticed that two planes were missing."

Thus, only Jack Hendry and John Doherty can say exactly what happened to them that day. Most of the air war is a blur to Jack Hendry now, but the climax of his Eighth Air Force career is indelibly imprinted in memory.

"All of a sudden the German fighters popped out of a thin cloud bank, and they were sitting right in front of us. There were between five and ten of them. I couldn't really identify them at first, but they got closer in split seconds and I saw they were Me-109s. I was mildly surprised because I was sort of expecting they would be our own fighters.

"Instead of coming head-on, like the Germans normally did, this group dove a little bit and then came up shooting. There wasn't any action I could take. I couldn't turn sideways because it would take too long to get a wing up. I couldn't kick the rudder, because that wouldn't do any good, either. The only thing I could do was pull up on the control column. That would work if you pulled up just at the time they commenced firing—you could judge it pretty well and take evasive action just when you figured they would open fire, about 1200 yards out. But with them coming up from below me, at close quarters, it didn't change the deflection very much and it didn't do any good.

"One of the 109s just raked us from stem to stem, dead center. I never saw him; he was hidden by the nose. There wasn't much noise when we were hit—it was hard to hear with the four engines going and the earphones on—but I knew we were hit very severely because of all the vibration from the hits. The nose was broken where a 20mm came through, and it hit Webster, the bombardier, in the chest. His flak vest saved him, but he got a badly broken leg."

Back in the tail, the fact that they were hit was about all that John Doherty knew. As he remembers now:

"Somebody called out fighters at eleven o'clock. In the tail I was unable to see this. A few seconds later, a 20mm exploded and hit me in the leg. It was just shrapnel, not the main force of the shell. I was on my knees and so I didn't know how bad I was hurt.

"I saw the plane that shot us. I gave him a short burst, but he was too far away and at too much of an angle for it to do any good. I was trying to see how bad my leg was hit but I couldn't see through my coveralls. I put my finger in the hole and could feel blood. But I didn't feel as though I was hit too badly and I was able to move around because I was on my knees the whole time.

"But at the same time the shell took my intercom system out. I turned around and I had this hole in the side of the plane from where the 20mm exploded. It was right along behind me to my left, where the intercom box was supposed to be. I tried to call the crew and I could get no answer."

Fire was the next thing the pilot and tail gunner both sensed. As Hendry recalls, "The shells started a fire in the gas tank of our left wing, inboard of No. 2 engine, right outside my cockpit window. I could look down and see holes in the top of the wing. There were a number of them about three to four inches in diameter, and I could see flames underneath in the tank.

"The plane was still flying all right, and looking out I could even see England. It really hurt when I realized I didn't have a chance to fight the plane back. I felt like cutting the engine and gliding back. You could have cut off all four engines, and that wouldn't have bothered me one iota. I felt I could take that B-17 and land it anywhere. I wasn't afraid of ditching either.

"But with that fire I knew we had to bail out. I had seen too many other planes where fires would break out in the fuel tanks, with the flames going back to the tail, and the wing finally just broke off. I didn't want to sit on top of a burning fuel tank. So I told everybody, 'We got fire in the gas tank, so bail out.'"

There was no way that Doherty could hear this in the tail but he saw the flames soon enough. "I pressed my face real hard against the Plexiglas and looked back over my right, on the plane's left over the stabilizer. I knew we had been hit on that side because it was my right leg that was hit. And these tongues of flame seemed to be coming out over the aileron or someplace.

"It wasn't a great big flame but you could see tongues of flame coming back. I presume the wind was blowing it. It was coming from under the wing or out from the end of the wing. That's when I decided it was time to get out of there. Something told me it was time to get out."

Up in the front of the stricken B-17 the men were doing just that. As Hendry recalls, "I gave the crew about a minute and a half to bail out, and then I went myself. The main spar on the B-17's wing was pretty heavy, so it would take a while for the fire to get to it and cause the wing to fold up. Since the gas tank hadn't exploded, I figured an explosion wasn't imminent, and that I had a little time.

"The bombardier and the navigator went out the nose and Harper, the copilot, did the same. He went out after the bombardier and navigator, but the engineer later told me that he saw Harper falling with his chute trailing behind. We wore parachute harnesses, and we had little chest packs which were tucked under our seats. I figure bullets must have hit his chute and damaged it. He died. I didn't know him that well, but he came from Jacksonville, like I did.

"I grabbed my parachute and clipped it on. When I bailed out, I doubled up and dived headfirst through the forward escape hatch. I knew I was at around 25,000 feet, and I was falling with legs and head up and rear end down. I didn't see the plane at all after jumping out. I looked over my shoulder at the ground, and went through several layers of cloud. I tried to judge when I was at five or six thousand feet, and pulled the ripcord. The chute opened nicely.

"I came down on the corner of a Spanish-type house with a flat roof that had a little parapet wall around the edge. I stuck out my foot to hit the corner of the roof and sprained my ankle slightly. This was in a little village near Lille on the border between France and Belgium called Vaenciens. The local people hid me in the church half a block away. I was in a little room in the church. But there was a German fighter base on the south side of town and there were quite a few Germans in town. They picked me up in about 15 minutes, and I was sent to a POW camp in Sagen, Germany."

Meanwhile, John Doherty was making his own exit from the burning B-17.

"I checked all my snaps on my parachute, and I hooked my regular shoes on my Mae West—tying the shoes on our Mae Wests was something we figured after we had been in combat a while. We used to discuss these things in the barracks. Before I went I took about three or four big gulps of oxygen. Then I went out the little escape door under the right stabilizer.

"I went out headfirst. In training they told us to be careful not to pull the ripcord of the chute too soon leaving the tail because the chute would come up and catch in the stabilizer. They told us to count to ten. In the tail we were using a back-type chute, which they had always told us not to pull out at a 90-degree angle, but to pull up, so we wouldn't bind the cable and would be sure the chute would open. On the end of the ripcord was a steel ring that you could put your hand right through. When I pulled my ripcord, I pulled it so hard that I struck myself in the chin with this iron ring, and knocked myself out. That ring took all the skin off of my throat under the chin.

"The first thing I knew when I came to was that my chute was open and I was in my stocking feet. My chute had snapped me out of my flying boots and my electrically heated felt shoes. I was drifting down, and I looked around to see if I could see the planes, our plane, or other chutes. I could see absolutely nothing, no planes, no chutes, nothing. The thought struck me that apparently I had bailed out on the crew. I had no instructions to bail out, and this bothered me very much, believe me it did.

"But there was nothing I could do about it then, and here came two Me-109s, and the first thing I thought was that they were going to shoot me out of the parachute. At this time they had been shooting parachutes and setting them on fire, and then of course you would just drop. So I was just waving back and forth as much as I could, trying evasive action in the chute. They didn't fire, but they did try to tip my chute. They buzzed me, trying to tip the chute so it would lose its air. They made two passes and left. So I drifted on down to the ground, and when I got to the ground I could see people following me."

Sgt. Doherty was on his way to a rendezvous with the Belgian resistance, but while he floated down, the Group's gunners were adding a couple of yellow German parachutes to the white American ones by getting two of the Me-109s, plus a "probable" and a "damaged." Moments later, the P-47 escorts engaged the retreating Germans. But none of this compensated for the loss of two B-17s with 21 men aboard.

As the Wing headed for home, the raid continued to go badly.

Clouds and improper coordination among the Wing's groups kept any bombs from getting on target. Major Shayler wrote:

"I called on VHF to the Combat Wing Leader for instructions and received no reply. I fell into position in Combat Wing formation and followed the lead Group. Without warning, the lead ship fired a red flare and announced over VHF 'turning on IP.'

"I took interval and instructed my Bombardier to look for the target. We could not determine whether the target was going to be Woensdrecht or Gilze-Rijen. My bombardier picked up the airdrome at Gilze-Rijen and we started to make a bomb run on it. The lead Group fired a red flare and dropped a phosphorus bomb before the bomb release line but I was unable to see any other bombs drop from the lead Group.

"My bombardier asked if he should drop his bombs and I instructed him not to unless he was sure that it was the target. We did not drop our bombs. The formation made a left turn, we fell in behind the 379th, and began descent at 170 IAS and 500 FPM [feet per minute]. The lead Group pulled away from us during the let-down and we returned home practically abreast of the 384th Group without difficulty, with the 379th Group out ahead."

Hullar's crew described the outcome more prosaically. Elmer Brown wrote that "When we got almost to the English Channel, the wing commander decided to try to hit another target. So we turned into Holland to Gilze-Rijen airport. We could not see the target so we brought the bombs back." However, he mentioned the loss of Hendry's crew and wrote that "Webster, the bombardier, was in Squadron 81 with me at Santa Ana."

Bud Klint noted the losses too, but tried to put a good face on the failed mission by observing: "We hauled our six 500-pounders and six bundles of incendiaries back to Molesworth rather than scatter them over the fields of the occupied Low Countries."

The crews of the 358th Squadron were far less charitable in their assessments. Captain Don Gamble confined himself to writing about the "piss poor landing" he made, but his crew commented at interrogation: "Can't imagine why bombs were not dropped—they should have been dropped."

Lt. McSween was even more scathing in his notebook: "Aborted from original target, turned north and flew 90 degrees over Gilze-Rijen. Haze bad. Lead never dropped bombs. Turned around and brought bombs home. Somebody with head up and locked."

Bill Fort believed: "We were piddling around like we were on a Sunday afternoon drive. They put the bomb bay doors down, figuring they were going to bomb an airfield down below us. Then they pulled up the bomb bay doors and we went home."

The bad feelings from this raid festered in the Wing for a few days, finally coming to a head in a meeting called by General Travis. Bill Fort was one of those who was there:

"There was a big hullabaloo about it in a couple of days. Travis got everybody together in the theater building, and really raised some hell. Everybody was talking. There were colonels blaming each other for it, and a lot of them were lying, especially one from the 379th. He was lying like a dog.

"Travis saw they weren't getting anywhere, and got up to talk. He said, 'Listen, let me tell you all something for the next time you're over there. If you're over France, you just bomb military targets. But the next time you're over Germany, you *leave those damn bombs over there*. Don't bring those bombs back even if you have to bomb [an outhouse]. That's all.'

"And nothing else was ever said about it."

The *denouement* of the day came in stages for John Doherty, as he gradually learned what happened to the airplanes and people he had searched for so earnestly after coming to in his parachute.

"At one of the houses I was taken to, I was told I was in the Underground and they were moving me out to Spain. I had a fake passport which said I was deaf and dumb; I had papers identifying me as a deaf and dumb shoemaker, and was to use these papers in my travels.

"I was taken to another house without anybody being able to see me. I was in the back of a truck and they backed it right up to the door. When I walked into this room in this house, there sitting on a chair was our Squadron photographer [a Sgt. Ralph Moffett].

"When I saw him, I asked him what he was doing here and he said, 'The same thing *you're* doing here.' He told me he was flying with the crew that was flying wing on the lead ship, and that them and our crew had got taken out on the one pass with the fighters.

"And I asked him what happened to our plane, and he told me that the plane blew up. And I asked him if he saw any chutes go out of it and he could only account for four chutes. That's all he saw.

"That gave me things to think about, but it did relieve my mind from the fact that I bailed out and left the crew. From that standpoint I felt a little better.

"Later we were to catch a train and start going to Spain. We met a girl that was about 25 years old. Our instructions were to stay about 30 feet behind her, and she would be our guide. Wherever she went, we would follow.

"So we went on the train and started going to Brussels. There we got off and followed this girl, she took us right through the Brussels railroad station—I walked within *five feet* of armed German guards on duty with fixed bayonets, helmets, the whole regalia—we walked right by them to what appeared to me to be the front door of the railroad station in Brussels. It was at night and it was pitch dark, no lights. And instantly this girl disappeared, she just disappeared, which wasn't too hard in the dark. And at this point three men took over, and the deal was that we were going to stay in Brussels that night. And they took us out, me and the Squadron photographer, and they put us into a 1929 Ford. We drove for a considerable amount of time, and we came to this big house that was surrounded by steel fencing.

"It was a real fancy place. We stopped and got out and walked up to the door and knocked on the door and somebody opened it. And when the door was completely opened, right in front of me on the full length of the wall was a picture of Hitler. And as we walked in, the guy that opened the door said, 'You are now in the hands of the German Gestapo.'

"About a month later I arrived at Stalag 17B prisoner of war camp. Upon my arrival there I found all the other members of the crew except Brown. I was the last one to get to the prison camp. I got there on Thanksgiving Day of 1943. Biddle and Hargrave were elated to see me, because they thought I was dead, so it was kind of a happy reunion even though it wasn't a happy time.

"I was told by the crew that Brown had been killed. Hargrave told me that he and Brown was standing at the waist door of the plane, ready to jump. Brown always had this idea that maybe the door wouldn't come off of the plane, but Hargrave said that the door was open—the hinges to the door were on this cable, and you pulled it and the two pins on the hinges would come out and the door would drop off the plane. The door was off, but Hargrave told me that as they were standing in the door the plane exploded. He was blown out and he also supposed that Brown was blown out with him, as they were both standing together at the door.

"He came to in the air and opened his chute. Whether Brown was killed in the explosion or whether when he was blown out of the plane, he didn't come to in time to open his chute, I do not know. But a piece of me has always stayed with Brown in Germany."*

There was a final, bitter irony to the events of October 20, 1943. Within days of John Doherty's arrival at the POW camp, the article "'Tail-End Charlie' Has System For 'Bringing 'Em Back Alive'" appeared in newspapers all over America.

The truth was harder to take. In the ETO, there was no "system" or sure-fire way to survive.

---

*The death of Sgt. James Brown on this mission provides telling evidence of the toll that war takes on families, too. According to Mrs. Henrietta S. Duke, Brown's niece, his widowed mother never recovered from his death. Nor did his fiancée, who lost all interest in life when she learned that he wouldn't be coming home. She drifted into alcoholism and died young, "a very unhappy woman."

17

A Week at the "Rest Home"

October 22–27, 1943

WITH THE OCTOBER 20, 1943 mission to Düren, Hullar's crew had finished 12 of their 25 missions, and had earned a well-deserved respite from combat.

Elmer Brown recorded in his diary: "October 22–27. Went to the Rest Home in southern England," and the enjoyment the crew had during their stay is beautifully described by Bud Klint's own diary entry. On October 27th he wrote:

"Back home again after the most glorious four days I ever hope to experience away from home. The rest home is wonderful. I only hope we get to go back for seven more days after we have a few more missions under our belts.

"The house itself is a tremendous mansion built in 1870 by a 'vinegar king.' It was later turned into a hotel, then taken over by the RAF, and finally leased to the Air Corps. It is a gray stone building, and inside and out looks like something out of the movies. It is located in the most beautiful section of England I have yet seen, and the grounds of the estate, which must cover several miles, are heavily wooded.

"The Red Cross runs the place, and they have everything for amusement that you could ask for. Football, baseball, tennis, badminton, archery, cycling, horseback riding, hunting, checkers, chess, cards, pool, volleyball, and many other things. The staff is lovely. They really make you feel like guests. They do everything they can to make you comfortable and happy.

"The butler, Michael, is a typical butler. He woke us every morning at nine-thirty, once by serving us a cup of pineapple juice. There are four Red Cross hostesses there, and they couldn't have picked better girls for the job. Our whole stay was like a glorious dream.

"The one thing that I shall probably remember best was our first morning there. We started out in a recon to watch a fox hunt. We missed the hunt and wound up on a sightseeing tour of Lady Arundel's estate. We climbed all through the ruins of the old castle built in 800 and partially destroyed by Cromwell after Blanche Somerset, Lady Arundel the VIII, with only nine men, withstood the attacks of Cromwell's 1500 soldiers for nine days.

"Then we went to the new castle. Lady Arundel the XV received us like we were her own children, and gave us a personally conducted tour of her home, showing us all the art treasures and historic relics therein. The new palace, built in 1750, is the most wonderful place I have ever seen. It is something like the Field Museum in Chicago. Some of the wallpaper is original. The ceilings are all beautifully painted and trimmed with gold carvings. All the doors are made of oak from boats of the Spanish Armada.

"She had numerous paintings, including portraits of all the Arundels from Lord Arundel the I to her late husband. Some of the pictures are original van Dykes. She has a bed rescued from the royal palace before Charles I was beheaded, in which he slept, and it is just as it was when he slept in it, original canopy, spreads and all. Many of the furnishings in the palace are from the old castle and are lovely antiques.

"I could write books on all the things we saw that morning: famous old books, papers, including an original letter from Queen Elizabeth I, a holy Grail, old china, et cetera, et cetera. It was the most amazing thing I have ever seen. There is even a complete chapel in the palace, one of the loveliest you could imagine. It is entirely Italian, and the altar alone cost 20,000 pounds. Her son, Lord Arundel the XVI, is a prisoner of war near Leipzig, and she has turned over one entire wing of her castle to refugee children. She was truly a wonderful lady.

"That was only one glorious morning of our stay, and all of it was just as amazing and wonderful. We wore civilian clothes, lounged, laughed, sang, rode, and played. We felt like human beings again instead of like automat killers. We forgot about war and its horrors for the most part."

George Hoyt and Merlin Miller also remember their stay at the rest home. Miller recalls:

"Ours was in the same vicinity as the officers'. It was a big estate out in the country, not too far from a small town. It must have had a good 40 to 50 rooms in it. There at the rest home we could play golf, go horseback riding, shoot skeet, play tennis, shoot archery, and play softball. It was a good place just to go and relax."

"I remember one night they had a famous English classical pianist. She played classical pieces for us, and also played a lot of requests from the GIs. There were two Red Cross girls there, one called 'Mississippi' because of her Southern accent. They were very nice and easy to get along with, and went out to play golf with us a number of times. Those golf matches were classics, seeing as most of us had never played golf before. I never saw anybody besides Marson hit a golf ball straight up in the air so that it came down behind him.

George Hoyt remembers the rest home as "a large stone edifice resembling a medieval castle. The building had a tall, three-story turret with a music room on the third floor. It had a record player with a good supply of classical recordings, and I spent an hour or so each day there. I love classical music, and I relished each moment.

"On one side of the manor building was a first-rate tennis court, and one day I accepted a challenge from the two Red Cross girls to play. I enjoyed it so much I played all day with only a break for lunch. It had been two or three years since I had last played and the next morning, much to my dismay, I could hardly get out of bed due to 'charlie horses.'

"The girls were very sympathetic, and served me breakfast in bed three times! Ah, for the life of a combat crewmember! Our time there really was a great morale builder. The people at Bomber Command definitely knew what they were doing in setting these places up."

Unfortunately, it was impossible to forget the war even at the rest home. Upon his return Bud Klint thought not only about the good times he had had, but also some other experiences the rest home provided. He concluded his diary entry on a somber note:

"I ran into Luke there. He went through cadets with me. He is at Chelveston [home of the 305th Bomb Group] and was riding in the only ship in his squadron which came back from Münster. We also met some boys from the 100th Group. Two ships from their entire Group were all that returned from Schweinfurt. Stories like these make me wonder if I'll live through my 25. We lost 150 ships on four raids. We can't take much more of that."

Back at the base there were other grim reminders of the toll that even noncombat operations could take. On the second day of the Hullar crew's stay at the rest home, Lt. L.E. Jokerst and Sgt. W.H. Stephen, his engineer, were killed along with six others in the crash of *Miss Patricia*, B-17F 42-29930, on a local night flight. They had flown with Major Shayler on the Düren mission, which had been their 15th.

Hullar's crew now had 13 more missions to fly, and only time would tell how hard they would be. What the crew could not know was that the very darkest period of the Eighth's air war was already past. The next raid would signal more of a false dawn than a definite turn of the tide, but it would demonstrate without doubt that a change for the better was on the way.

It featured something new: an escort of P-38s.

18

"Those Lightnings Were Beautiful That Day"

Wilhelmshaven, November 3, 1943

Sure, we're braver than hell; on the ground all is swell
In the air it's a different story;
We sweat out our track through the fighters and flak,
We're willing to split up the glory.
Well, they wouldn't reject us, so Heaven protect us,
And, until all this shooting abates,
Give us courage to fight 'em—one other small item—
An escort of P-38s!

—from "Lightnings in the Sky"
Anonymous

The P-38s that some of the Hell's Angels had seen on the Second Schweinfurt journey were better equipped to help the "Big Friends" when November arrived. At this time the Eighth had two groups of Lightnings in the ETO; the 55th, located at Nuthampstead, and the 20th, based at Kings Cliffe. Only the 55th Group was fully operational, but the 20th Group had enough aircraft to serve as a "fourth squadron" for its sister unit. Thus, the Eighth's planners were able to put 45 long-range Lightnings into the air to complement the 333 P-47 Thunderbolts from the seven groups it had in England.

For this first mission of November 1943 Pinetree had planned a maximum effort. More than 566 heavy bombers were to attack Wilhelmshaven, an ideal objective for another new weapon the Americans had: H2X, code-named "Mickey," a U.S. version of the H2S "Stinky" radar the Eighth had borrowed from the British for its early PFF missions against Emden.

All First Division B-17s available were to take off to form four combat wing formations. The 40th CBW would lead; a Composite Wing drawn from all of the Division's groups would make up the second combat box; the 1st CBW would be third in line; and the 41st CBW, led by the 303rd, would bring up the rear.

The 482nd PFF Group, based at Alconbury, was tasked with supplying H2X ships to each of the Division's wings: two each to the 40th CBW and the Composite Wing, one each to the 1st and 41st CBWs. The Second and Third Divisions were to follow the First into the target at five-minute intervals.

Though the total number of bombers was impressive, the emphasis was on the fighter escort. The naval dockyards of Wilhelmshaven were the primary target, but the field order stipulated that the secondary target could be "any German industrial target which will not disrupt fighter support."

Two groups of P-47s were to escort the First Division on the first inbound leg; a second P-47 group would overlap on the way into the target. The Lightnings of the 55th Group and their 20th Group contingent would escort the bombers into and out of the target area, and a final group of P-47s was to take the B-17s home. A "special note" to the First Division field order stated: "There will be additional groups of American fighters with 2nd and 3rd Divisions which together with 1st Division fighter support will put several hundred friendly fighters on this route within approximately a 30-minute period." All in all, the mission amounted to a major demonstration of American airpower.

The 303rd's contribution was a force of 23 B-17s. Six bombers from the 358th Squadron—including Captain Gamble's crew in *Sky Wolf*—would form the low squadron of the Composite Wing's low group. The balance of the 303rd would take up position two wings back at the head of the 41st CBW.

Captain David Shelhamer's crew was leading the high squadron in *Vicious Virgin*; Hullar's crew was flying second element lead of the high squadron in B-17F 42-29823, and the Group was headed by a 482nd PFF ship with the Wing Air Commander, Lt. Colonel Eugene A. Romig, aboard. Romig was no stranger to the 303rd. He had commanded the 359th Squadron during the 303rd's early days and he would go on to a distinguished career as Commander of the First Division's 351st Bomb Group.*

The 303rd got off between 0940 and 1000 and began to assemble above Molesworth. At 1035 the PFF ship joined the Group over the base at 8000 feet. Wing assembly was next, and in his Group Leader's Narrative Colonel Romig commented: "The Combat Wing assembly was effected smoothly and this was about the best coordinated mission I have ever seen." The 303rd climbed to its assigned altitude of 22,000 feet without difficulty, and before long it and the rest of the Wing were over the enemy coast.

*The ranks of the 351st would soon include another early 303rd veteran. In January 1944 Captain Ken Davey, who had ridden as tail observer in Major Lyle's aircraft during First Schweinfurt, also joined the 351st.

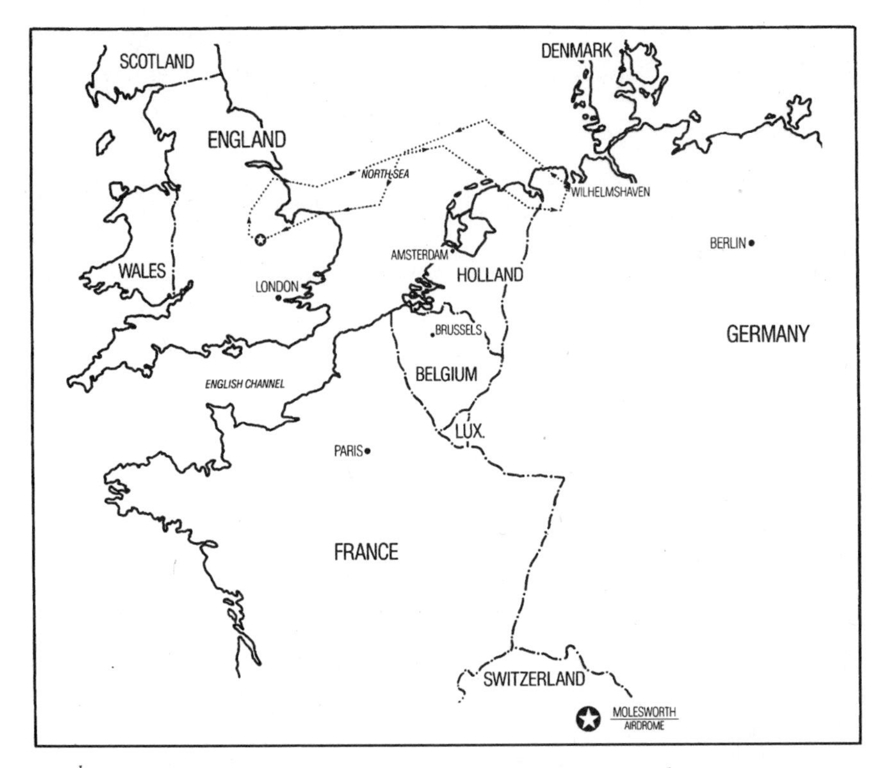

303rd Bomb Group Mission Route(s): Wilhelmshaven, November 3, 1943.
(Map courtesy Waters Design Associates, Inc.)

The P-47s assumed their normal escort formation, crisscrossing above the bombers at 27,000 feet. Looking to the rear from the radio room, George Hoyt did not see them, but about 15 minutes into enemy territory he did spy another group of aircraft:

"I was looking at the vast sky behind us devoid of any other planes. My eyes picked up a large formation of planes approaching us from five o'clock level. They were very far out, and from that distance looked like a group of bombers. I called Bob Hullar to inform him, and he said, 'Keep an eye on them, because there are not supposed to be any more B-17 groups behind us according to the briefing.'

"Slowly these planes gained on us, in a tight formation that looked like one of ours. Finally, about 1200 yards out, they were close enough for me to identify. They were P-38s! They throttled back and stayed about that distance behind us. It sure felt good having them there."

The P-38s were employing a new "close escort" tactic, staying right with the bombers to ward off enemy fighters. Two wings ahead of Hullar's crew, Captain Gamble also noted their proximity: "Get high P-47 protection at coast and on in. P-38s come in close."

The tactic soon bore fruit, all along the bomber stream. Lt. Bill McSween wrote: "Just before the IP I saw approximately 25 fighters break through the clouds below. I called Don and he called the P-38s."

Don Gamble wrote: "All 25 or 30 enemy aircraft break out of the clouds below. I call the P-38s and they come right down to protect us from these bandits. They fly in close and only one bandit gets through but he doesn't attack the formation."

McSween happily concluded: "They came right in and definitely saved our hides. Jerry stayed well out of range. There was very little flak at the target, 10/10 cloud cover, the bombs dropped, and we drove out unscathed."

Another group of Germans popped up through the clouds in front of the 41st CBW. In the nose of No. 823 Elmer Brown was with Sgt. Royal Plante, a "togglier" whose primary job was to drop the bombs when they went from the lead PFF ship. As the Germans formed for a head-on attack, Brown could not resist the temptation to take the front seat and fire from the .50-caliber in the Plexiglas nose.

"I acted as both bombardier and navigator. The only time I got a good shot at enemy fighters was when we were on the bomb run. I had to keep one hand on the toggle switch, but I managed to fire the machine gun with the other hand."

In the radio room Hoyt "heard Brownie call 'Bandits at twelve o'clock!' which meant they were attacking us head-on. I knew it would only be seconds before they would be barreling through our formation towards the rear. I shot at them as they zipped over the radio room hatch and then those gray 109s really ran into a surprise!

"I saw the P-38s suddenly break formation like a swarm of angry hornets. An intense dogfight ensued, with a P-38 on the tail of a Messerschmitt everywhere I could see. Directly behind us at six o'clock an Me-109 climbed almost straight up with a P-38 right on his tail. I saw 109s pull this maneuver many times to shake off P-47s that could not climb as fast, but that twin-engine Lightning closed right in and parked right behind this 109 at what must have been less than 50 feet.

"The P-38 let go with his nose guns, four .50-caliber machine guns, and a 20mm cannon, and pieces flew off the German fighter right and left. Then he started down in a blaze, leaving a trail of black smoke. It all happened in a matter of seconds."

From the tail, Miller saw another kill: "I remember a P-38 with a 109 on his tail. He was maneuvering around, and pulled up into a climb straight up, which he could do pretty well with his twin engines. The 109 tried to follow him and stalled out. Just as he did, the P-38's wingman came in and chopped him in two."

Bud Klint saw the same combat from the cockpit: "The Lightning's big advantage seemed to be its rate of climb. I clearly remember seeing one P-38 with an Me-109 on its tail. The Lightning pilot stood his plane up on its tail and the 109 tried to follow him. The Kraut was hanging on his prop and looked about to stall out when a second '38 came roaring in and cut him to shreds. I watched four Jerries come out of dogfights on the losing end that afternoon. Those Lightnings were beautiful that day."

From *Vicious Virgin*, David Shelhamer was similarly impressed: "The P-38 pilots were very much aware that their primary duty was to protect the bombers. I'm prejudiced, but I always felt that the P-47 pilots just loved to make pretty contrails about 5000 feet above us while the Germans were shooting the living hell out of our formations. The P-38s came in quite close to us, and the moment they were aware of enemy activity, they were right on the German fighters. In my book those P-38 pilots were just great."

The dogfight between the P-38s and the German fighters—about 30 Me-109s and FW-190s in all—took place near the IP as the 303rd turned left and began its bomb run through meager, inaccurate flak and dense cloud cover. The fight lasted less than five minutes, but it was not all one-sided—Captain Shelhamer's crew saw a Fortress ahead of the Group go down two minutes before the 303rd dropped its bombs at 1314. It exploded above low clouds at about five to six thousand feet. Once "bombs away" was called, the Group made another left turn and headed for home.

In one sense the flight back was uneventful; there were no other aerial combats and only a few bursts of flak for the bombers to avoid. But for George Hoyt there was a sight that has stayed with him all these years as a symbol of what took place that first day the Group had an escort of P-38s:

"After we left the French coast I looked out the starboard window of the radio room at the B-17 flying next to us. Under the protection of the bomber's wing, almost glued to its ball turret, was a P-38 with its right engine out and its propeller feathered. I went back to the waist to get a better look from Marson's large open window and as I did the P-38 pilot

P-38 escort. [Photo courtesy National Archives (USAF Photo).]

looked at me and gave me the 'thumbs up' sign. I answered with a 'V for Victory,' and felt a surge of emotion which brought tears to my eyes."

The 303rd landed back at Molesworth between 1530 and 1600, bringing the mission to a close without a single lost ship. But the sense of elation and victory that George Hoyt and the others had spilled over into the interrogations the crews underwent. Crew after crew registered their praise for the job the Lightnings did. Bob Hullar summed up the escort in a word: "Beautiful!"

Lt. "Woodie" Woddrop's crew judged the "P-38s very good, especially so."

Lt. G.S. McClellan's crew saw that they "gave crippled Forts excellent coverage."

Lt. Carl Fyler's crew said, "Good cover, especially the 38s."

Lt. W.C. Heller's crew reported that "P-38s were right in there giving us personal attention. Very good support."

And lastly, Lt. Ambrose Grant's crew—the one that had flown the crippled *Cat O' 9 Tails* back from Schweinfurt—spoke for everyone when they said: "More P-38s—they're eager."

The P-38s and the mission as a whole made a positive impression on more than the B-17 crews. The numbers looked good. The score was seven B-17s and two P-47s against 16 German fighters: 13 claimed by the P-47s and three credited to the P-38s.

The American press reported favorably on the operation, describing it as a "thousand plane raid," which it was when all the heavy bombers, fighters, and twin-engine B-26s flying diversions were taken into account. Thus, the mission did much to restore a measure of public confidence in the Eighth's ability to bomb the enemy without extreme loss.

The reality was more complex. The damage done by the PFF bombing was difficult to assess, and much of the P-38's success in safeguarding the bombers flowed from its novelty. The history of war is filled with cases where a new weapon momentarily throws an enemy off balance, and so it was with the Lightnings and the Luftwaffe. In time, German fighter pilots adjusted their tactics to the twin-engine fighter's weaknesses—inferior diving ability, ongoing mechanical problems with engines and superchargers, and poor high-altitude performance—and the odds evened out.

Moreover, the Eighth's leaders realized that the P-38, despite its advantage in range over the P-47, still did not have legs long enough to escort the bombers to *all* critical targets in the Reich.

19

The Silence before the Storm

November 4–25, 1943

TWO DAYS AFTER the Wilhelmshaven raid the 303rd took part in another PFF attack under the protection of P-38 and P-47 escorts. The target was Gelsenkirchen in the Ruhr, home of Happy Valley's notorious flak.

The Group sent 19 B-17s out of 374 from the First and Third Divisions, while 118 B-24s from the Second Division launched an attack against Münster. All 303rd ships made it to the target, dropping their incendiaries with unobserved results due to industrial haze and a smoke screen. "Intense and accurate flak" damaged 15 Group ships and one, *Ramblin Wreck*, B-17F 41-24565, fell to fighters. She was a veteran 359th Squadron Fort on her 28th trip and was flown by one of the crews who had praised the P-38s for their work on the previous mission: Lt. Ambrose Grant's, now on their eighth mission.

Lt. Carl Fyler saw Grant's ship go down: "I flew No. 483 [*Red Ass*] leading the second element in the top squadron. There was lots of flak. The ship on my right wing had been hit. The trailing edge of his left wing was on fire. Red flames ate along the wing towards the gas tanks. I waved for the pilot, Lt. Grant, to leave formation and bail out. He was determined to get to target. They never made it."

Ramblin Wreck fell out of formation on the bomb run and followed 400 yards below and behind the formation on the return trip. It was last seen under fighter attack about forty miles from the enemy coast at an altitude of 12,000 ft. Four parachutes were reported.

Sgt. Ed Sexton, the radioman who made the sign of the cross so often on Black Thursday, was wounded in the foot by shrapnel. Killed on his third raid was Sgt. J.J. Hauer, the replacement right waist gunner for Sgt. "Woody" Greenlee, who had been hit in the face by the exploding 20mm shell on October 14th. Shot down on his 23rd trip was Sgt. H.A. Kraft. a fill-in at the left waist gun for Sgt. Robert Jaouen, who was grounded for tonsillitis. (Jaouen ultimately completed a tour of 30 missions, flying his last operation on D-Day.) The others in Grant's original crew spent the rest of the war in POW camps, but all came home unharmed.

Shortly after this mission the 303rd and 41st Wing were taken off operations to test a new weapon that Bud Klint remembers "was supposed to revolutionize daytime bombing. We received a shipment of so-called 'glide bombs' for a secret program called 'Project Grapefruit.' They were built like miniature P-38s, with a 2000-pound bomb as the fuselage. They had a glide ratio of nearly six to one and were supposed to permit us to hit heavily defended areas from outside of flak range."

The Group tried the weapons, officially known as the GB-1, against some small uninhabited islands in the Irish Sea. At the time Klint quickly concluded: "These glide bombs aren't infallible. Their accuracy is pretty poor, and the added difficulties they present to formation flying have to be ironed out," and he recalls now that "When they were released they went in all directions, some spinning in, some doing loops, and very few going anywhere near the target."

The way they were supposed to work. This photo shows GB-1 "Glide Bombs" being launched from a B-17E during weapons development and testing at Eglin Field, Florida (officially known as the Valparaiso Bombing & Gunnery Base) in 1942. Operational conditions in the ETO in 1943 doomed "Project Grapefruit" (the code name for use of these weapons) to failure. It is believed that this is the first time this photograph has ever been published. [Photo courtesy National Archives (USAF Photo 32348 AC).]

Merlin Miller remembers them as "just one big fiasco. I felt we were in more danger getting hit by one of those things than we were flying through flak on a mission." Ultimately the whole program was cancelled—to everyone's relief.

The 303rd returned to the mission roster on November 11th, when Pinetree organized a strike by the First Division against Wesel, Germany, and a simultaneous attack on Münster by the Third Division.

The Münster mission got through, but the Wesel operation was abandoned. Elmer Brown wrote: "We started out on a raid to Wesel. To be pathfinder at 27,000 feet. We were recalled within two miles of the enemy coast. We were at 29,500 feet and our lead and low groups were still heading into dense clouds."

Two days later the Group went out on another effort stymied by weather. This time Brown wrote: "We started out for Bremen, but could not rendezvous the Wing, so we were recalled at the English Coast."

It was not until Tuesday, November 16, 1943, that the 303rd was able to get another operation off. The mission was a long overwater flight to Norway, a country the Eighth had bombed only once before—on July 24, 1943, when the B-17s were sent to attack a nitrate plant at Heroya and the ports of Trondheim and Bergen. This time the Third Division was to hit a hydroelectric plant at Rjukan suspected of providing power to a German "heavy water" nuclear research facility. The Second Division was to bomb an aircraft repair depot at Oslo-Kjeller, and the First Division was to strike a molybdenum mine in southern Norway near the town of Knaben. The mine was a pinpoint objective on the side of a mountain, whose importance lay in the fact that it produced practically all German supplies of this vital steel-hardening mineral.

The 41st CBW led the Division, and the 303rd was low group in the Wing with the 384th as lead and the 379th high. Major Ed Snyder led the Group with Captain David Shelhamer in *Mr. Five by Five*. For Shelhamer this trip was special; it was his 25th raid and the end of his tour. With him was Lt. Paul Scoggins as lead navigator.

Hullar's crew took the No. 4 slot in the lead squadron, and today they were a mixed bag. Mac McCormick was not aboard, nor was Bud Klint, who had been bumped out of the copilot seat by a Lt. Colonel Culbertson visiting from VIII Bomber Command. Klint "tried to talk Operations into letting me go as a waist gunner in place of Marson, who was in the hospital with his wrenched knee, but no luck. This put me one mission behind most of the other crew and two behind Brownie."

Klint didn't miss much. The only thing novel about the raid was the ship the crew drew: *The Flying Bitch*, B-17F 42-29795. She was one of

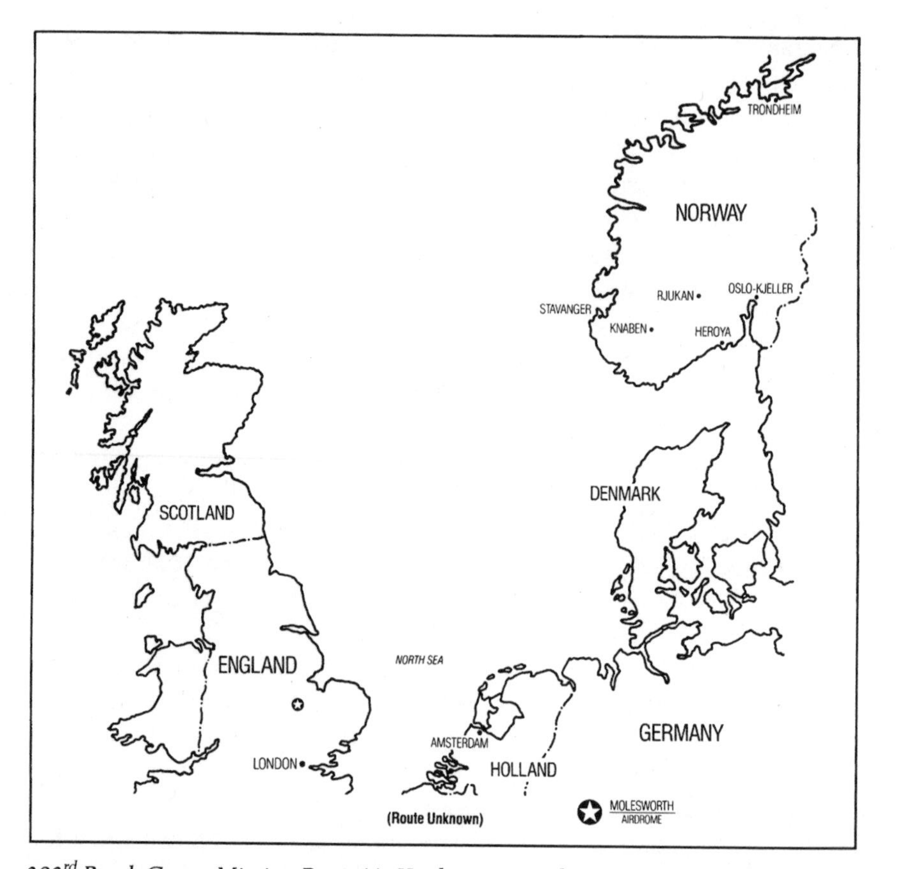

303rd Bomb Group Mission Route(s): Knaben, November 16, 1943. Route unknown. (Map courtesy Waters Design Associates, Inc.)

the few B-17Fs equipped with the new Bendix chin turret (which was standard on the B-17G), and she was the first they had ever flown with this armament.

Mr. Five by Five got off three minutes behind schedule at 0718, but the 303rd soon caught up with the other two groups of the 41st Wing, falling in trail of the 384th Group and climbing to 20,000 feet to get over the weather and clouds while crossing the North Sea. Hullar's crew settled down to what Sampson remembers "sure was a long, long, long trip," with Brown noting that "we had to go up to 22,000 feet about halfway to Norway to get above some clouds."

The one most agitated by the flight was George Hoyt, who recalls that "On the way our No. 2 engine was throwing oil very badly. I reported this to Bob Hullar, and he said to keep an eye on it. The stream of dirty black oil was running back over the top of the engine nacelle from the cowl flap openings, back over the top of the wing and off the trailing edge into the slipstream. We were losing oil way too rapidly from this leak, and my anxiety mounted by the minute."

After nearly four hours the Group crossed the Norwegian coast. Hoyt "held my breath waiting for flak and fighters as we approached the jagged, mountain-scarred coast of Norway at our briefed altitude of 12,000 feet," and Sampson "wasn't happy about going in at this altitude at all," but nothing happened.

Lt. Scoggins couldn't fix his position. There were, he reported, "cloud layers at the Coast and scattered...cloud over the snow-covered hills...the shapes and sizes of the Fjords were hard to distinguish and there were too many lakes for accurate pinpointing of any one."

As a result, Brown wrote, "We flew all over southern Norway trying to find the target. We split up into groups and there were B-17s all over the sky trying to find the target."

The mission became a hit-or-miss proposition for all the bomber formations. Both the 384th and 379th Groups found the target and bombed it successfully, as did most of the First Division groups. But the Division's 91st Group ran into the same kind of navigational problems the Hell's Angels encountered, and the Second Division's B-24s were unable to locate their target at Oslo-Kjeller. Three Liberator groups joined with the Third Division's B-17s to make a successful strike against the hydroelectric plant at Rjukan, and 12 more B-24s bombed a chemical works there, causing major damage. Only two heavy bombers were lost.

Shortage of fuel forced the 303rd to abandon its search and head west. The Group crossed the Norwegian coast above the port of Stavanger, which gave Lt. Scoggins a chance to get his bearings, but "we didn't have enough fuel to return to the target area after our position was established." Then the enemy briefly came to life.

"At this point we noticed a couple of bursts of flak from a couple of ships there," Brown wrote, but the only thing they achieved was to provoke one of the few disagreements Shelhamer and Snyder had during their service together in the war.

"The harbor was just loaded with ships," David Shelhamer recalls. "We were fairly low, at about nine or ten thousand feet, and I debated whether we should open the bomb bay doors and drop the bombs, even though we didn't know how many of the ships were German and how many were Norwegian. I wanted to drop the bombs anyway, but Snyder wouldn't let me. I've always felt those ships were a target of opportunity, and that we blew the opportunity."

Ed Snyder responds: "Our objective was to hit a pinpoint target in a friendly country, and we didn't dare drop the bombs just anywhere. On the way out I remember seeing those ships in the harbor, and Dave Shelhamer wanting to bomb them, because there were definitely German military vessels down there. But I wouldn't let him because I felt there was too good a chance of us creating an incident with our Norwegian friends. We weren't briefed for this target, and I didn't think it was worth the risk."

Elmer Brown added that "We met enemy fighters at this point also, and they attacked us for about 30 minutes out to sea." But he noted "only about 10 fighters, and like the few bursts of flak, they did not create much excitement."

As Merlin Miller recalls: "They made only half-hearted passes at us. Just about the time they got to within firing range they broke away, circled around, came back, and did the same thing all over again. They were either real green pilots or afraid of a B-17 formation."

George Hoyt remained the one most exercised by the mission:

"After the fighters left, my thoughts returned to that leaking No. 2 engine of ours. I sat at the radio room table and kept staring at it. Then my heart skipped a beat when I began to realize that the wing was accumulating ice on the top surface. 'That's all we need,' I thought. The ice buildup could create so much additional weight, and alter the shape of the airfoil so much, that we could lose lift and not be able to stay airborne.

"I called Bob again, and he said, 'Yeah, I've been looking at it, and it's not good.' The flight back seemed endless."

Happily, the icing on *The Flying Bitch*'s wing got better as the 303rd made a slow descent during the return. Shelhamer reported that the bombers "proceeded back to England, letting down from 11,000 feet to about 1000 feet, coming back over the water under the weather as a single Group.

"At the English coast we encountered a bad snowstorm and broke up into Squadrons and climbed up through the low clouds. Due to gas shortage, I landed at Grimsby and the Squadron proceeded back to base."

Elmer Brown wrote: "We hit the U.K. at Sunburn, which was quite a distance north. I believe it is in Scotland. Our lead ship had to land at an emergency field near there for fuel and I had to bring the Group home. It was a long mission as we were in the air eight hours and 40 minutes."

Back on the ground, Hoyt remembers, "Bob Hullar had the ground crew pull the engine cowling off and check it out. Dale said there was less than two gallons of oil left in the 35-gallon oil reservoir. This engine was a license-built Studebaker Cyclone, and they were notorious oil-throwers. I wished we were back with *Luscious Lady*, but as things turned out we never got to fly her again."

Paul Scoggins was understandably sensitive about his performance. Two days later he confided to his diary:

"I was lead navigator and I'm thoroughly ashamed! We flew all over Norway and didn't find the target, because of a little weather and thousands of little lakes and fjords all covered with snow and looking exactly alike. I'm not feeling too badly about it—no one else knew where we were either."

For David Shelhamer all such concerns were academic. His tour was over and he didn't quite know how to take it.

"My last mission was just one big exercise in futility. I was rather mind-boggled for a few days as to whether or not to volunteer for five more, but the fact that I had a wife and a child back in the States who I hadn't seen yet ended up being the determining factor."

But Shelhamer did not leave right away because Major Snyder kept him on at Molesworth for 90 days ground duty, during which time he and Scoggins shared a room. Paul Scoggins now had no one left from his original crew. Two crewmates had gone down on an early mission with another crew; one had been killed in action; four had finished their tours and gone back to the States, and Captain Jake James got a transfer with the rest—"new boys, mostly" Scoggins noted—to the 482nd PFF Group.

There were many "new boys" entering the Eighth's ranks at this time, for the Army Air Force was embarked on a huge buildup of Eighth Air Force strength designed to double the number of aircraft a bomber group could deploy. Eaker and Anderson were still committed to the offensive, and the next series of missions had as their objective a wearing down of Luftwaffe strength through massive PFF attacks on targets within P-47 and P-38 range. In short, a battle of attrition was in the offing: a series of no-quarter encounters between equally matched enemies in an arena without parallel for physical danger—the frozen air in late autumn at high altitude over northern Germany.

These missions were the closest one could come to a Stalingrad in the sky, yet to this day they bear no name. They deserve one that will place them on a par with the other significant phases of the European air war—"Blitz Week," "Black Week," the "Big Week" of February 1944, and RAF Bomber Command's "Battle of Berlin." "The Battle of Bremen" is apt, since this German port was the Eighth's prime target for PFF bombing, with no fewer than five attacks launched against it between November 26th and December 20, 1943.

One of the 303rd crews destined to contribute greatly to these battles was Lt. John F. Henderson's, a replacement crew assigned to the 358th Squadron, who arrived in the ETO in early October 1943. Because of the role they play in this book, brief introductions to them are now in order. The best way of meeting the ball turret gunner, Ed Ruppel, a 27-year-old from Passaic, New Jersey, is through his memories of what happened after the crew reported in at Molesworth:

"We were supposed to be replacements for the Schweinfurt raids, but we didn't put any combat missions in right away. They took us by truck to gunnery training at a place called 'The Wash' on the coast of England.

"When we got there they took us into this big room, and the man said: 'Everything you've ever learned in the States, forget about it! You don't know anything! We're going to teach you now what it's all about.'

"This guy says, 'First of all, we're going to give you how long you're going to live.' And he went through each gun position, and he told us how long that guy would live. I forget how long he said the man in the ball would live, but it was very short. The two most vulnerable spots were the tail and the waist guns. A waist gunner had something like two to three minutes of combat. They said that combat could be three seconds, one pass and it was all over for you. It was all very hard-core.

"At the end they explained the reason why they were telling us this. The man said, 'All of you that don't want to go into combat, step over on the side. Nobody's going to holler at you, nobody's going to knock you down or anything, I just want to know now.'

"So somebody raised his hand and said, 'Why the hell are you so interested in that?'

"The man said, 'I want to know *now* that you're going to quit. I don't want you to quit when you're upstairs, and mess nine other people up. Understand?'

"Then they started to teach us what it was all about. We were out there for days, going through various stages. In the last stage we fired .50-calibers at various silhouettes that were set up. Then we went back to Molesworth to fly practice missions with instructors to show us what we were doing wrong. When we finished, they figured we were ready for combat."

There was one member of Lt. Henderson's crew who took the gunnery instructor's warning to heart. The crew's youngest member was another New Jerseyan, 19-year-old William Simpkins from Egg Harbor. As he recalls:

"Our tail gunner quit after our first mission. He just up and said he didn't want to fly anymore. We got George Buske as a replacement. He was an experienced gunner who had flown before, and he hung around with Ralph Burkart, our right waist gunner. Burkart was from Columbus, Ohio. Our left waist gunner was a guy named Stan Moody, from Maine. He was a real gun expert, a gunsmith or something. He hunted in the woods a lot and used to fool with guns all the time. Our bombardier was Woodrow Monkres from Oklahoma—small but tough. And our navigator was Warren Wiggins, from Long Island. He could really get you around."

The crew's radioman was Forrest L. "Woody" Vosler, from upstate New York. Ed Ruppel remembers that "Vosler was 20 years old, and as cocky as the day is long. He acted like he had the world by the tail.

"Our copilot was a guy named Ames, who had washed out of fighter pilot training. But he wasn't with us for the first five missions. Instead, we had Captain Hungerford as our Instructor Pilot."

The crew's pilot is the man the others remember best. As Forrest Vosler describes him:

"Henderson was 29 years old and had many, many hours in the air, much more than the average pilot. He started out flying civilian aircraft when he was 16 years old. He had been in the RCAF, and then he transferred into the Army Air Corps, where he had to go all through flight training again. By the time he joined us he was a seasoned, tremendously meticulous pilot."

As Bill Simpkins recalls: "We first met Henderson when he was a staff sergeant. He had been in the Canadian Air Force, and then they made him a flight officer. Before the war he and his brothers had an airport in California. He had flown for years, and he was a great pilot."

Henderson's crew was slated for the next maximum effort planned by Pinetree for November 26th, as were Hullar's crew, Captain Don Gamble's, Lt. Carl Fyler's, and Lt. Bill Fort's. After the Knaben raid the Eighth had launched a number of small-scale attacks, on the 18th and 19th of November, and the 303rd had not been scheduled for them. This next operation would make up the difference; it was the mission Norman Sampson remembers as "the most scary of them all."

20

The Battle of Bremen Begins

November 26, 1943

THE MORNING OF November 26th saw the Eighth mounting its largest mission ever—633 heavy bombers were aloft, escorted by 353 P-47s and 28 P-38s. The majority of the bombers—505 from the First, Second, and Third Divisions—were heading for a PFF attack on Bremen; 128 more from the Third Division were en route to strike an industrial target in Paris.

The 303rd launched no fewer than 37 B-17s from 0835 to 0859, filling both the lead and high group positions in the first of nine wings bound for Bremen. General Travis was leading the Eighth in a PFF ship flown by Captain Jake James, and Captain Gamble's crew was flying as deputy lead in the No. 1 position of the lead group's high squadron aboard "old reliable," *Sky Wolf*.

To Gamble's left in the No. 3 slot of the high squadron was Lt. Bill Fort in *Star Dust*, and below Fort in the No. 5 slot of the lead group's lead squadron flew Lt. Carl Fyler and crew in *Dark Horse*, B-17F 42-9498. Hullar's crew led the high squadron of the high group in No. 823, the B-17F they had flown on the November 3rd Wilhelmshaven raid. With them again as copilot was Lt. Colonel Culbertson, who had bumped Bud Klint a second time, and since Mac McCormick no longer flew with the crew regularly, Elmer Brown was in the nose with a new bombardier on his second raid, Lt. E.L. Matthews. Below them in the No. 3 slot of the high group's low squadron flew Lt. John Henderson's crew in *Hell's Angels*, with Captain Merle Hungerford as Instructor Pilot.

Lt. Bill McSween's notebook contains a good description of the journey into the target area:

"We took off early and climbed to 10,000 feet. We took the usual route out just north of No. 4 splasher. The target course wasn't followed exactly

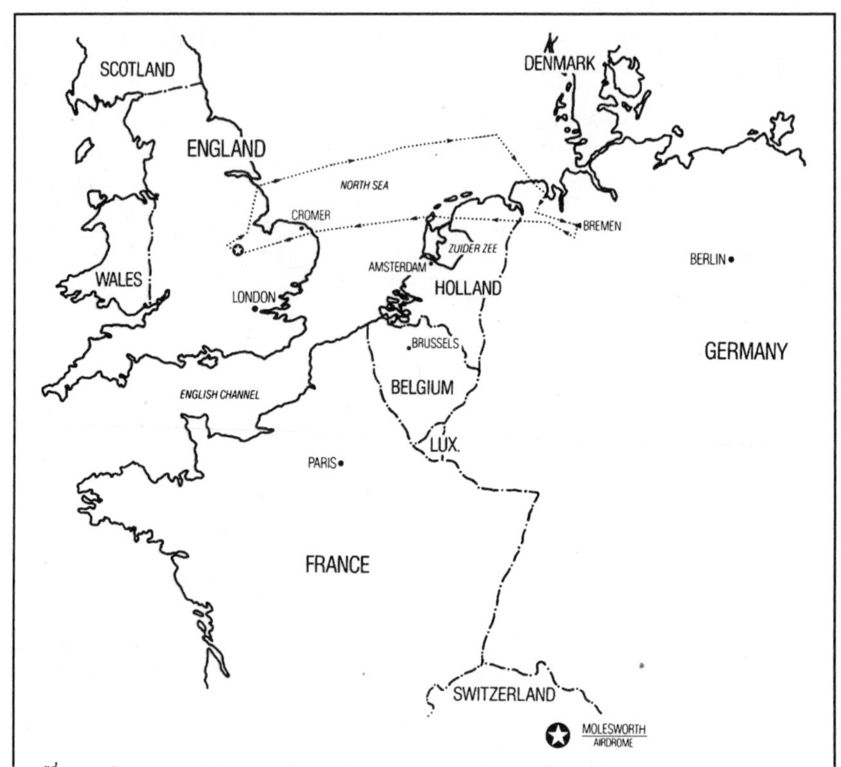

303rd Bomb Group Mission Route(s): Bremen, November 26, 1943.
(Map courtesy Waters Design Associates, Inc.)

as briefed over the North Sea. Due to high clouds our climb was up to 27,500 feet...Our DR turning point was passed by about 15 to 20 degrees before turning to the enemy coast, where again we were left of course, hitting an island to left of course.

"After flying back west for a few minutes we made a left turn on to the bomb run at 100 degrees, magnetic heading. There was a 10/10 undercast, and vapor trails were the worst I ever saw...The temperature was at least -50 degrees C. and possibly lower...Ice on windows made visibility practically nil."

Captain Gamble commented on this too: "Wag, Ralph, and Bill's windows frosted up bad. Was about 50 to 55 degrees below Centigrade."

It was in these terrible conditions that the Luftwaffe pounced. The P-47s were "too high and too late" according to many disgruntled bomber crews, and they were unable to stop the enemy from hitting the 303rd hard. Its two group formations were assaulted by about 50 FW-190s, Me-109s, and rocket-firing Me-110s.

The Germans began with head-on attacks against the lead group in an effort to knock out the mission leaders, namely Captain James's PFF ship and Gamble's. Don Gamble wrote that "Ten to 15 fighters hit us just as we crossed the coast. Made definite passes at lead squadron and us. Had to do violent evasive action. One fighter—109—passed close overhead," and it must have been this incident that Ralph Coburn was referring to when he commented that evening about "one nose-on fighter attack that was a lulu."

Lt. McSween noted that "several close passes were made at the General," with one fighter actually trying to ram Captain James's ship.

McSween took the attempt to get *Sky Wolf* personally, writing: "Jerry attacked in formation, coming in four abreast. The sweep made to get us (me) got Fort." *Star Dust*, badly damaged, fell out of the formation while McSween noted: "The only trouble we had was No. 4 engine shaking lead in the nacelle."

Star Dust's fall had an even greater impact on Lt. Carl Fyler in *Dark Horse*, who was flying just below Lt. Fort. Fyler was on his 24th mission, and the incident was, perhaps, just too much like the close call he had had with Lt. Crockett's stricken ship on the August 27, 1943 mission to Watten:

"The lead ship in front of me took a direct hit in the nose section. The front of his aircraft came clear off...His plane mushed down on top of me. I had to leave formation to avoid a collision. Pieces of metal and such littered the sky. I was shook! Too many trips and then this. I was having such a hard time coping that even flying the plane had become a challenge. My nerves were so shot I dared not rejoin the formation. One by one, our guns froze up and we were all alone. We had become a straggler and so vulnerable, as if we were just begging to be shot down. At last I managed to move up parallel with the group, and we flew back to England that way."

This first sweep occurred at 1130 and the fight lasted over an hour, running from the enemy coast on into the target and back out, the Germans hitting the bombers around the clock but concentrating on nose and tail attacks.

It wasn't long before other gaps were torn in the 303rd's ranks. From the 359th Squadron Lt. H.S. Bolsover's crew was flying B-17F 42-5117 in the No. 6 slot of the high group's lead squadron. They were attacked by Me-109s that knocked out two engines and so badly riddled the bomber that she was later described as "almost completely shot to pieces." Lt. Bolsover's tail gunner was hit in the shoulder and seriously wounded. His left waist gunner was badly hit in the shoulder and head. His radioman, Sgt. R.K. Roberts, was fatally wounded. It was Roberts's 24th mission.

Lt. John Henderson's crew was also having a rough time. The two waist gunners, Moody and Burkart, passed out during the battle due to an oxygen system failure. From the radio room Sgt. Vosler saw that they were out and rushed back to revive them with portable oxygen bottles; he earned an Air Medal for his actions.

In the meantime, Sgt. Ed Ruppel was fielding attacks from the ball turret. He later commented that "They tried to hide in the vapor trails left by the Forts and sneak up without being seen. I saw three or four of them try this, but it didn't seem to work too well. The gunners drove them off as soon as they came out in the open."

Hullar's crew was fighting the same kind of battle in No. 823; they were getting beam attacks and a large number at the tail. Merlin Miller remembers one in particular where Norman Sampson's help was critical.

"We were leaving heavy contrails. Sammy called from the ball turret to say he saw a fighter sneak into the contrails way back behind us. So I pointed both tail guns straight back, just above the vapor trail, and waited. Sure enough, that fighter popped right up out of there. I fired, and I'll bet that German pilot still had a surprised look on his face when St. Peter opened the Pearly Gates for him."

Up forward, Elmer Brown and Lt. Matthews were getting nose attacks from one o'clock high, but he and Lt. Matthews weren't able to fire with effect because their nose guns were in poor condition. Hullar later insisted that better firing pins be put in them.

The Group finally got on the bomb run where flak gave the bombers an equally rough time. The 303rd's B-17s were unable to see the Bremen industrial area through the clouds, but this didn't prevent the German radar flak from seeing *them*.

Lt. McSween wrote: "Flak through the overcast was *damn* accurate, but moderate. We got a few flak holes, and some awful close bursts of flak."

Captain Gamble also commented on the "Fairly intense and very accurate flak [that] came up through 10/10s clouds. Best I've seen when they fired by screening us. Tracking fire several minutes into target..."

Elmer Brown likewise wrote that "The flak was quite heavy over the target," and it was this deadly antiaircraft fire that almost finished the Hullar crew's career.

Brown described what happened next this way:

"Right after the target we lost oil pressure on No. 2 engine, and the pilot couldn't feather it. The engine started running away. He cut the engine, but the propeller continued to windmill at a terrific rate of speed, and turned the engine over. This caused a violent vibration of the aircraft, and danger of tearing the wing off.

"The pilot said, 'Put on your chutes and prepare to bail out.' There we were over Germany heading for Holland, which was 60 miles away."

Bob Hullar recorded the incident with a coolness that matched his conduct: "Lost No. 2 over target." He might have been "too busy to be scared," but the others waited nervously to see what was in store.

Norman Sampson remembers that "The aircraft shook and shuddered so bad I thought it was going to fall apart. Hullar told us to prepare to bail out. The only thing I could think of was, 'My parachute, how tight is it on me?!' I made sure I had my escape kit. I was going to walk back home. At least that's what I thought."

Merlin Miller was reluctant to give up the ship: "We were still over Bremen when all of a sudden the plane started shaking to beat hell. The motion was magnified back in the tail and I was rattling around like a pea in a pod wondering what was going on. Hullar came on and very briefly explained the situation, and told us to get our chutes on, get ready to bail out, and he actually rang the alarm button. Then the intercom went dead.

"I couldn't talk to anybody and I figured everybody jumped. I crawled to the little escape hatch in the tail ahead of the tail gunner's compartment. It had a handle on it which you could pull and kick the whole door off. But I was a little cautious. I didn't do that because I wasn't sure if I was going to jump or not. I pushed the door open with my foot, and sat there with my feet hanging out for a minute looking down.

"I could see the fire and smoke down there in Bremen, and I decided, 'To hell with this. I'll be safer in the airplane.' With what we were doing to the people down there, I figured they weren't going to be too friendly to anybody who landed in their midst. So I crawled back to the tail guns and started looking for fighters, just waiting, under the impression that I might be by myself because the plane was still bobbing around quite a bit."

While Miller waited, up in the radio room George Hoyt and Dale Rice were playing out another drama over the decision to jump or stay. As Hoyt recalls:

"The alarm bell rang, which meant 'Get ready to bail out.' I got my chest pack off the floor and snapped it on my harness. I saw the two waist gunners at the waist door ready to go. Then Dale came back to the radio room. I knew there was a good chance of the whole outer wing panel shearing off from the vibration, but I refused to accept that this was how it was going to end. I said my prayers and believed very strongly, 'We're not going down. There's no way.' We had a tour of missions to finish, and we were going to get back to the base and were going to finish them. And the worse the vibrations got, the stronger my belief became.

"Dale was yelling in my ear, 'We need to go out the bomb bay!' He opened the radio room door, and I said 'No, close the door, we're not going to have to go!' He kept shaking his head, saying 'We're going to have to bail out.' So I gave him the thumbs-up sign, and I kept giving it, but he opened the radio room door a second time, and now he had his foot on the catwalk above the open bomb bay doors.

"I could see that he was ready to jump. So I gave him forceful thumbs-up gestures once again, trying to convince him that we would make this thing, we'd beat it, we weren't going to go down over German territory."

Something had to give, and after a number of minutes that seemed like a lifetime, it did. Elmer Brown wrote: "Fortunately, the propeller connection snapped loose from the engine. It continued to windmill, but as the engine was not being turned over, the vibration ceased."

A semblance of order returned and Merlin Miller heard "the interphone click and Hullar ask, 'Is there anybody back there? Crew check.' So we checked through and we were all still there."

Everyone shared the one, overriding emotion Norman Sampson expressed: "What a relief!"

But the crew was still in a bad way. As Elmer Brown described it:

"The mechanical danger was past, but we had fallen way behind our formation. Our two wing planes had stayed with us, and a plane from a strange group filled in our No. 4 position. There we were, a four-ship formation, an easy target for enemy fighters. Other troubles we had were that the temperature was down to −50 degrees C, and all our guns froze up except the nose guns. When we lost our No. 2 engine we lost our heating system also."

It didn't take long for the German fighters to arrive. Merlin Miller remembers that "Right about the time the prop shaft snapped, our little formation was jumped by a bunch of German fighters, and things started to look bad." It is a point of pride to Miller that "I managed to keep at least one of my guns going at all times," but even so it was a sticky situation.

As Hoyt recalls: "We got several very intensive attacks at the tail from FW-190s, but we received great help from our wingmen, who covered our rear with all the guns that were able to fire behind blazing away. Bob had ordered the wingmen to get back to the main formation earlier, but they stayed with us at very great risk to themselves. I saw some true heroism that day."

Not surprisingly Brown wrote: "We were very happy that P-47s soon joined us and carried us to the Holland coast. We had to cross 134 miles of the cold North Sea. I didn't have a Gee box, [a radio navigation system very similar to present-day Loran], the radio compass was erratic, and the flux gate compass was inaccurate. I brought them home by pilotage, and was very glad to get home."

While Hullar's small formation was crossing the North Sea, the main body of the 303rd began landing at Molesworth beginning at 1420. Lt. McSween wrote that "We flew 270 degrees out to the Zuider Zee, then to the coast south of Cromer to the base. Don did a good job of flying." Gamble commented: "Had to go around on landing. Made a nice three-point landing."

When the mission was finally over, 33 of the Group's ships had returned to Molesworth, three landed away, and one didn't come back at all.

The missing 303rd bomber was *Mr. Five By Five*, flown by Captain Adele A. Cote of the 427th Squadron at the head of the lead group's low squadron. He was on his 24th raid and was last seen at 1245 over the Zuider Zee with his wheels down, under control and turning left at 10,000 feet. It was believed at the time that Cote went down over water while trying to return to the Continent, but the fate of *Mr. Five By Five* and Cote's crew remains a mystery to this day.*

For the Eighth as a whole the November 26th mission was an expensive proposition. The Bremen bombers and fighters claimed a total of 70 German aircraft but the cost was high—22 B-17s plus three B-24s and four missing P-47s. The raid to Paris cost four more B-17s and proved fruitless due to cloud cover over the target.

For most of the bomber crews there was plenty more to come, since the "Battle of Bremen" was just beginning. But any account of this day would be incomplete without the stories of two men who returned from this raid, never to fly again. What Lt. Bill Fort and his bombardier endured is a tale that deserves a separate chapter. It is a hymn of perseverance, faith, and hope utterly without equal in the Eighth's air war.

---

*After the war, captured German documents revealed that the body of the crew's bombardier. Lt. John W. Hull, was discovered on the north beach of Terschilling Island in the Frisian Islands Group on February 13, 1944. It was Hull's 24th raid. Among those who remain missing is the crew's copilot, Lt. Clarence C. Bixler, who had been in the ETO since May 1943 and who had volunteered for this, his first mission. The author is indebted to Lt. Bixler's sister and brother-in-law for providing this information.

21

Star Dust

Bremen, November 26, 1943

Out of the depths I cry to you, O LORD;
LORD, hear my voice!
—Psalm 130

The story of *Star Dust* is a tale of two journeys—one in a badly crippled B-17, the other into the inner reaches of a man's soul. It is a story that is best told by three individuals, William Fort, Grover Mullins, and Charles Spencer, but some introductions are necessary to place their accounts in proper context.

Lt. Fort was now the captain of his own ship, having been left by Lt. Calder Wise, his Instructor Pilot, two missions before. This was his eighth raid and his third trip with his crew's regular copilot, Lt. MacDonald Riddick, of Beaumont, Texas. Lt. Fort's navigator this day was Lt. Harold J. Rocketto, of Brooklyn—this was his first mission. The engineer was a regular member of Lt. Fort's crew, Sgt. Grover C. Mullins, of Windsor, Missouri. And Lt. Charles W. Spencer from Peoria, Illinois, was the bombardier. He had been with the 358th Squadron before Fort's crew had arrived, and he had flown a number of raids with them. He and Bill Fort were friendly even though they were not particularly close to one another.

How Spencer came to fly this mission and the events leading up to the first fighter attack are things he can best describe. From here on he, Fort, and Mullins will relate what took place; Spencer recalls the preliminaries this way.

Spencer: "I had returned from R&R not in the best physical condition because I had the flu the last few days. I was there so I spent it in bed. So when I came back to the Group and to the 358th Squadron and talked to Captain Hungerford, he said. 'Well, I'd like to have you go on a mission, but I don't know whether to have you go since you're getting over flu.'

"And I said. 'Well, I think I feel good enough to go, so why don't you put me down.'

Lt. William C. Fort, Jr.'s crew. Bottom row, L-R. Sgt. Grover C. Mullins, engineer; Sgt. James Pleasant, waist gunner; Sgt. Howard Zeitner, ball turret gunner; Sgt. Bernard Sutton, tail gunner; Sgt. James Supple, radioman; Sgt. John Viszneki, waist gunner. Top row, L-R. Lt. Calder L. Wise, Instructor Pilot (not on 11/26/43 mission); Lt. Fort, pilot; Lt. John Nothstein (not on 11/26/43 mission); Lt. Charles W. Spencer, bombardier. (Photo courtesy William C. Fort, Jr.)

"So I signed up for the mission. It was going to be a maximum effort and that meant that everything that could fly would fly. The only thing I didn't like was that we were going to fly in *Star Dust*, which was not the best airplane on the field and was usually thought of as a 'Hangar Queen.' But that didn't make too much difference to me. I was looking forward to finishing up my missions now that I had 15, so I was ready to get my 16th mission in.

"I remember the morning of the 26th being awakened early and having a nice regular breakfast of real eggs, having the briefing and having the time to get to the dispersal area and to get all ready for the flight.

"We took off about 0800 and it was a cold November morning. It was dismal, it was dreary, and it was overcast. We flew through a cloud layer to get up and out and on our way and I was flying with a navigator that I just met. His name was Rocketto and he was eager, just like I was on my first mission. We went over Eyebrook as usual and then headed out over the coast, headed for the Dutch coast on up the Channel and the North Sea and headed on for Bremen.

"It was a morning that was kind of ominous. There was something about it that I didn't like. I don't know if it was that I was flying an airplane I didn't like and having this strange fellow that I didn't know much about and all that, but it seemed to portend something.

"And my gun was sluggish. When I test-fired that gun I knew it wasn't going to fire real good. I tried to fix it by checking the headspace, but I left my knife in my room and I couldn't take the plate off so I wasn't able to remedy it. I tried to shake it off and kind of had a little banter as we went on.

"As we got near the target I began to realize that they were an awful long time on the run and I realized maybe they had overshot. I realized that when they began to turn back and then they turned in. So everything seemed to be just a little bit *off* on this. As we headed in towards the target I began to have some kind of qualms. I suppose it was because everything seemed different this morning.

"As we looked for our fighter escort, we found the P-38s very easily, because they would throw their wings up and show their twin tails. I certainly saw the 38s. But I couldn't pick out the Thunderbolts.

"And then I saw the Focke-Wulfs. They were coming in from the right. They were slightly high and they had plenty of contrails; it looked like there were plenty of them and then they seemed almost to turn in *en masse*, and I just picked out one to start working on. As I got off my bursts, I could hear Rocketto's gun firing at the same time. We were both firing. My gun was sluggish, but I fired it."

Fort: "I could see contrails about six miles straight ahead of us, and about the same distance 90 degrees to our right. I figured the Germans were the ones ahead of us, and our fighters were off to the right. But our fighters never seemed to get any closer to us, while the Germans did. They came barreling right on through, firing, and we got hit."

Spencer: "That's when the big 'splat!' hit the nose, and that's all I can describe it as, a big 'splat!' It seemed as if someone had thrown a wet mop against the nose and that's all I remember."

Fort: "They blew out the Plexiglas nose. Both the navigator and the bombardier had been firing, but they only got off a very few shots, just one little burst. The navigator was under my feet in the nose, and out the cockpit window I saw his gun barrel suddenly go straight up—it had weights on it to counterbalance it—so I knew something had happened to him. It was the first German fighter he saw, and he didn't fire more than a dozen shots before he was hit. I called on the intercom to him and the bombardier, but I didn't get an answer."

Spencer: "The next I remember is the fact that I was looking into a brilliant red, and then I realized I was looking into my own blood. It was covering my chest chute pack and I usually kneeled on it. There was all this red. And then I felt, to see if I could feel Rocketto, and I couldn't feel him.

"Then I began to get a little weak and I began to realize the cold, the terrible blast of cold that was coming in there. As I felt the movements of the plane I began to wonder if we were going down. The helmet that I usually wore over my leather helmet I removed, and since I didn't hear anything through the earphones, I took them off. I took off the flak pack that was covering my chest and clipped on my parachute, and got ready to hear the bell to get out when the bell would sound."

Fort: "We fell out of formation, and things were quite confused for a few minutes. The Germans knocked out part of our oxygen supply, running down the right side of the plane, and the cold air was blowing up into the cockpit through the floor hatch to the nose, even though it was closed. It was like a wind tunnel. With the oxygen system out the copilot passed out pretty quick, and the engineer kept putting walk-around bottles on him to keep him going. The guys all down the right side of the plane back to the waist gunners had the same problem, and were using walk-around bottles. Mullins did quite a job helping them out."

Mullins: "I got all the crew that was passed out from lack of oxygen on portable oxygen bottles, and got the copilot woke up and out of his daze from being passed out from lack of oxygen."

Fort: "I wasn't getting any oxygen, either. The plane had the old type constant-flow system, and my mask froze. I had to pull it up to get any air, but we somehow got back into another formation. We went on into the target, but didn't drop any bombs.

"I sent my engineer down to the nose. He came back up, saying the navigator was dead and the bombardier was dying. Mullins couldn't get an oxygen mask on the bombardier because his face had swollen up so much in the cold. It was swollen up the size of a basketball. Mullins like to froze to death himself up there."

Mullins: "I went down into the nose compartment and got Lt. Spencer up into the cockpit area, but he wouldn't stay and went back down into the nose section, which was blown away, in an attempt to use his guns to ward off fighters in order for us to get home."

Spencer: "I began to realize too that the plane was more or less under control. It was in a struggle of some kind, I recognized that. But I began to search for the rheostat to turn up my heated suit. I had a heated suit finally, but I had no gloves or boots to hook onto it, but the heated suit was on medium and I wanted to put it up to high because I knew I needed all that extra heat.

"So that thought was in my mind. And then a great darkness seemed to begin settling over me, with pressure like a thousand vises crushing in. I began to realize that maybe this was it for me. Maybe I was dying."

November 26, 1943. Lt. Charles W. Spencer at the forward gun of Star Dust *after shells from a head-on German fighter attack shattered the plexiglas nose, killing the crew's navigator, Lt. Harold J. Rocketto. Badly cut in the face and removed from the nose by the crew's engineer, Sgt. Grover C. Mullins, Lt. Spencer returned to man his battle station despite freezing air at a temperature of −60°C blasting through the compartment at over 125 m.p.h. Horribly injured by frostbite in these terrible conditions, Spencer was awarded the Distinguished Service Cross for his determination to protect his ship and crew at all costs. Black and white reproduction of an original 1997 painting by Geoff Pleasance, part of the "Heroes of Molesworth" Series commissioned by Brian S. McGuire for display at the Joint Analysis Center Headquarters Building, RAF Molesworth, England.* (Reproduced with permission of Brian S. McGuire).

"It was then that I really thought about my life, and how I hadn't accomplished too much in the 24 years I had lived. Then I thought of my loved ones back home. And I saw their faces, I saw their anguish, I saw their sorrow, I saw their tears, I saw their heartache, all these things had begun to build up in my mind's eye.

"And whether I cried out in actual words I do not know, but I did cry out unto the Lord to save me, to spare me, not to let me die because of the loved ones and their hearts, and their receiving news that I wouldn't return. This was tearing me apart in my mind. And certainly I asked the Lord's help.

"And certainly we know that He *stepped in there* and preserved me during those hours that I lay in that cold air until we got back to the English Coast. But it was a terrific experience, and I must have blacked out again, for I don't remember anything until we got back to England."

Fort: "I forget how far we had gone on the way back when No. 3 engine quit. It just died. So we could have done one of two things. We could have pulled up out of the formation and hit the deck. We might have made it back, and might not have. But as a rule they always said stay with the formation, so that's what we did. I don't know if it was the right thing or not the way it worked out.

"The engine was wind-milling pretty good, but it wasn't causing any real problem except for the drag on it. We were running the other three engines at almost the maximum power. The guys, a lot of them, were crying, they wanted to go down and get oxygen, but I figured if I wasn't getting it they could do without it too. We were at 26,000 feet. They say you can't live at that altitude, but you *can*, don't let them kid you. That bombardier did."

Mullins: "I remember some of the crew yelling in the intercom, 'Grover, take us down, we're dying.' I remember yelling back in the intercom, 'SHUT UP, we're not going down, we're getting home somehow.'"

Fort: "After we finally saw water, we knew we'd be safe. We decided to go on down and take our chances. There was a lot of clouds, and if we got shot at we could slip into them, maybe, and hide for a few minutes. My hands were hurting me pretty bad at the time; I hadn't been wearing my regular flying gloves, just the silk liners, and they were badly frost-bitten from my having them on the controls with the cold air in there, so I told the copilot to go down lower. He put the plane into a dive like we were going to dive-bomb something.

"That's when the prop to No. 3 started racing like mad. I took the wheel and pulled the plane up level to slow the prop down, but it didn't slow down much. We started vibrating bad enough to shake the wing off in a couple of minutes. And then we heard something give. The shaft broke loose. The prop was still windmilling, but it wasn't connected to the engine anymore.

"So we got down over the water, and I salvoed the bombs. I didn't do it over land because if the Germans had seen us they would have figured we were already hit, and they went after stragglers. Then we had to worry about gasoline. I knew we were going to hit England; I figured we couldn't miss it, but I didn't know where."

Mullins: "I tried to salvo the bombs that wouldn't drop. The pilot's hands were frozen; I tried to thaw them out, transferred gas from the engines that wasn't running to the tanks where the engines were still running, manned my guns when I got a chance, helped feather the props, and helped adjust the speed on the remaining engines that were running. The pilot's hands were frozen so bad he could do very little."

Fort: "Quite a ways out over the North Sea we ran into an air sea rescue B-24 flying around. We tried to get in touch with them to give us a heading to the nearest landfall, 'cause that would help us out quite a bit. But their reception was bad, and we couldn't get anything out of them, and eventually they just flew off and left us.

"So then we just kept going. We were down to about 500 feet over the water, and we made landfall. Luckily, this little English grass field was close by. So I was trying to go around a downwind leg to come in to land, and No. 4 engine quit. With the copilot's help we eventually got the plane turned around, and came back to this little field and landed. Within an hour of landing, myself and the severely injured bombardier were flown to a British hospital for treatment."

Mullins: "I remember helping the pilot and copilot trying to land as we only had one or two engines left running at that time. All and all it was a very stressful mission, but then in a few days, the ones of us that were able, were right back on a plane on another mission."*

Spencer: "I didn't remember anything until someone was shaking me, someone was trying to rouse me, and then I heard voices that were definitely English, 'Limey' we'd say, and they said, 'This one's alive. This chap's alive.' And the next thing I knew they were filing on my wedding ring, they were dipping my hands in the water, and they were doing various things to me that I was just half conscious about."

Lt. Bill Fort's landing marked the end of his Eighth Air Force career. His hands had been so badly frostbitten by the frigid air in *Star Dust*'s cock-

*Mullins extraordinary efforts on this mission earned him the Silver Star, though others on his crew thought he deserved the Medal of Honor and tried, unsuccessfully, to put him in for it.

pit that he spent the next 13 months in military hospitals having them treated and operated on. A number of joints on his fingers had to be amputated, and he was mustered out of the Army Air Force on a permanent disability in January 1945.

While he never regained full use of his hands, Fort feels he has compensated for this well over the years, and he has enjoyed a full working life. When asked about the events of November 26, 1943, he still wonders, in view of what occurred, whether he made the right decision by staying at altitude after *Star Dust* was hit. But Bill Fort is reconciled to the realization that "I'll never know."

Lt. Charles Spencer lost far more physically. All his fingers and all of his toes had to be amputated, making the way in which his wedding ring was removed all but incidental. And his face was virtually destroyed by that merciless blast of freezing air; the Army's plastic surgeons labored for years creating a new one for him. Charles Spencer also lost an eye, frozen in its socket during that flight, and the vision in his other eye was permanently impaired.

But what Charles Spencer lost in his body, he gained in his soul. In *Star Dust* he begged the Lord to spare his life for the sake of his loved ones; and his wife, Jeanne, has never left his side. And while his vision is now fading even more with the years, those hearing this man speak immediately sense in him the Spirit still burning with a fierce, internal light. For three of the past four decades Charles Spencer was chaplain to a Soldiers' Home in western Kansas, and he is, even now, tremendously active in his calling as a Baptist minister. As he states his case today:

"I praise the Lord not only for what He did for me in 1943 but for what He's done for me ever since. I praise the Lord for a good wife who stood right with me, and I praise Him for two fine boys who did their part serving Uncle Sam in the Service. Life is good and I am thankful, indeed, that the Lord spared my life, that I might be used somewhat in this world for Good. Amen."*

*Lt. Charles Spencer was awarded the Distinguished Service Cross, the Nation's second highest award for military valor, for his extraordinary conduct in returning to *Star Dust*'s shattered forward compartment to man the nose gun despite his terrible wounds. Like Mullins, there are *many* who feel that Spencer deserved the Medal of Honor for his actions.

22

Sawicki's Sacrifice

Bremen, November 29, 1943

THE EIGHTH'S LEADERS pressed their assault on Bremen again on November 29th. This time the plan called for blind bombing by 360 B-17s from the First and Third Divisions with 314 P-47s and 38 P-38s providing escort. The 303rd contributed 20 B-17s from the 358th, 359th and 360th Squadrons, so Hullar's crew got to stay home. It was just as well for them, for this trip was to prove every bit as rough as the mission three days before.

The 384th Bomb Group was leading the 41st CBW, with the 379th flying high group and the 303rd flying low. As on some prior missions, lack of coordination among the groups in the Wing would again make a big difference in how the 303rd fared.

The 303rd formation was led by the 358th Sqdn., with Major Kirk Mitchell as Group lead in No. 865, B-17F 42-30865. The 359th Sqdn. was flying low, and the 360th Sqdn. was high. Flying as Deputy Group Lead at the head of the 360th Sqdn. was the experienced crew of Lt. Pharris Brinkley, aboard B-17F 42-5859. They are introduced by Guy Lance, the crew's left waist gunner, this way:

"I was assistant engineer and left waist gunner. This was our twelfth mission...Because we were flying lead our regular bombardier was replaced for this mission by a lead bombardier [Lt. F.T. Clark]. The other nine men aboard had been together since our flight training began in Ephrata, Washington about six months prior to this jaunt. [They] were: Pharris Brinkley, pilot; John Parrott, co-pilot; Sylvester "Red" Becker, navigator; Arthur Worthington, engineer and upper turret; Richard Snyder, radio; Charles Ferguson, right waist; Harold Reid, ball turret; and William Rein, tail gunner."

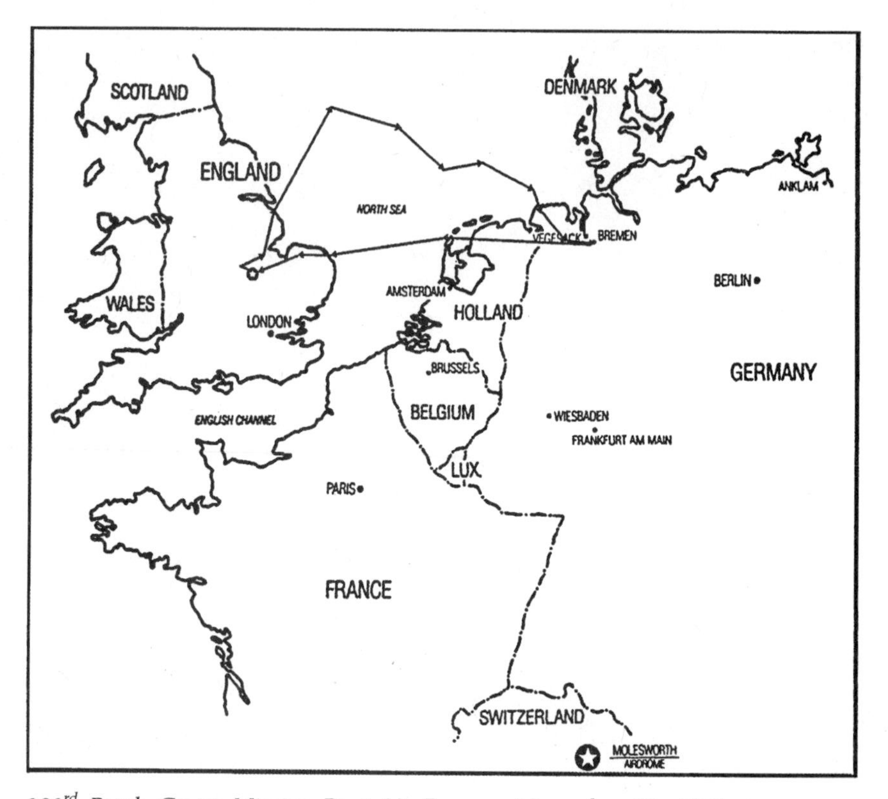

303rd Bomb Group Mission Route(s): Bremen, November 29, 1943. (Map courtesy Waters Design Associates, Inc.; mission route added by author.)

Flying above and to the left of Brinkley's crew in *Dark Horse* was Lt. Carl Fyler and crew. As Fyler recalls, "This was to be my 25th mission. I was so battle weary that I hardly functioned, but I acted. I did what I thought was right. Maybe, subconsciously, I knew it would be over, one way or another."

In addition to having on board Lt. John Petrolino, the regular bombardier on Brinkley's crew, and an aerial photographer, Sgt. N.P.S. Egge, Fyler's men included many regulars who had been with him from the beginning: Sgt. Bill Addison, engineer and top turret gunner, on his 25th raid; Sgt. Ray Ford, ball turret gunner; and Lt. George Molnar, navigator, both on their 24th raid; and Fyler's two veteran waist gunners, Sgt. George Fisher with 23 missions at the right waist and Sgt. Martin "Marty" Stachowiack with 21 missions at the left. Also aboard was an experienced copilot, Lt. Robert Ward, on his 13th raid, and a new radioman, Sgt. Raymond O'Connell, on his second mission. Last but not least there was Sgt. Joseph R. Sawicki, a Polish American from Detroit with 14 B-17 missions to his credit who had joined Fyler's crew after transferring to the 303rd from the RAF's Free Polish Air Force.

Flying off Fyler's right wing in the No. 5 high Sqdn. position was the rookie crew of Lt. F.B. Brumbeloe, in *Red Ass*, all of whom were on their third raid.

Major Mitchell's Group Leader's Narrative chronicles the mission up to the enemy coast: "Took off on time at 1116 hours. The Group assembled over the field at 1500 feet. Departed field at 1500 feet for Eyebrook. Rendezvoused with 384th Group over Molesworth and the 379th Group fell in on us 10 minutes after leaving Molesworth. Went to Splasher #4 and began climb as scheduled. The climb was terrifically fast. For 35 minutes I indicated from 155-158 MPH at 400-500 F.P.M. [Feet per Minute].

Lt. Pharris Brinkley's crew standing before the 360th Squadron's Red Ass, *B-17F 42-5483, PU✪F, lost with Lt. F.A. Brumbeloe's crew on the 11/29/43 Bremen mission. Note "kicking jackass" nose art under cheek gun window. Pictured L-R, Top Row are: Lt. John Petrolino, bombardier (shot down with Lt. Carl Fyler's crew on 11/29/43 Bremen mission); Lt. John Parrott, copilot; Lt. Sylvester "Red" Becker, navigator; and Lt. Pharris Brinkley, pilot. L-R, Bottom Row: Sgt. Richard Snyder, radioman; Sgt. Harold Reid, ball turret gunner; Sgt. Charles Ferguson, right waist gunner; Sgt. Guy Lance, left waist gunner; Sgt. William Rein, tail gunner; and Sgt. Arthur Worthington, engineer and top turret gunner.*
(Photo courtesy *Hell's Angels Newsletter.*)

Lt. Carl J. Fyler's crew late in their tour posing in front of the 360th Squadron's Alley Oop, *B-17F 42-5854, PU✪C. Note the unconventional cheek gun and window just behind the aircraft's nose. Pictured L-R, Top Row are: Sgt. George Fisher, right waist gunner; Lt. George Molnar, navigator; Sgt. Bill Addison, engineer and top turret gunner; Lt. S. Gibson, bombardier (not on 11/29/43 mission); Lt. P. Tibbits, copilot (not on 11/29/43 mission); and Lt. Carl J. Fyler, pilot. L-R, Bottom Row are: Sgt. Joseph R. Sawicki, tail gunner; Sgt. G. Crowder, radioman (not on 11/29/43 mission); Sgt. Ray Ford, ball turret gunner; and Sgt. Martin Stachowiack, left waist gunner.* (Photo courtesy Carl J. Fuler).

Crossing the North Sea and still were dropping behind. When the lead Group turned into the enemy coast, we cut them off on the turn and closed on them indicating 155 IAS at 26,500 feet...We had dense, persistent 'con' trails from 17,500 feet on up over the North Sea and on over the continent. East of us at this time lay altostratus clouds up to about 23,000-24,000 feet. Below this layer was a deck of strata-cumulus of 7/10-9/10ths. We closed into low Group position about the time we crossed the enemy coast at 26,500 feet and after catching up to the lead Group we were indicating about 140-145 and overrunning. The P-47s picked us up about the time we crossed the enemy coast. Our course took us right along the top of a thick but not too dense cloud layer and at least twice we went through wisps of cirrus tufts of cloud at 27,500 feet. We were flying just above the cloud layer and if we had been in proper Low Group position on the Lead Group, our formation would have been down in the clouds."

The Group suffered a 25% loss in strength from aborts before reaching the enemy coast, due in part to the "terrifically fast" climb to altitude Major Mitchell reported. The first abort occurred in the 358th Sqdn., some 30 minutes after leaving the English Coast. Capt. Merle Hungerford took *Connecticut Yankee*, B-17F 42-29629, out of the No. 6 slot in the lead squadron after becoming sick to his stomach following takeoff. "I went on to midchannel hoping to get over it."

The low 359th Sqdn. lost its No. 5 and No. 7 ships—B-17F 42-3448, piloted by Lt. H.J. Eich, Jr. and *Nero*, B-17G 42-39807, piloted by Lt. W.M. Goolsby—three minutes off the enemy coast and right off the enemy coast when both "couldn't catch formation" because "during the climb, the last element of the Low Squadron was forced to enter the clouds, causing them to string out their element and also due to the Group Leader losing contact with the Wing formation and using excessive MPs [manifold pressures] and RPMs in an attempt to catch up with the Wing."

In the high 360th Sqdn. Lt. W.M. Cavanaugh took B-17G 42-39781 home when he "was unable to maintain place in formation without using excessive manifold pressures and RPMs." Excessive vapor trails which caused a heavy buildup of ice on the pilots' windshield and on the ship's turrets also contributed to his decision. Lt. W.C. Heller took *Alley Oop*, B-17F 42-5854, out of the No. 7 slot and back to Base after the No. 4 engine started losing oil pressure and began to smoke badly. "Excessive manifold pressures had been used and there was every reason to assume the trouble indicated."

Had the Wing's progress across the Channel been better, at least four of these five B-17s could have continued on. How much of a difference their defenses that would have made can never be known.

The next leg of the trip went better. Major Mitchell wrote that: "We continued on to the I.P. without experiencing any flak or fighter attacks having excellent fighter support with us." However, "'Con' trails were very persistent. Windshields and nose iced up and guns froze up. At the I.P. a left turn was made and bomb run was set up."

It was at this point that the Group formation suffered another loss, causing Carl Fyler to believe that "as I approached the target there were only two of us left in the top formation." Guy Lance explains what happened, and the narrative will follow Brinkley's crew and ship to the end of their mission before returning to the main formation.

"Flak was heavy from the IP. Goering had his bunch up in force. The combination knocked down several of our B-17s in the immediate target area. As we approached the target we lost an engine which the pilots feathered. Because we could no longer keep up to the formation Brinkley turned the lead over to the deputy leader and we headed for cloud cover en route back to England.

"Except for occasional flak bursts we encountered no opposition until we broke into the clear near the Frisian Islands. Several FW-190 fighters were waiting for us and they hit us from 12 o'clock. The B-17 fell into a near vertical dive from about 22,000 feet.

"The pilots and the engineer together, hanging onto the columns, finally persuaded the airplane out of the dive. As they fought for control I lay on my side facing the waist door. I vividly recall looking at my chest pack laying about three feet away. Centrifugal force kept me glued to the floor. Even if I had reached the chute I would have been unable to overcome the gravitational pull to bail out.

"As we were leveling off the fighters pressed for the kill, coming in on our tail from between four and seven o'clock. Worthington had not yet reentered his turret when a 20-mm exploded between his twin 50s disabling his gun position. My gun was now inoperable. Ice and frost had built up on the receiver and I could not budge the charging handle, which left me helpless to assist in our defense. Ferguson at right waist and Rein at the tail were getting most of the action. The plane was being torn apart by 20-mm cannon and 13-mm machine gun fire from the FW-190s.

"In the initial attack Rein was hit in the left shoulder. During the dive his gun sight glass frosted over and he continued the fight without a sight. He was then hit in the left leg and left arm. Meanwhile Ferguson was bleeding profusely from the nose and mouth. He had taken 20-mm fragments in the face but kept firing throughout the ordeal. I was hit in the left forearm by 20-mm shell fragments and had taken a 13-mm machine gun slug in the left thigh. Snyder at the radio was hit in the wrist as he worked at his position. The ball turret glass was obscured with oil and Harold Reid could not see to fire, which was rather inconsequential, because the attacks were from the high positions. Reid spun his turret and opened the hatch into the waist just as a 20-mm exploded with shrapnel hitting him in the arm. He ducked back into the relative safety of the turret and snapped the hatch shut.

"The fighters repeatedly attacked until suddenly we saw an orange flash behind us and the attacks ceased. In all probability Rein blew an FW-190 out of the air and the others, for whatever reason, decided to go home.

"Our B-17 still flew but her ordeal was not yet over. As we flew over one of the Frisian Islands at almost ground level, German machine and flak gunners opened up on us. Rein returned the fire. The plane had so many holes in it by this time we had no idea whether the ground fire did further damage.

"We were now over the North Sea free from enemy fire. I went back to the tail and assisted Rein to the radio room where we were all now assembled with the exception of the pilots. Those of us who could still function did what we could to administer first aid, tearing up a parachute for bandages and compresses. Ferguson, though badly wounded, took over the radio to send SOS messages. Assessment indicated that cables were damaged which made control difficult. The life rafts were destroyed. Number 1 engine had been lost over Bremen. Number 2 was maintaining power. Number 3 caught fire and the prop was feathered. Finally the fire went out. Number 4 engine was running rough requiring constant throttle manipulation. The pilots dared not feather it because whatever assistance it could offer was badly needed. We were fighting for altitude.

"Except for the radio everything that could be pried loose was thrown overboard to lighten the plane. That included guns, ammunition, oxygen bottles, parachutes, flak suits, helmets, and boots. Gathered in the radio room, we sat braced against the forward wall in ditching position. It was now almost totally dark. Following Brinkley's instructions, Reid and I moved back to the waist windows to watch for a friendly ship we might ditch near. We passed within a few hundred feet of a freighter but could not identify it and Brinkley opted to go on. The seas were rough for ditching and with less than two good engines it would have been extremely hazardous to attempt turning to look over the freighter.

"When finally we cleared the English Coast, we were directly over a new air base at Ipswich. Worthington had to crank down the main landing gear. The tail wheel refused to come down. The pilots made a two-wheel, near-perfect landing. As the plane stopped we pushed and pulled one another to safety as quickly as possible, fearing the plane would burst into flames.

"John Parrott kissed the ground. Reid noted, 'This is enough to make a man think about giving up flying.' Every gunner flying to the rear of the bomb bay was wounded and hospitalized. All recovered to resume flying status...Our original bombardier, John Petrolino, was not as lucky as we were. He was flying with Carl Fyler."

Fyler's ship was flying with the rest of the Group formation into Bremen's "intense and accurate flak" and then into a hornet's nest of German fighters, who were aided by particularly dense, persistent contrails. The Group met between 50 and 125 enemy aircraft—FW-190s, Me-109s, Me-110s, Me-210s, Me-410s, Ju-88s, and even Ju-87 Stuka dive bombers, firing machine guns, cannons, and the air to air rockets the B-17 crews called "rocket guns." Carl Fyler picks up the narrative from here:

"As we dropped our bombs on the target the flak hit us [and *Red Ass*]. The copilot, Bob Ward, was wounded in the face. Another burst hit the tail section. My ship lurched up violently. Both waist gunners had been thrown against the roof, and then slammed down on the floor, which was littered ankle deep in empty shell casings. The two engines on our right wing were damaged. Bob shut them down and feathered the props. I did not know that the right stabilizer and part of the right wing were gone. Later, the third engine caught fire. I was flying on only one engine, on the left wing. The ship nosed up, turning to right. I ended up putting both feet on the control to hold the nose down. I knew we were in deep because I could not steer to the west, and home.

"Then the FW-190s hit us. One came past very close, and I thought he had hit me. He was still firing as he passed our left wingtip. S/Sgt. Bill Addison, the top turret gunner, was hit in the thigh by a 20mm shell. He was slammed out of his turret and onto the flight deck. He looked dazed, as he lay bleeding on the floor alongside me. Other shells passed through the cockpit and the navigator and I were hit. There was blood running down my back and puddling in the bucket seat. I did not know the condition of the rest of my crew because the intercom had quit working. I motioned to Bob and Bill to get out. They went...I continued to try to fly west. No luck. I could hear one gun in the nose still firing, though it was firing very slowly for a machine gun. By now the ship was in a downward spiral to the right. I could see things on the ground, rotating. At last I decided it was time for me to go!"

Carl Fyler was on his way to a rough landing in Nazi Germany, where he was to spend the rest of the war in very difficult conditions of captivity. Bombardier Lt. Petrolino parachuted safely and escaped injury, ending the war as a POW. Navigator Lt. Molnar recovered from his back wound and also survived as a POW, as did engineer Sgt. Addison. Copilot Lt. Ward survived too, though he lost an eye from the 20mm wound to his face.

Less lucky were the men in the middle of the aircraft. Though Sgt. O'Connell, the radioman, reportedly bailed out the bomb bay, he is listed as KIA, as is Sgt. Egge, the photographer. Sgt. Ford, the ball turret gunner, was killed by gunfire at his battle station: His remains were found in the wreckage of *Dark Horse* where it crashed 25 miles SW of Bremen. The same was true of Sgt. Sawicki, the tail gunner, whose body was found with his left arm missing from the elbow down.

Therein lies a tale of true heroism, which Carl Fyler learned of first hand from Sgt. Marty Stachowiak, the left waist gunner, when Fyler met him at a repatriation camp at the end of the war. For it was only due to Joseph Sawicki, who looked first to save his comrades rather than try to save himself, that Stachowiak and George Fisher, the other waist gunner, did not die in the aircraft along with the ball turret gunner. The story of what happened is well related by Robert A. Hand, Sr., a 303rd veteran and artist who interviewed Carl Fyler for a painting of the incident and who also prepared an account of it under Fyler's guidance. Part of Hand's account is quoted below:

"A burst of flak hit the plane, tearing off the right horizontal stabilizer, part of the right wing, and kill[ing] both right engines...In the tail section, S/Sgt. Joseph Sawicki was struck by the flak burst that tore away his left arm below the elbow and that also inflicted mortal wounds to his mid-section. Bleeding profusely and in unimaginable pain, he crawled forward to the waist section to find both waist gunners, Sgt. Fisher and S/Sgt. Marty Stachowiack, wounded and dazed on the floor of the aircraft. They had suffered multiple wounds and each had a broken arm.

"With his last ounce of energy, he managed to buckle a chest pack chute on each and drag them to the waist door. Pulling the hinge-pin cable, he kicked out the door and wrestled both gunners to the exit, literally booting them out of the faltering aircraft into the minus 50°C air outside. They were able to pull their own ripcords and safely parachuted into enemy territory. Sgt. Sawicki collapsed from his wounds and went down with the flaming Fortress."

The end of *Red Ass* was similar, but less well documented. The ship was damaged by flak and finished off by fighters, and came down near the town of Renslage, Germany. Four of Lt. Brumbeloe's crew were KIA, and the rest became POWs.

Major Mitchell's Group Leader's narrative ends this mission account: "After we bombed, the 384th Group dropped down 1,000 feet and were in the top of the overcast which would have put us down in it. We went down into it but since the fighters were working us over, I climbed back up on top of the overcast. I did not see the Lead Group again but came on home letting down over the North Sea without difficulty and arrived back at Base at 1717 hours. The friendly fighters stayed with us until we left the enemy coast."

It had not been a good mission, for the Group or the Eighth as a whole. The 303rd claimed three German fighters destroyed, one probable, and two damaged, but the lead navigator believed that the Group's bombs "fell about 12 miles west of Bremen after the formation had passed over the City." Less than half the total bomber force—a mere 154 B-17s—managed to unload at all. Many groups were stymied by cloud tops running to

November 29, 1943. Despite loss of his left arm at the elbow and grave abdominal wounds, Sgt. Joseph R. Sawicki, tail gunner on Lt. Carl J. Fyler's crew, boots the two wounded and disabled waist gunners, Sgts. George Fisher and Martin Stachowiack, to safe parachute landings over Nazi Germany. Sawicki then collapsed of his wounds and was found dead in the wreckage of Dark Horse, *the crew's B-17. Up to the present day Carl Fyler has been unrelenting in his efforts to secure a posthumous Medal of Honor for Sawicki's extraordinary actions "above and beyond the call of duty" on this mission.* Illustration reproduced from an original color painting by Robert A. Hand, Sr. Copyright © 1997 Robert A. Hand, Sr. (Used with Permission).

29,000 feet that seriously disrupted their formations, and when the day was done there were 13 missing B-17s.

The fighter escort fared even worse. The 55th Group was savaged by the loss of seven Lightnings, and nine Thunderbolts fell. German losses were put at 30.

The Battle of Bremen was proving to be a bloody affair.

23

The Battle of Bremen Continued

December 1–December 16, 1943

SOMETIME AFTER THE late November Bremen missions Bud Klint wrote in his diary: "I think I'll be getting my own crew soon... Our Group lost Captain Cote and his crew...That puts Bob in the Squadron lead."

Hullar's men were now on the brink of a completely new career: After 16 missions they were the most senior crew in the Squadron, with a leadership position of prime importance in the Group. Hullar would now work hand-in-glove with the mission leader as the pilot responsible for taking his formation through the many course and altitude changes required by the mission plan and the tactical situation. The whole crew would share his responsibility, each man helping in his own way: The navigation team would guide Hullar to the target area; the lead bombardier would get the Group's bombs on the target; Hoyt would serve as the vital communications link; and the other gunners would keep Hullar informed about developments in the formation.

December 1st was the day Hullar's crew had their first chance to lead a mission. The target was Solingen, a famous metalworking city in the Ruhr well suited to PFF attack. It possessed the largest nonferrous metal casting foundry in Europe, which produced thousands of aircraft engine castings. It had many other plants that manufactured steel castings and other parts for airframes and engines.

The plan called for two combat wings from the First Division to lead the bomber force. They were to attack Leverkusen, near Solingen, with the help of H2X ships from the 482nd Group. Behind them, two more First Division wings were to strike Solingen with a mixture of H2X and H2S ships. A total of 215 B-17s were to fly, with the 41st CBW in the rear. Following would be 78 B-24s from the Second Division, under orders to

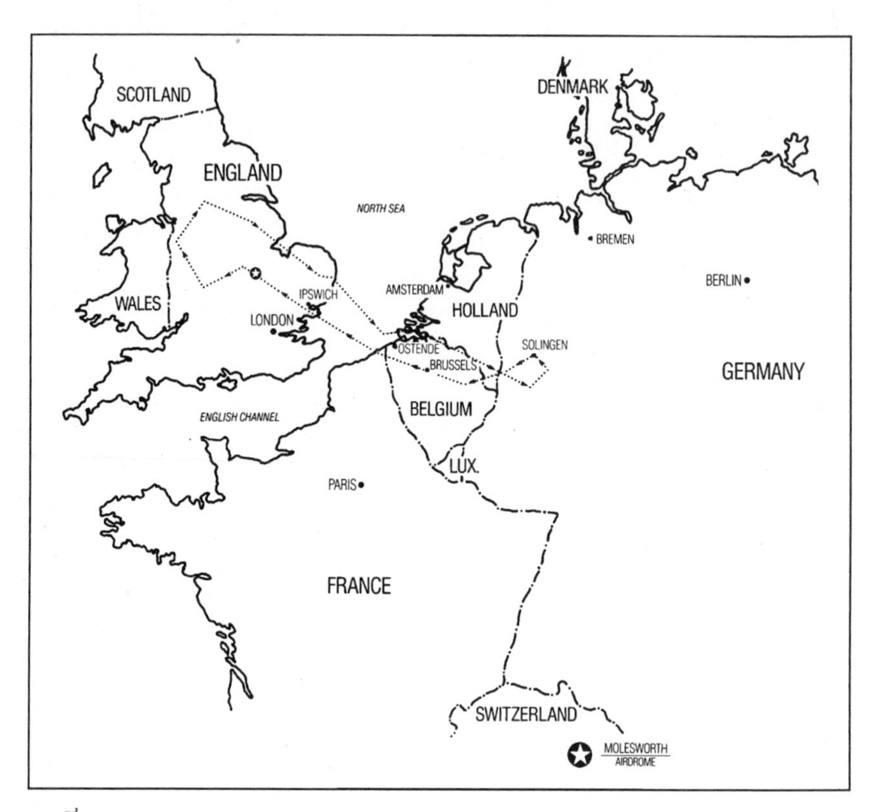

303rd Bomb Group Mission Route(s): Solingen, December 1, 1943.
(Courtesy Waters Design Associates, Inc.)

trail the 41st CBW as closely as possible, dropping on the sky markers left by its PFF aircraft. The bombers would be escorted by 374 P-47s and 42 P-38s with eight RAF Spitfire squadrons providing withdrawal support.

In the 41st CBW, the 303rd was high group. Since they were now the Squadron Lead, Hullar's crew was assigned to one of the 427th's regular lead ships. She was Captain David Shelhamer's old favorite, *Vicious Virgin*, a Queen who would prove a worthy successor to *Luscious Lady*. Flying with the crew were Major Ed Snyder; Lt. Paul Scoggins; the Squadron Bombardier, Captain Byron Butt; and Bud Klint, who was "in the process of being checked out as a first pilot" and who "rode as an observer in the tail as part of my training." Elmer Brown was assigned to Lt. "Woodie" Woddrop's crew in *The Flying Bison*, B-17G 42-37875, behind *Vicious Virgin* at the head of the lead squadron's second element.

The *Virgin* climbed into the air at 0802 and as she cleared the end of the runway, George Hoyt rejoiced because "Now that we were Group lead, we had no one ahead of us to lay dangerous currents of propwash right in our flight path."

Group Assembly and the cross-Channel flight were fairly routine, Hullar noting that "We crossed the Dutch coast south of course behind the Lead Combat Wings and got some flak over the coastline." After three aborts, the Group proceeded to Solingen with 17 ships, flying south of course all the way.

The 41st CBW Wing Leader, Major Marcus Elliott of the 379th Group, explained that "I crossed in the rear of several other Combat Wings from the First Division and lined up to the right and rear. Both deputy Pathfinder crews aborted prior to the enemy coast, and shortly after passing inside Germany we definitely ascertained that our Pathfinder equipment was completely out."

The PFF equipment was also out in the First Division wings assigned to attack Leverkusen, and soon Major Elliott saw more than the expected number of wings ahead of him. He reported:

"I placed our Combat Wing to the right of the three lead Combat Wings and just in back of them. I had another Combat Wing to my right and in front and planned on bombing on whichever flare we could see. The bombardier dropped as we passed over a quite visible flare, presumably dropped by the Division Leader...The high group was straggling a bit on the bombing run."

Hullar reported it differently: "Before the IP, the Lead Group speeded up and pulled away from us. We indicated 155-160 IAS and cut the corner at the IP to close on Lead Group. Flares were not fired at the IP by the Lead Group. The code word was not announced over VHF, but a left turn was made. The bomb run was made on a magnetic heading of 338 degrees and bombs were released at 1203 hours from 26,700 feet. We were slightly in trail of the Lead Group, but bombed on the Lead Group's red parachute flares. [A] smoke bomb was also seen."

From the nose of *The Flying Bison*, Elmer Brown noted: "We flew way south of course the whole way in, and after a whole lot of turning and fooling around, we dropped on Pathfinder through 10/10s."

Major Elliott "believed that none of the ships in the Combat Wing were lost due to enemy action," but the 303rd's trailing position forced it through heavy flak.

"Over the target there were two barrage-type boxes of intense and accurate fire," Hullar wrote. "We flew down a lane between the two boxes."

The antiaircraft fire got the No. 5 plane of the low, 360th Squadron, B-17G 42-39781, flown by Lt. G.W. Luke, Jr.'s crew, and it also gave Woddrop's crew a rough time.

Brown commented: "We hit quite a bit of flak. One piece came through and hit my instrument panel and another piece put a hole in the fuselage between the pilot and me."

In contrast, only one to five enemy fighters were seen. As Elmer Brown observed, "We had wonderful fighter support all the way in and out. We had so many escorts, Jerry didn't even make one pass at us. Mostly P-47s. At one time we had P-47s above us and P-38s below. Spits also escorted us on the withdrawal."

Major Snyder commented: "We had very good fighter protection and I didn't see even one enemy plane."

This was just as well, for in the *Virgin*'s tail, Bud Klint was having no end of trouble with the guns.

"We had excellent fighter coverage—so excellent that the only occasion I had to fire the tail guns was to test them as we crossed the English Channel. When I did try to fire them over Germany, they were frozen tight. It was probably my own fault, for shortly after we reached altitude I turned my electrically heated suit up high enough to keep me comfortably warm, only to find that that caused moisture to condense and immediately freeze on everything in the tail compartment. Finally, I had to turn the rheostat down to a minimum and open the windows long enough to clear the glass of frost so that I could observe the formation. The episode taught me what an uncomfortable time the gunners spent while we were at altitude. If such a thing were possible, I respected them more from that day forward."

On the way back, the Group had more to worry about than frozen guns. This was one of the first raids the 303rd had flown with B-17 F and G models in one formation, and there was a real difference in the fuel consumption of the two aircraft. The G's chin turrets meant more weight, more drag, and less fuel economy, and they all had long-range "Tokyo Tanks" to compensate. But on this mission they hadn't been filled, so for Elmer Brown the return trip was "a milk run except that we were sweating out a gasoline shortage all the way."

The problem was aggravated by air turbulence. Bob Hullar reported that "Considerable propwash was encountered on [the] route out," and the other crews complained that "Any unexpected conditions would force A/C down."

By the time the Group crossed the enemy coast, the wing formation had split up, Hullar observing that "Apparently ships were running short of gas. We continued let-down and about halfway across the Channel we took out as a group of seven A/C for base."

Among the missing was *The Flying Bison,* which pressed on with a fellow 427th Squadron orphan, B-17G 42-31243, flown by Lt. A. Eckhart's crew. As Brown described their return:

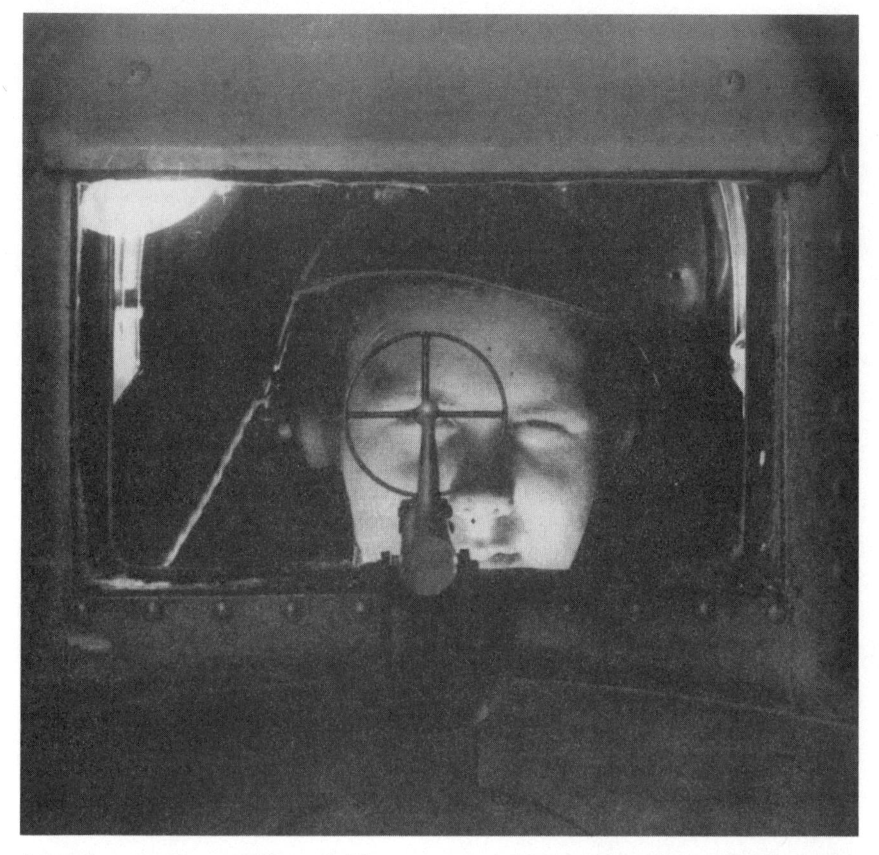

A good man to have in the tail. This wartime publicity photo depicts Merlin Miller peering through the primitive "post and ring" sight early B-17Fs had instead of the more sophisticated optical sights that graced later F and G models. In actual combat, Miller would have worn a leather helmet instead of the baseball cap. (Photo courtesy Merlin D. Miller.)

"We left the formation at the Holland coast (Ostende) and I gave the pilot a heading for the nearest English airdrome, Manston, ESE of London. It was a fighter drome. As we approached the field, we saw a B-24 circling the field explode and go down in flames. No runways, so we had to land in the grass. Several other Forts landed here with the same trouble—no fuel. Eckhart followed us, but couldn't quite make it and had to ditch in the bay right off the coast."

All ten of Lt. Eckhart's men were saved, but no one in the exploding B-24 survived; it came from the 389th Group based at Hethel. The total price the Eighth paid for the mission was 19 B-17s, five B-24s, two P-38s, and five P-47s. German losses were put at 24 aircraft.

Meanwhile, the main 303rd formation had landed back at Molesworth. *Vicious Virgin* touched down at 1419, and by 1440 twelve Group ships had made it home. Three Group ships landed away, leaving the 303rd with two lost aircraft but only one lost crew.

For Hullar there was the satisfaction of a job well done on a flight that posed special challenges. He merely recorded: "Lead Group for first time," but the other crews commented that "leadership and evasive action were excellent."

Something of what Hullar must have felt is expressed by the man who flew next to him in the copilot seat, Ed Snyder: "It was considered a privilege to be able to lead a mission, a real feather in your cap. You felt good about it, but you also realized you had a lot of responsibility riding on your back. You never could do it exactly by the book. There was always something—clouds, somebody late—that made you have to figure out what to do. It was a hard job, but it was one that gave you a great deal of satisfaction. I'm sure Bob Hullar felt the same way."

The Solingen mission ended with an incident involving George Hoyt, which shows that Hullar's leadership was not limited to the cockpit. As Hoyt recalls:

"After landing from this mission, I felt quite tired and skipped cleaning my gun, which was our standard routine. The next morning I ran into Bob near the Squadron HQ and he said, 'George, a little bird told me that you did not clean your gun when we got down yesterday.'

"I said, 'Yes Sir. I'll go over and clean it right now.' He could have chewed me out, but he smiled in that dignified, quiet way he had and then returned my salute. I couldn't help but respect and admire this man. He was a real leader."

As December wore on, Hullar's crew labored under a busy schedule while Elmer Brown found himself chafing under a long period away from the mission roster. Bud Klint was back in the cockpit with Hullar on December 5th, when the Eighth's bombers were launched to hit targets in France. The 303rd's objective was the Luftwaffe airdrome at St. Jean D'Angely and the mission "didn't sound very easy" to Klint at the 0500 briefing:

"We were to go deep into southwestern France—a round trip of seven hours and ten minutes. We were to go in at an altitude of 19,000 feet until we were about 150 miles from the target and then descend to 11,000 feet for our bomb run. That would put us in range of not only the usual heavy ack-ack guns, but also the more plentiful medium caliber defenses."

Since the 303rd was low group, Hullar's crew was particularly exposed. They were low squadron lead in *The Flying Bison*.

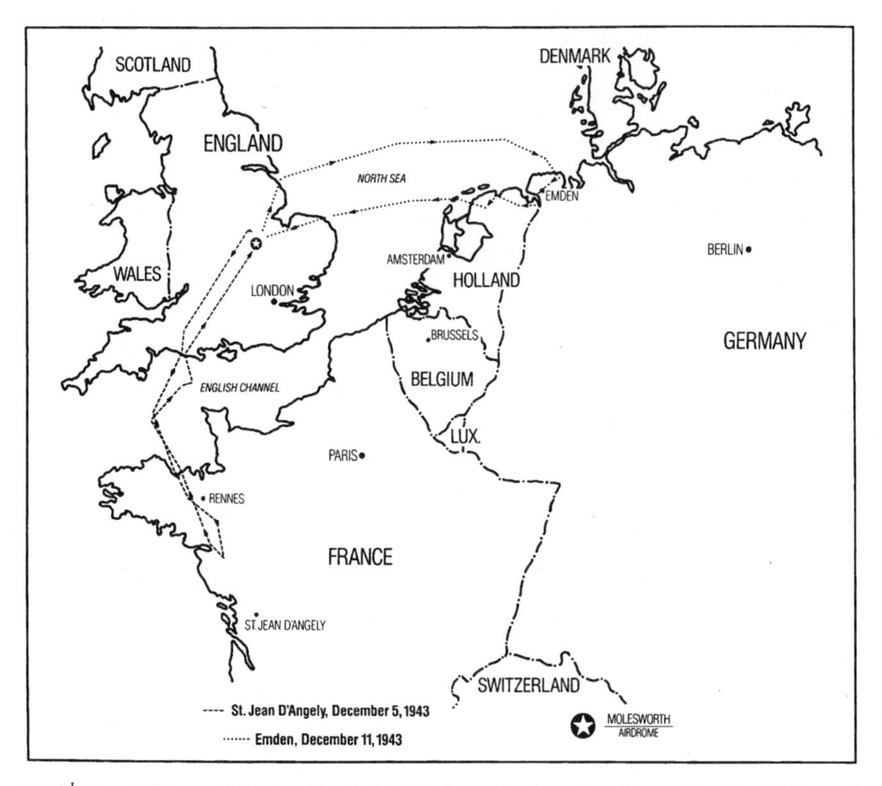

303rd Bomb Group Mission Route(s): St. Jean de Angeley, December 5, 1943, and Emden, December 11, 1943. (Map courtesy Waters Design Associates, Inc.)

The bombers got as far as Rennes before 10/10 clouds forced the formation back in a turn that nearly brought disaster to Hullar's squadron. He reported that "We found ourselves on a collision course with another group [the 92nd] to our left at our altitude, which forced us to make an excessively steep turn to avoid collision. In spite of our maneuver, we were unable to avoid them and they flew through our formation."

Nor was this the only excitement that the trip home contained. Hullar also reported: "On the return course while still over France, one aircraft from the Lead Group opened bomb bay doors and dropped its load through our formation, ahead of and between the No. 1 and No. 2 men of the Low Squadron."

Understandably, that evening Hullar described the raid as a "rough do," but he also wrote that the "mission itself was a Christmas present from 8th BC. No flak, no fighters, no target."

Klint concurred: "We always considered this one our Christmas present from the Eighth Air Force."

Nine bombers failed to return, but the mission was noteworthy as the first in which P-51 Mustangs were used to escort the heavies: 36 of the new long-range fighters were borrowed by VIII Fighter Command from the Ninth Air Force's 354th Fighter Group to augment the P-38s and P-47s.

The next mission, on December 11th, was going to be tougher. Hullar wrote that the target was "Emden, Germany. Second trip to that town."

It was another raid on which Klint "rode with the old crew while waiting for Base Operations to assemble a crew for me," and he recalls that "at the briefing, we were told 'The Eighth Air Force will erase Emden from the map of Germany.' General Eaker had made this promise to the RAF and our job this day was to help fulfill that pledge." The Eighth was sending 583 heavies escorted by 388 fighters to get the job done.

The bombing was to be visual rather than PFF, and Mac McCormick was to lead the 303rd in the last of 12 combat wings headed for Emden. Hullar's crew was to fly *Vicious Virgin* as deputy group lead on the second raid in a row Elmer Brown was forced to sit out on the ground.

To Hullar's right in a Fortress named *G.I. Sheets*, B-17G 42-39786, was Lt. James F. Fowler, a new 427th Squadron pilot flying under Lt. Robert W. Sheets, his Instructor Pilot.

Fowler was a 22-year old from Mullins, South Carolina, who found the prospect of a military flying career far more exciting than his prewar job as an assistant grocery store manager. He volunteered for aviation cadets in January 1942, but he and his crew encountered one frustrating delay after another during training, and they had been on the base since November 16th waiting for this, their first raid.

The trip went without incident up to the German coast, where the Group was met by P-47 escorts. Then Hoyt had a problem with his electric suit outlet, and Hullar sent him up to the *Virgin*'s nose, where he had a ringside seat as the bombers neared Emden.

"I saw us flying right smack into the middle of a flak barrage that filled the sky with ugly black explosions. It was incredible!"

Fresh to combat, Lt. Fowler saw nothing surprising about the flak. In a notebook that he kept, he wrote: "The flak was light," adding, "We got one small flak hole about as big as my thumb about two feet back of the tail gunner."

Klint and Hullar had a greater appreciation of the danger. "It was extremely accurate antiaircraft fire," Klint wrote.

Hullar commented: "Flak saw us and was good. Rough after so much PFF flak. The boys pleaded for evasive action."

Hoyt peered out of the *Virgin*'s nose and "to my right I saw a B-17 on its back that had just broken in half at the radio room, with the two pieces plummeting crazily downward. I saw no chutes. Later on I saw giant bursts of phosphorus flak with streaming tentacles that seemed to reach out for our plane."

Others in Hullar's crew described them as "two spiraling trails...which burst into several white balls."

The flak didn't stop the Group from getting to Emden, but the enemy's last line of defense frustrated the Group's effort to get bombs on the target. Hullar's crew was one of many that reported "a very effective smoke-screen" blowing "lots" of smoke across the target, and it made McCormick's job all but impossible. He had to synchronize on railroad yards in the dock area rather than the briefed aiming point, and he later reported that "the bombing run which was as briefed was directly into the sun, which cast sort of a sheen over the smokescreen around the target so that details of the target area could not be identified until the A/C was right over the target." He dropped, but the Group's 229 500-pound GP bombs exploded harmlessly in open fields to the right of Emden's outer harbor entrance.

The Group's crews reported three B-17s in distress on the way back to England, but all the 303rd ships returned safely. The Eighth lost 15 B-17s and two B-24s. However, much damage was inflicted in the northwest quarter of Emden.

December 13th dawned with VIII Bomber Command launching what was described as "the largest single force of heavy bombers ever to participate in a daylight attack." No fewer than 710 B-17s and B-24s were flying in 12 wings to hit Bremen and Kiel with 394 fighter escorts: 322 P-47s, 31 P-38s, and 41 P-51s from IX Fighter Command's 354th Group.

Hullar's crew fit into the mission plan leading the second 303rd formation the Hell's Angels were sending. Their "Group B" was the high group in the second Bremen wing, prompting Hullar to record in his notebook: "Bremen (again! 3rd time)."

Lt. Paul Scoggins bumped Brown out of the navigator's slot, and since the Group Lead position called for an officer to observe from the tail, Chuck Marson was left behind in favor of Lt. Bernard W. Rawlings. He was Lt. Jim Fowler's copilot, but had stayed behind during his crew's first raid two days before. His initial meeting with Bob Hullar, and how he came to fly with Hullar's crew, are memories he treasures:

"I was a 23-year-old Second Lieutenant more anxious than just about anybody to get into combat. When we first joined the 303rd Bomb Group and the 427th Squadron it seemed like there was an interminable time of training before we were considered combat-ready. It was a trying time for me, and I think for all of our crew. I was dying to get into the war, to be

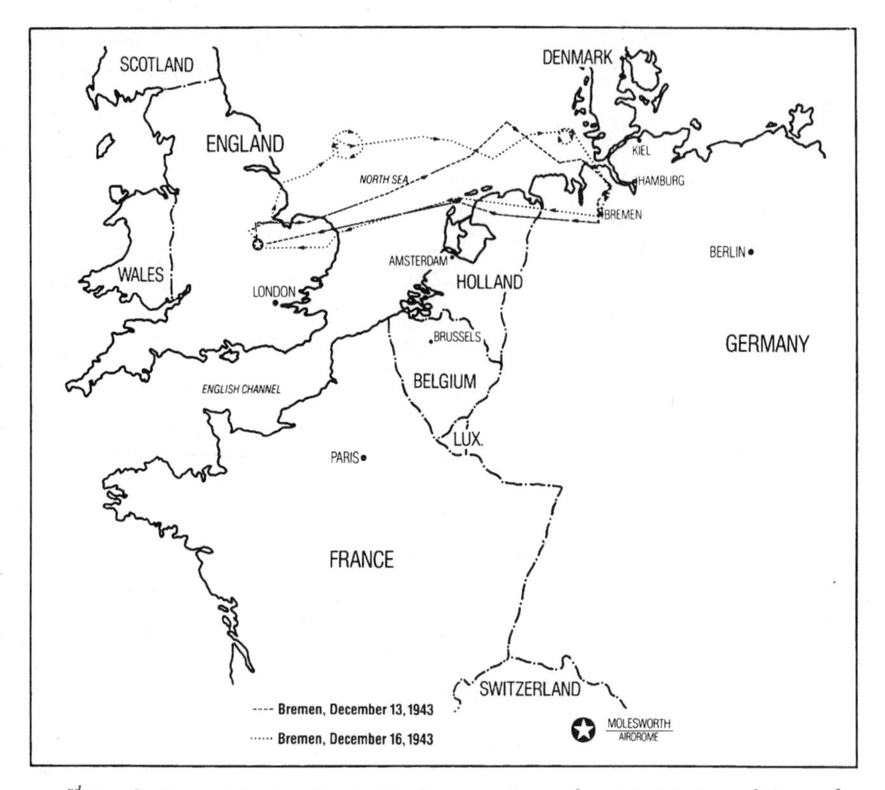

303rd Bomb Group Mission Route (s): Bremen, December 13, 1943, and December 16, 1943. (Map courtesy Waters Design Associates, Inc.)

a hero, and it seemed they were going to horse around with us forever before letting us go fly a mission. I found this time very difficult.

"I first met Bob Hullar when I'd been in the 427th Squadron a couple of weeks. I got up early in the morning and went over to the mess hall to get some coffee. I had some project—going out to the airplane to sit in the cockpit blindfolded and memorize where everything was once again, something like that. I was sitting there drinking my coffee and the famous Lt. Bob Hullar came over and sat down at the table with me.

"I knew him by reputation already. After we had been there a few days, some of the older crews started to accept us and talk to us a little bit, and it seemed to me that Hullar was esteemed by them just below Lew Lyle for his abilities as a pilot, for the way he ran his crew, for his maturity and courage, and for everything, really, that was important about a person in a combat situation.

"I didn't know what to say. It was like sitting there next to God. But I got up my courage and introduced myself.

"'I'm Lt. Rawlings,' I said. 'I've been here about two weeks, and they haven't let us fly yet. We're anxious to go.' I told him I knew he had about 17 missions in, and I said, 'I suppose after you've flown a few missions, it gets easier.'

"He gave me a look of despair, almost, like he wasn't even going to try to explain the facts of life to this dumb kid who had just joined the Group. But he was kind about it. He didn't say, 'You dummy,' or anything of that sort. He was a very controlled and yet warm individual, quiet but competent. I was really impressed. He was the guy I wanted to be.

"It wasn't too long after this that I was scheduled to fly a mission with his crew. I had been devastated when I found out I couldn't go on the Emden mission. I had come over here to win the war, and they wouldn't let me fly! Then a verbal request went out to the pilots in my Squadron that if anybody wanted to volunteer to fly tail gunner the next day, he could get credit for a mission. Instantaneously I signed up for that. It would give me a start towards winning the war and being a hero and all. I found myself in the lead ship with Bob Hullar in the left pilot seat and Major Snyder in the right."

The ship was *Vicious Virgin* and off to her right was Lt. Rawlings's crew in B-17G 42-31233. Lt. Fowler's Instructor Pilot was Lt. Grover Henderson, Woddrop's regular copilot. To the *Virgin*'s left, Bud Klint was serving as Instructor Pilot to another new crew, Lt. D.L. Barnes's in *Flak Wolf*.

The 303rd got more than 40 B-17s in the air by 0848. Once aloft, Rawlings began to get a first-class education in how an experienced lead crew operated.

"The regular tail gunner, Miller, was shifted up to the right waist gun, but the understanding was that if we got out of formation, I was to get my rear end out of there and let a real tail gunner get back there. I was there on the theory that an officer was better qualified to keep track of the formation."

"I was just fascinated by my admiration for this crew. It was really something to hear them on the intercom, and to listen to Hullar's responses. As we were going over, I picked up, 'Skipper, left waist gunner here. There are a couple of fighters three o'clock high. I can't really tell for sure, but I think probably they're ours.'

"Hullar responded, 'Roger waist gunner; keep me advised.' I could tell there was a really high-class operation going on here. I was proud of my own crew. I thought we were a good crew, but we were nowhere close to the professionalism of this bunch. I was very aware that this was an outstanding group of people. The quality really came through."

It was Pete Fullem who had spied the Group's fighter escorts; the B-17s enjoyed what Klint felt was "beautiful close coverage by both P47s and P-38s." The 303rd's formation proceeded to the IP free from fighter attack, encountering only "scattered meager and inaccurate fire" from flak along the route.

Lt. Scoggins, riding with Hullar's crew again as lead navigator, wrote: "It was PFF mission and we dropped through 10/10th cloud cover. There were no enemy fighters and the flak wasn't too rough, so I consider it an easy one."

Even so, the bombers faced the ever-present hazards of high altitude. The 303rd formations were at 28,000 feet and the temperature was −43 degrees C. The cold and the lack of oxygen came close to claiming victims in both Fowler's and Klint's B-17s.

In Fowler's ship it was Lt. Henderson who nearly gave up the ghost. Fowler wrote: "Henderson passed out twice from lack of oxygen. Had on a borrowed mask." But Fowler could readily see Henderson's condition and got life-saving oxygen to him.

In *Flak Wolf*, the man in peril had a much closer call. It occurred on the bomb run, as Klint explains:

"We had to fly through dense contrails and fight propwash all the way from the IP to the point of release. To top that off, one bundle of incendiaries got hung up in the bomb bay. The gunners finally kicked the bombs out about 10 minutes past the target. By that time we had dropped behind and below the rest of the formation, but under the protective 'mothering' of four of our fighters.

"Our troubles still weren't over, for now the bombardier was unable to close the bomb doors. The engineer started to wind them up manually, but in the process exhausted his walk-around oxygen bottle and passed out. Fortunately, I noticed the man in time as he lay draped across the step leading to the open bomb bay. After giving him a fresh supply of oxygen and cranking the bomb doors to the closed position, we rejoined the formation."

The engineer, Sgt. W.T. Sparks, owed his life to Bud Klint's quick action.

Another crew on their first mission also averted disaster, but only by taking an action which would not have been possible had there been heavy fighter opposition. Lt. Vern Moncur's men, of the 359th Sqdn., were flying *The Dutchess*, B-17F 41-24561, in the A Group's No. 2 lead squadron position when problems occurred in the waist. That evening Moncur wrote:

"We had a few minor troubles which could have been disastrous had prompt action not been taken. Sgt. Baer, left waist gunner, passed out from lack of oxygen soon after we left the target. The intake valve on his

Lt. Vern L. Moncur's crew from the 359th Squadron posing before The '8' Ball Mk. 2 (The Eight Ball), *B-17F 41-24635, BN✪O. Pictured L-R, Top Row are: Lt. Vern L. Moncur, pilot; Lt. James Brooks, navigator; Lt. Billy A. Cunningham, copilot; and Lt. David K.S. Chany, bombardier. L-R, Bottom Row: Sgt. Robert L. Rosier, engineer and top turret gunner; Sgt. James S. Andrus, radioman; St. Thomas J. Dickman, left waist gunner; Sgt. Walter E. Hein, ball turret gunner; Sgt. Richard K. Baer (tail turret gunner; and Sgt. Leonard L. Wike, right waist gunner. Moncur's crew went on to complete a tour as the first and the only regularly assigned crew of* Thunderbird, *B-17G 42-38050, BN✪U, a veteran of 112 missions and one of the 303rd's most well-known B-17s.* (Photo Courtesy Gary Moncur).

mask had frozen solid, preventing him from getting enough oxygen. We broke away from the formation and descended rapidly to 11,000 feet. He soon recovered and felt all right. Sgt. Dickman, right waist gunner, also passed out temporarily from lack of oxygen as he attempted to revive Sgt. Baer. S/Sgt. Andrus, radio operator, experienced the same thing because of a frozen oxygen mask. The temperature was 50 degrees below zero."

Unfortunately, the top turret gunner in another 303rd ship wasn't as lucky. Flying in the No. 3 lead squadron slot just off Moncur's right wing was the 359th Squadron's *Baltimore Bounce*, B-17F 42-29894, piloted by Lt. L.E. Daub. When his engineer got into trouble, his difficulty wasn't discovered in time. Sgt. J.J. Barrett died of anoxia.*

---

*Jon Krakauer's *Into Thin Air: A Personal Account of the Mount Everest Disaster* (New York:Villard 1997) is strongly recommended for the insight it provides into the extreme difficulty humans have functioning at 28,000 ft. even *with* oxygen.

All the 303rd's ships made it back to Molesworth, landing between 1353 and 1430.

"When Hullar brought the formation back over the field and we peeled off and made our landing," Rawlings recalls, "I had this feeling of absolutely wonderful pride and satisfaction that I was part of this magnificent effort."

The Eighth had hit Bremen with 171 bombers, Kiel with 352, and Hamburg with 116 as a secondary target. Only five bombers were lost against "exceptionally nonaggressive" enemy opposition.

Just 12 hours later, Hullar's crew numbered among hundreds roused out of bed for one of the most ominous briefings of the war. As Klint recalls:

"December 14, 1943, started out to be a very memorable day. The CQ shined his flashlight in our eyes and awakened us at 0230 with the announcement: 'Breakfast at 0300, briefing at 0330.' This meant predawn takeoff, which was in no way appealing.

"As it turned out, that was the least of our worries. When the Intelligence Officer uncovered that briefing map and revealed that black tape which was to be our course—a direct line from Molesworth to Berlin—I think every heart in that room sank. When they told us the details of the planned attack, we were all convinced that the Eighth Air Force was about to be completely wiped out.

"Reconnaissance planes had reported a heavy overcast over most of Germany—an overcast heavy enough, they believed, to ground all of the Jerries. This, Headquarters figured, was the opportunity they had dreamed of for a surprise attack on the German Capital. We were to take off at 0535, attempt to rendezvous over the field, but in any case, after a given time, to proceed on course. They estimated that it would be light when we crossed the Belgian coast.

"At that point we were to try to join enough other Forts to make up a suitable formation. If unable to join other ships there, we were still to hold our course and go in to the target area in groups of two, three, or even as single ships. Pathfinder ships were to have preceded us to the target and were to be circling outside the flak zone. As the bombers arrived, the PFF ships were to gather them into a group and lead them over the target. Each PFF ship was to take two groups over the target. We were to climb on course from the field to an altitude of 25,000 feet and the round trip was to take just seven minutes short of seven hours. It sounded like suicide to us.

"We went to our planes and were greatly relieved when, at 0500, just 10 minutes before start engines time, they sent up the two red flares indicating that the mission had been scrubbed. They called us back to the briefing room and told us that later reconnaissance flights had revealed that the clouds were starting to dissipate, and that, as an alternative, we were to hit Bremen that morning.

"So we went through the whole briefing procedure again and returned to our ships, only to have this one scrubbed, too, just 10 minutes before we were scheduled to get underway."

Two days later, the Eighth's planners decided once again to attack Bremen. The mission was another all-out assault: 631 B-17s and B-24s were dispatched in 12 wings from all three bomb divisions. They had a full complement of fighter escorts: 31 P-38s, 131 P-47s, and 39 P-51s.

The 303rd fit into the plan at the head of the last combat wing. Hullar's crew led the low squadron in B-17G 42-31241, the *City of Wanette.** The crew included Klint and Marson, but two new officers were aboard as navigator and bombardier. Occupying the No. 5 slot of the Squadron was Lt. Fowler's crew, with Lt. Rawlings finally flying in the copilot seat.

As George Hoyt recalls: "We sat down in the pre-dawn in the briefing room with an officer up front. He had a long pointer, and the words he uttered were a wartime classic.

"'Gentlemen, three days ago we destroyed Bremen, Germany. Today, we are going back and destroy it again.' Laughter erupted all over the room."

The less humorous side to the renewed assault was the Germans' obvious awareness of the greatly increased attention the Eighth was giving this city. The Luftwaffe did not always rise to meet the bombers, but the flak was ever-present—and growing in strength. As Bud Klint observed:

"During the early days of the European air war, the famous subpen town of St. Nazaire, France, had come to be known as 'Flak City.' Now, since the Eighth Air Force had switched its efforts to German industrial centers, that title had been transferred to Bremen. This city was supposed to have something like 600 heavy AA guns."

The Group launched more than 20 ships between 0906 and 0918 and crossed the enemy coast at around 1300.

"Our escort kept the enemy fighters off our necks," Klint wrote, but as the Group approached the IP, "it seemed like we ran into every one of those guns. Every plane in our Squadron had at least one hit. We took a blast in the nose that broke a hydraulic line and another which tore a good-sized hole in the waist compartment."

In the No. 5 Squadron slot, Lt. Fowler was in B-17G 42-31233, and also got multiple hits. He "counted 10 flak holes in ship—one through

*This aircraft was named after the small city of Wanette, Oklahoma, which is located southeast of both Norman, Oklahoma and Oklahoma City. It was also known as the *Spirit of Wanette*.

side of radio compartment cutting the muff at the radio operator's foot." It didn't bother Lt. Rawlings—"I didn't have enough experience to feel vulnerable yet"—but it made Hoyt jumpy, especially when he observed a new type of antiaircraft fire.

"This time I saw vapor trails rising up and streaking straight on through our formation to explode with good-sized detonations. They were a strange sight, later diagnosed to be ground-to-air rockets, and they got on my nerves."

Even so, the 303rd bombed without loss and returned to England without further incident.

Over Molesworth there was a final obstacle to face. Bud Klint remembers:

"When we returned to the base, we found it socked in. We had to make low visibility approaches and instrument let-downs. They were nearly as nerve-wracking as another combat mission. All these planes were milling around in the soup, and we never knew when one was going to come streaking past.

"Of course we had a system, planned to reduce the risk to a minimum, but in order to get all of the ships down in the least possible time, no system was completely effective from a safety standpoint. Even with oil drum flare pots lining the runway, it wasn't always possible to land on the first attempt and planes circling for subsequent attempts added to the hazard."

Nonetheless, all the Hell's Angels got down safely. The Eighth lost 10 B-17s and one P-47.

The weather over England closed in completely during the next three days, and Hullar's crew had some time off. Bob Hullar and Bud Klint went to London with two other officers, but Elmer Brown remained at Molesworth in the hopes of flying with another crew and catching up on the four missions he had missed.

Brown was, Klint recalls, "a real eager beaver at this time. Elmer wanted to get his missions over with and get back to Peggyann in the worst way, and he was ready to volunteer for just about anything."

Thus, on the next clear day, Elmer Brown would find himself flying with another crew on yet another attack against Bremen. This one would differ from the ones preceding it in one significant respect: This time the Luftwaffe would be out in full force, determined to stop the American bombers at all costs.

24

"Half a Wing, Three Engines, and a Prayer"

Bremen, December 20, 1943

BY THE EVENING of December 19th, the weather picture had improved enough for VIII Bomber Command to lay on a raid for the 20th. It had been a long time since Elmer Brown had been on the mission roster—his 17th and last mission had been to Solingen on December 1st with Lt. Woddrop's crew—but when the Group's flying personnel were awakened in the frigid, predawn hours of the 20th, Elmer Brown found himself slated to go with Woddrop's crew again.

This assignment wasn't just a matter of chance. On the Solingen raid, Brown's talents had been strongly tested when Woddrop was forced to leave the group formation over Ostende due to a serious fuel shortage. Had Brown's navigation not been flawless, the airplane and crew might well have ended up in the Channel instead of the RAF field near the coast where they came down.

Brown's performance had impressed Woddrop, who, though still a First Lieutenant, was a senior enough pilot in the 427th Squadron to make his preferences known.

As his copilot, Grover Henderson, puts it: "Woodie liked Elmer, and he had a lot of confidence in his ability as a navigator. So when we were supposed to fly lead and our regular navigator wasn't available, Woodie would ask for Elmer to be assigned to us. That's why, I think, he wound up flying a good number of missions with our crew."

Brown's ability would be put to an even greater test on the December 20th mission, but this operation was to prove a supreme challenge for everyone on Woddrop's crew. Special mention must be made of Sgt. Charlie Baggs, the tail gunner, and three others who particularly distinguished themselves:

Woddrop himself; Sgt. Bill Watts, his top turret gunner and engineer; and a new radioman on his first combat mission, Sgt. W.S. O'Conner. While little is known of O'Conner but his deeds, for the others, an in-depth view into personalities and backgrounds is needed. What these men were truly shaped the events that brought all of them home.

The 303rd's wartime history, *The First 300*, contains an intriguing reference to Woddrop and Baggs: "The 427th has had some outstanding personalities among its airmen...The RAF gave them Captain Edward M. Woddrop, pilot, and T/Sgt. Charlie Baggs, tail gunner, who made a hot team on the old *City of Wanette*." Though Woddrop's favorite ship was *Flak Wolf*, the two men were truly soldiers of fortune.

Ed Snyder remembers Woddrop as "kind of a fat, roly-poly guy who sported a typically British waxed mustache. When he came to us initially he had kind of a superior air because he had been in the RAF. But he realized he was coming into a different type of outfit and that he had to make a place for himself. Which he did. It took a little getting to know each other on the part of him and the organization, and then he seemed to fall in pretty well.

"He was eager to get into flying. He was on my tail all the time about getting a crew. He had come to us from the RAF alone, without a crew of his own, and eventually we got that set up. He got to know the B-17 quite well, and there was no question that he knew how to fly. He was pretty much on the cowboy side at times. He kind of reminded me of a hyperactive kid."

Lt. Edward M. "Woodie " Woddrop. (Photo courtesy Mrs. Lorraine Shelhamer.)

David Shelhamer also knew Woddrop well: "Woodie was a crazy kind of a person, but we got to be quite friendly. He said he was an heir to the Singer family sewing machine fortune, and he had been in England long enough, and he had enough money, to belong to practically every private club in London. These all opened at different hours. One would open at 10:30 and close at 2:00; another would open at 12:00 and close at 3:00 and so on. It was his contention that one day we would get a 48-hour pass together, and he would drink me under the table.

"We finally got into London, and hit his first club at about 11:30 in the morning. From that time till about 2:00 the next morning, it was no food and nothing but drink, drink, drink. We were in a club called The Coconut Grove when I finally had to pick him up off the floor with the help of one of the attendants."

Though Snyder and Shelhamer were well acquainted with "Woodie," no one knew him better than Grover Henderson did. Henderson was a big, friendly soul from the small rural city of Greenwood, South Carolina and their first meeting was an experience he will never forget.

"I got to the ETO in the early part of August 1943. I had been separated from my crew right after my arrival because of a personality conflict between me and my first pilot. He had been an enlisted man who had been made a Flight Officer, and he deeply resented the fact that I outranked him as a Second Lieutenant. We fought practically the whole flight across the Atlantic, and when we got to Scotland he insisted I be put off the crew, which was just fine with me. He and the rest of my original crew went on to another group, and I learned later that all of them, except the bombardier, had been killed on their first mission in a midair collision. I went on to Molesworth, the luckiest day of my life.

"When I got there I was put into the 427th Squadron and was assigned to Woddrop as his copilot. He had already flown a number of missions in the right seat, and was just being given his own crew. He wasn't your typical pilot. I was 23 years old and thought of him as an 'old man.' He was in his early thirties, and had a good-sized beer belly, but he was very agile and quick on his feet. He stood about five foot ten and had chestnut-red hair and a red beefy face.

"His appearance was very distinctive, but the impression you got from looking at him changed the minute you heard him talk. He had a very high-pitched voice that he was real sensitive about. One of the first things he said to me was that he didn't know why the Almighty had given him this 'girl's voice,' but he wanted me to know that he had nothing to do with it.

"I was just getting to know him a bit when he said we should take an orientation flight around the Molesworth area that afternoon. We had a flight surgeon in the Group who was very unpopular. He was afraid to fly, but he forced himself to because of the extra flight pay that he got for flying four hours every month. He would only fly with a pilot who was highly experienced, and he would put on a parachute and sit right up front near the escape hatch in the navigator's compartment the whole time. On my first flight with Woodie, this flight surgeon came out and climbed on the airplane to get his flying time in so he could draw his 50 percent flying pay.

"We climbed up to about seven or eight thousand feet, and on reaching that altitude Woodie turned to me and said, 'Feather the engines.'

"I said, 'Which one?'

"And he said, 'All of them.' I looked at him quite surprised, because I had always been told you didn't do that in a B-17.

"So I said, 'You're not supposed to feather all the engines at the same time. You might not be able to get them back.'

"He answered right away: 'I said feather *all* the engines,' but I refused.

"Woodie went ahead and feathered all of them himself. He was quick as a cat, pulling back the mixture controls, cutting the throttles, punching the propeller feathering buttons, and killing the ignition switches. All four engines were completely out of action inside of 30 seconds. All you could hear was the wind whistling.

"We could see the doctor through the trap door in the cockpit floor, but he was out of touch with the rest of the crew since he was not plugged into the aircraft's intercom system. The doctor raised forward into the navigator's area and looked out and saw the propellers stopped on one side, and then he looked on the other side and they had stopped on the other side. Then he made a mad dash back to the escape hatch. He already had his parachute on and was going to go.

"Bill Watts, our engineer, jumped down through the cockpit and he and another fellow started to wrestle with the doctor to keep him in the plane. The doctor was a big fellow, bigger than either of the other two, and he was really fighting to get out, and Woodie was hollering in the intercom, 'Let him go! Let him jump! When we get back to the base, we'll all swear we don't know why he went!'

"They wouldn't do it. They held on to him, and finally we had to start the engines again. When the doctor realized it was a joke, he got *real* mad about it, and he wouldn't speak to us the whole rest of the time he was there—about four or five months.

"After that flight, I had serious reservations about my new pilot's mental stability. He didn't look too stable to me, trying to make people jump overboard. I began to wonder what in the hell I was doing here in the first place. What had I gotten myself into with this guy?

"I learned soon enough. Woodie came from the town of Westfield, New Jersey, near New York City, and he told me he was the black sheep of his family. He spent a lot of time in New York, and he was very cosmopolitan. I saw some of this myself when I met up with him at a nightclub in Galveston, Texas, after we had finished flying with the 303rd.

"Les Paul and Mary Ford were there that evening. They were really big stars back in the '40s. He played the guitar and she was a singer, and when they were finished Woodie walked right up to them and started talking, and they recognized him!

"When I asked him about it, he said, 'Oh, you know, there were a lot of clubs up North on the Jersey side. I used to hang around in all those places.'

"Woodie left home as a young man and started bush flying in Canada and Alaska. He joined the RCAF as soon as the Germans marched into Poland and he instructed in Canada for a while, and then he wanted to fly combat. So he requested a transfer to England, and he flew twin-engine Wellington bombers for RAF Bomber Command as a Sergeant Pilot. Then he transferred to our Air Force and he was made a Second Lieutenant.

"I met him about halfway through his tour, when he was a First Lieutenant, and he went on to make Captain in less than a year. Woodie was on the wild side, he was reckless, but after flying with him I felt that he had enough ability so that he could be reckless and get away with it."

His fellow officers and crewmembers were not the only ones to see Woddrop's wild side. It was also apparent to Sgt. A.R. "Roy" Westfall, a member of the 427th Squadron's Personnel Flight Equipment Section. Westfall was responsible for custody and issue of flak suits and parachutes, heated clothing, and the like to combat crews, and always saw them as they left and when they returned from missions. He offers this story about Woddrop:

"One day early in Captain Woddrop's tour of duty he came in and was going through his flight bag, and as he emptied it on the counter I noticed his oxygen mask had a hole about the size of a nickel cut in one side.

"I said, 'My God, Captain, you can't fly with that. I'll get you a new one.'

"He said, 'No, I want it that way so I can smoke my cigar while I'm flying.'

"I was horrified. I said, 'Captain, with the raw oxygen in that mask, that cigar is liable to blow up in your face!' Well, Captain Woddrop flew 50 or 55 missions, my memory fails on that point, but when he checked in his equipment for the last time, he still had that oxygen mask with the hole in it."

What Woddrop's crew possessed in their pilot was an exceptionally skilled, absolutely fearless aviator who loved combat, thrived on physical danger, and always managed to pull his men through despite the dangers they faced. As Henderson puts it: "Woodie was at his very best in a crisis. That's when he could really rise to the occasion."

The other soldier of fortune on Woddrop's crew was Sgt. Clarence C. "Charlie" Baggs, whose attitude towards combat was best expressed by his comments after the raid to Knaben, Norway on November 16, 1943. As Hoyt recalls:

"Back in the barracks that night we compared notes with Bill Watts and Charlie Baggs. Watts said Baggs kept hollering over the intercom, 'Come on in, little FW-190, and I'll blow your ass off!'

"Charlie was on the bunk polishing off a fifth, and he said, 'Yes, that's right. I always cut my oxygen down some, and it gives me a cheap drunk. I don't give a damn about anything then.'"

Grover Henderson first met Baggs when he was in his late thirties:

"He was a Southerner, like me, from Colquitt County, Georgia, in the southwest corner of the state next to Florida and Alabama. He told me that he had been a Second Lieutenant in the U.S. Army, but he lost his commission in the late '30s in a matter connected with hard liquor. When the war came in Europe he wound up not just in the RAF, but as part of the RAF's Free Polish Air Force. He was in charge of a training program for aerial gunners. He even learned to speak Polish, and could speak some Russian too. He also completed a tour in RAF Bomber Command. He was a tail gunner in Wellington bombers.

"In the 303rd, Baggs was only good for one thing. He was a complete misfit on the ground. He was a heavy drinker and always got into trouble with officers. I had to talk him out of jams more than once. But in the air, there he was a master. Charlie Baggs was one of the two truly great aerial gunners I knew in World War Two."

The other gunner Henderson held in such high esteem was Sgt. Bill Watts, whose home was about 40 or 50 miles west of Atlanta. His prowess as a gunner was indisputable, despite a physical problem that should have barred him from any place in a gun turret. As Henderson recalls:

"Watts's eyes wouldn't focus. When you'd look at him, one of them would be looking at you and the other would wander off and be looking across the room. It gave you an uncomfortable feeling, and would break your chain of thought. How he ever managed to get into combat with that condition I'll never know, but he could shoot a machine gun better than anybody I've ever seen."

Other regulars in Woddrop's crew slated for the raid were the waist gunners, Sgts. W. Valis and H. Hoff, and Sgt. Royal Plante, the ball turret gunner. On the morning of December 20th, they all climbed aboard the B-17G named *City of Wanette*. Through one of the most remarkable entries in all of Elmer Brown's diary, it is now time to accompany them on "a most exciting day."

"December 20, 1943 (Monday)—Eighteenth Raid. A most exciting day. Half a wing, three engines, and a prayer. Target Bremen, bomb load 42 60-lb. incendiaries. Bombing altitude, 25,500 feet. Our position, lead ship, high squadron, low group. I was flying with Woddrop and crew: Henderson, copilot, Matthews, bombardier. Bombing to be either PFF or visual. It was clear at the target except for a slight haze, so visual bombing was done.

"We ran into contrails and clouds at our altitude just before the English coast and over the Channel, which compelled us to do a lot of circling and climbing. This made us six minutes late reaching the

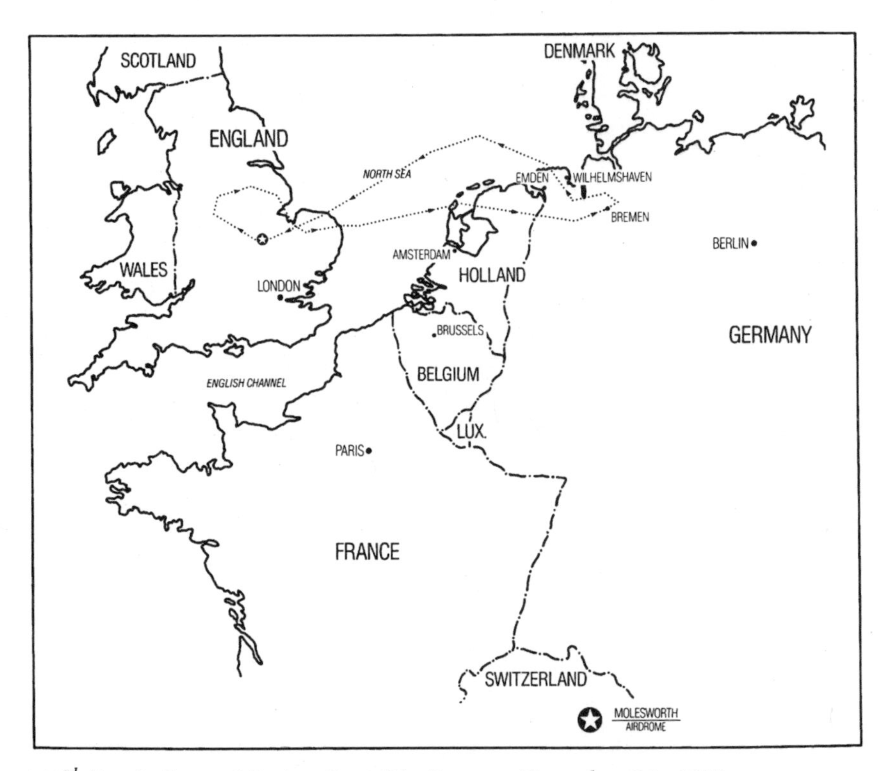

303rd Bomb Group Mission Route(s): Bremen, December 20, 1943. (Map courtesy Waters Design Associates, Inc.)

enemy coast—Holland. Despite this we met our fighter support. The fighter cover was wonderful going into the target. The P-47s were so thick it looked like our escorts had escorts. We also had a few P-38s just before the IP. When we turned at the IP on the bombing run, our escort left us.

"Then came the flak. Barrage-type, predicted fire, and tracking; they threw up the works. Looking on the ground I could see the flashes of the gun batteries for about eight miles around on all sides blasting away. In the sky around us there were bright red flashes followed by large puffs of black smoke that filled the air.

"We got severely hit by the flak. One explosion put a big hole through about three feet of our right wingtip. Another burst in the bomb bay and missed the control cables for the rudder and elevators by a hair. It did put a hole through the web of the pulley for these control cables. We got a hole in the gas tank for the No. 3 engine. Couldn't run the engine without gas and couldn't feather the propeller, which made for still more drag on our right wing and made it still more difficult to trim and fly the ship.

"One flak shell burst so close to the right waist window that it damaged the right waist gun and knocked the two waist gunners down. One turned to the other and said, 'I think I am dead.' The other said, 'Me too.' The plane had flak holes all over it except the nose. It only had one hole and that was in the tip of the glass in the nose.

"We bombed visually and laid them right in the center of the city. Bombs were away at 1143. We turned north and then west trying to stay south of the heavy flak at Wilhelmshaven and then headed northwest avoiding Emden, going out over the North Sea. Flak continued to burst around us, with a few intermittent letups, all the way to the coast. Some of our electrical wiring got hit and it knocked out my flux gate compass and radio compass and left me without a compass or any navigational instruments other than airspeed indicator and altimeter. I first noticed my compass was out when according to the sun we were going northwest and my compass pointed east. We had to buck a west wind and it took us until 1215 to reach the coast.

"We were without escort this whole time and Jerry really gave us the works. They were mostly twin-engine fighters—Me-210s, Me-110s, and Ju-88s and a few single-engine fighter Me-109s and FW-190s. A twin-engine got our No. 7 plane, Leve, with a rocket. The plane burst into flames and then exploded. They also knocked down our No. 2 plane, Alex.

"It was about −45 degrees C at altitude and when the radio operator tried to fix his jammed gun, he got a severe case of frostbite on his right

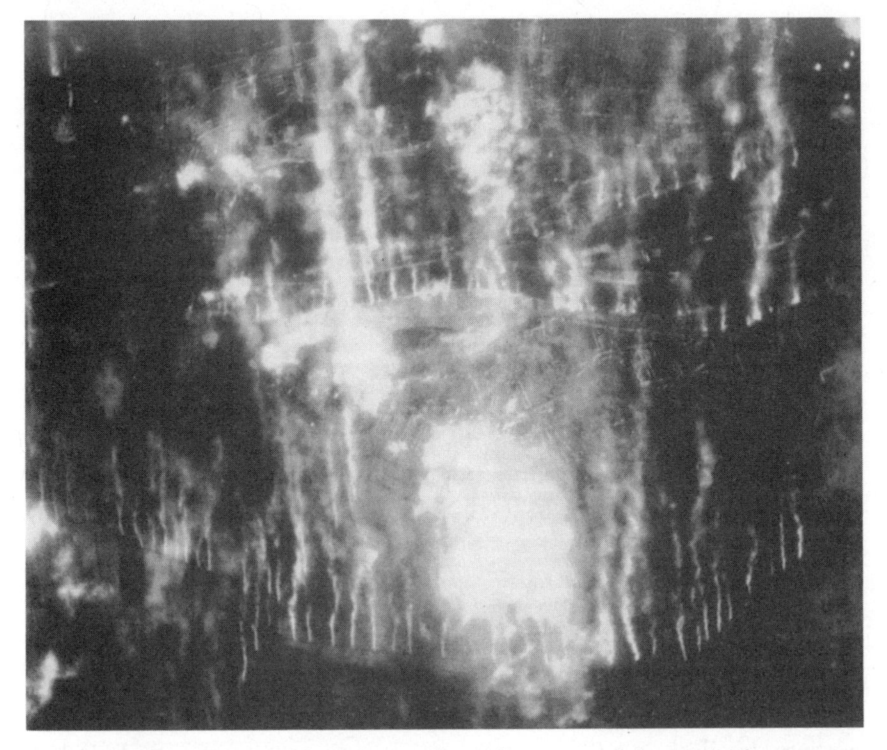

"We bombed visually and laid them right in the center of the city." The Germans tried to hide Bremen under a smokescreen, but the large explosions in the lower center of this photo confirm the accuracy of Elmer Brown's diary entry. This picture was taken by Sgt. R.G. Hunter from Wallaroo, *in the second element lead of the 303rd's low squadron.* (Photo courtesy Louis M. Schulstad.)

hand. All of our guns were either out or not working properly except our top turret. There was a big old Me-210 that attacked us from about eight o'clock high and he put three 20mm in our left side. One about six inches to the rear of the left waist window exploded right in the waist but didn't hurt either gunner. Another hit the left wing near the fuselage about 12 inches from the trailing edge of the wing. The third went through the fuselage heading right for the top turret gunner. It hit his ammunition cans and really scattered his ammo.

"This Me-210 then flew under us and came up at about two o'clock about 400 yards out. If I could only have gotten my right gun to feed, I am sure I could have knocked him down. He swung around and made another pass at us from one-thirty o'clock. Then he came in again at eight o'clock and this time when he came up at two-o'clock our top turret gunner poured lead into him. He smoked and then went down in a burst of flame.* The top turret gunner also knocked down a Ju-88 later that was attacking from the rear.

"In the heat of battle I heard someone say they were out of ammo. I said, 'I have a spare box.' The next thing I knew someone was tapping me on the leg and it was the copilot coming up in the nose. After he got in the nose, he moved the spare box of ammo I had back into the hatchway. He then took off his parachute and as he is a big man (about 6'1" tall and weighs about 200 lbs.), I figured he did it so it would be easier to carry the ammo back through that small hatchway. Then off came his Mae West and when he started to take off his flying suit I grabbed his oxygen bottle and saw that it was empty.

"I connected a fresh bottle for him and pushed the mask tight against his face. Well, he continued to try to undress—and at 45 degrees C below zero. I figured he was suffering from anoxia (lack of oxygen). So every time he would try to take his clothes off, I would pull them back on him and slap that oxygen mask on his face tighter.

"Finally he grabbed my pencil and wrote in my log, 'My electric suit is on fire.' Then he and I both started undressing him. His flying suit fit so tight we ripped it in shreds getting it off. Before he came down in the nose, the engineer had given him a shot of carbon tet from the fire extinguisher. It has a freezing action and really froze his old tail.

"While all this was going on, we were in the thickest part of our battle with Jerry. Our wing lead was very poor. Instead of indicating 155 mph like they should have, there we were in trouble and at times indicating

---

*During research for both the first and current editions of this book the author corresponded with *Luftwaffe* veterans and researchers in an effort to learn the identity of the tenacious Me-410 crew involved in this fierce combat with Sgt. Bill Watts. From correspondence in the 1980s with Dr. Karl Ries, a noted *Luftwaffe* historian and author, it appeared the mystery was solved when the author was informed that *Luftwaffe* records recorded the loss of only one Me-410 on December 20, 1943. This aircraft was "Black 10" of *Gruppe* II, *Zerstorergeschwader* 26 (Destroyer Wing 26) (II/ZG26) an Me-410A-1, *Werk Nr.* 420 087, with the Aircraft Code "3 U + S 10." "Black 10" was flown by Lt. Hans Joachim Rettig, Pilot, and Uffz. Heinz Zobel, Observer, and was reported lost SW of Cuxhaven, Germany, both crewmen being listed as "lost in action."

Hopes for certainty were dashed shortly after receipt of Dr. Ries' letter when another *Luftwaffe* researcher, the late Emil Nonnenmacher, wrote to confirm the loss of this single Me-410, but also to caution that many *Luftwaffe* records were destroyed at the end of the war, so that all *Luftwaffe* losses cannot be established with certainty. Thus the author could not be sure that "Black 10" was the aircraft involved in this combat.

For the present edition of this book another attempt was made to learn more about *Luftwaffe* daylight losses on December 20, 1943. In July 1998 the author corresponded with Herr Rudolf Tyrassek, a present day *Luftwaffe* researcher who was a nightfighter trainee at the end of the war. While it remains impossible to confirm that "Black 10" and her crew were Sgt. Watts's opponents in this fight, information provided by Herr Tyrassek leads the author to conclude, at a minimum, that the Me-410 in combat with *City of Wanette* did belong to II/ZG26, since this was the only *Luftwaffe* day-fighter unit equipped with the Me-410 operating in Northern Germany at this time.

Anyone having further information bearing on this subject is invited to contact the author.

Combat! This remarkable photo actually shows the 427th Squadron under attack by an Me-410 (called an Me-210 by Elmer Brown in his diary) on the December 20, 1943, mission to Bremen. Although the pursuit curve of the fighter is very similar to those made by the Me-210 Sgt. Watts got on this mission, the Fortress under attack here cannot be determined since the squadron formation is too badly scattered. This photo was also taken by Sgt. Hunter from Wallaroo. *Fortress in the center right is* Lonesome Polecat, *B-17G 42-31177, BN✪L, flown by Lt. W.A. Purcell and crew.* (Photo courtesy Mrs. Jean M. Rice.)

200 mph and still our formation was running away from us. We couldn't keep up so we started to lose altitude, figuring that flying 1000 or 2000 feet below the formation would provide some protection and would be better than trailing way behind.

"Since we only had one gun working, we all have Woodie, the pilot, to thank for pulling us through. Woodie kicked that ship around like a pursuit plane and it was his beautiful, violent evasive action that kept us from being blown to Kingdom Come by the many attacks Jerry made with his rockets, cannons, and machine guns. To illustrate the evasive action taken, when an Me-109 made a head-on attack one time, Woodie put that tail up and down so fast that the tail gunner hit his head on the roof and it knocked him out cold.

"We came back alone, the shortest and safest way we could. We dropped down to about 1000 feet altitude and flew over about 200 miles of the North Sea about 20 miles off the Holland coast. When we were heading home over the North Sea, we thought we might have to ditch so the radio operator tried to send an SOS through. He could not use his frostbitten hand, so he stuck a screwdriver between his thumb and first finger and beat out the code with the screwdriver and never stopped working.

"It was an uncomfortable feeling trying to navigate without a compass to look at, but the pilot's little compass seemed to be correct enough because I missed crossing the English coast at the intended place by one and a half miles. I got my wind direction and velocity by watching the ocean waves.

"We had sufficient gas to get back to our home field, much to our surprise. We were also surprised to learn we landed immediately after the last ship in our group. McClellan and Barker were the only two ships in our Squadron to come back with the Group. When we landed, an ambulance was waiting for us and took the radio operator and one of the waist gunners to the hospital. They took a piece of flak out of the waist gunner's leg.

"I could have been on a 48-hour pass today but I hadn't been on a raid since December 1st—I had missed four of them because I was doing special work—so I talked them into letting me go on the raid. There were moments when I wished I had been on pass.

"I learned this evening that my promotion came through. At last I am a First Lieutenant."

Grover Henderson's recollections of this mission are equally fascinating, and proof positive of the adage that no two men view the same events in exactly the same way.

"We got hit real bad by the flak going into the target. They put it right in our Squadron. The flak usually came up in a cluster of four, but this flak was so heavy it came up in clusters of 16, four different groups of guns firing at the same time. There were a hell of a lot of guns down there on the ground firing at us.

"It was a cluster of four shells that damaged our right wing so badly. The one that hit our right wingtip had to be a direct hit by a dud. All the G model B-17s had 'Tokyo tanks' in the wingtips, eleven interconnected fuel cells, and when the right wingtip got hit I saw a big shower of gasoline.

"Another burst in that string put a small hole under the right wing in the main fuel tank for No. 3 engine, and the gas started to drip out. Our intercom had been shot out, and when I saw us losing fuel out the tank to No. 3 engine, I got excited and tapped Woodie and pointed it out to him on the fuel gauge. He looked but he acted like nothing was wrong. So I slowly turned the fuel gauge knob—the B-17 had a six-position switch with one gauge that you'd turn in sequence to find out what was in each tank—to make sure he knew the situation, and it still didn't phase him a bit. He just kept flying the airplane like nothing had happened.

"Back on the ground, the only thing he said to me about it was, 'You had me worried there for a while. When you turned the switch, I thought you were trying to tell me there was no fuel in *any* of the tanks!'

"A third hit caused us to lose the supercharger for our No. 4 engine. There were some vacuum tubes that were part of the electrical controls for the supercharger located in the floorboards of the radioman's compartment and shrapnel got these. So with No. 4 engine not developing full power, we were actually worse off than Elmer knew.

"I remember the flak shell that exploded near the right waist window but I recollect the conversation between the gunners a bit differently from Brown. I believe both waist gunners were actually knocked unconscious for a while, and as I recall, when Hoff woke up, he said, 'Hey Valis, are we dead?'

"Somebody else, Plante I believe, piped up: 'You boys ain't in heaven yet. Get back to your guns!' I really don't know how the flak missed shooting us down.

"I thought that it was the flak that got Leve and Alex, but it could have been rockets, like Elmer said. I didn't see Leve get it, but I believe Charlie Baggs called out from the tail that he blew up.

"There was a period of just a few seconds between him and Alex, and I did see Alex get it. He was flying right off our right wing, and I was looking at his airplane at the time. The whole airplane went up in a red, red flash, and there was an enormous amount of black smoke. The only thing that looked anything like an airplane was a part of the main landing gear with a wheel flying through the air. In two or three seconds everything was gone. There was nothing left except the smoke blowing away.

"It was terrifically cold that day, probably 50 degrees below zero, and most of our guns did freeze up. They'd get hot when being fired and begin to sweat and this would turn to ice. After the mission I learned that in the middle of all the fighter attacks, the radioman took his mittens off to work on his gun. He couldn't do anything with his mittens on because they were too heavy and clumsy, and touching the gun was like holding a piece of dry ice in your hand for a long time.

"Well, that boy took the gun down from its bracket, put it on the radio room floor, took it apart, scraped the ice off with his knife, and then reassembled the gun and put it back in position and got it firing. By that time both his hands were shot, completely ruined. I saw them before he was taken to the hospital and it didn't look like he had any skin on his fingers at all.

"I never saw him again after he went to the hospital, and I don't remember his name. Woodie put him in for a Silver Star, and I understand he was awarded the DFC and was sent back home. He only flew that one mission, his career in combat lasted about six or seven hours, and he went home. Of course he should have.

"I do remember my problem with the electric flying suit. Actually, everybody called them 'blue bunny suits' because they were light blue in color. They were just like an electric blanket except they were formed like a flying suit. We wore regular underwear, and then the electric suit and then a wool flying suit on top of that. The blue bunny suits were brand new and I had never worn one before.

"Anyway, smoke started to accumulate in the cockpit, and I got concerned about it. The B-17 cockpit had a lot of electric wires and hydraulic lines and we were always worried about the danger of fire. I called the smoke out to Woodie's attention and he called Watts down from the top turret and told him to find out where the smoke was coming from.

"Watts looked around and came over and tapped me on the shoulder and said, 'Lieutenant, the smoke's commin' out of yo' collar!'

"At that point I felt a burning hot sensation in the seat of my pants, and I didn't know what it was except it felt like my clothes were on fire. I got out of the seat and examined the seat of my pants with my hand. What happened was that a fragment of a gun shell or some shrapnel had wedged itself between the seat armor plating and my back so that it cut across the seat of my pants and short-circuited the wires in my suit.

"Watts had a hunting knife in his boot and he split the seat of my wool flying suit, got a CO_2 fire extinguisher, and stuck it in the hole he cut and set the fire extinguisher loose in my pants. I got some relief from that hot spot and sat down and started flying the airplane again. But in a few minutes it began to get hot again, so I got up and Watts gave me a second application with the fire extinguisher.

"This was in the middle of the fighter attacks we were getting, and on that second occasion when I complained about getting hot, Woodie said, 'Go down in the nose and get Matthews or Brown to get you some help out of that flying suit and see what's on fire.' So I went down to get one of them to help, and that's when Elmer thought I had lost my mind, and I was trying to pull the suit off and he was trying to put it back on me at the same time. I couldn't make him understand that I was burning inside.

"When I did make him understand, the two of us got me out of my wool flying suit. It was smoldering and we opened the hatch in the bottom of the airplane and threw the suit out. Only later did I realize that I had my billfold in a pocket of that suit—we weren't supposed to carry it in combat—which had a considerable sum of money in it, a few five pound British notes and some Belgian and Dutch and French money. Afterwards we joked about some German farmer finding it in his field.

"The *real* funny part about this incident is that I could have solved my problem just by pulling the plug on the blue bunny suit. That's all I needed to do. I just had to pull the plug. It even had a rheostat control I could cut it off with, but the blue bunny suits were brand-new and I hadn't worn one before and in the excitement I just plain wasn't thinking.

"Woodie really did fly the B-17 like a pursuit plane that day, and he did knock Charlie Baggs unconscious for maybe ten minutes swinging the tail around trying to stay out of the way of the fighters that were making passes on us. He was very flexible and loose with the airplane, and never did believe what they tried to teach us about the stable firing platform, always hold your airplane real steady and still so the gunners can shoot at the fighters. He felt that when the fighters started shooting at you, get the hell out of the way. In formation he actually preferred the tail-end-Charlie position because it was easier to maneuver around and get out of the way of fighters in case they made a pass at our airplane.

"I do remember the Ju-88 that Watts got. This happened later on, when we were on our way home, completely separated from the formation and slowly losing altitude. We were at about 15,000 feet. I was swiveling my head and saw two German fighters out about 2000 yards to our right, slightly higher than we were and slightly behind us at about four o'clock.

"They were bluish-gray night fighters with hornlike radar antennae in the nose and rockets under the wings. They had slowed down to what they were cruising at the same speed we were, and they just stayed out there for some time. We knew they had something in mind for us. I think they were trying to figure out why we were not shooting at them, but of course our guns were frozen. At that time our gunners were still working to clear the frozen guns up.

"Finally one of the Germans started edging in closer and closer. I could tell he had turned in and had taken a heading that would eventually work him in right up close to us. He slowly came in, he was very cautious, and I watched him, and when he was about 800 yards away he fired. I saw the smoke from the two large rockets he carried under the wings. It billowed out under the rocket pods and I could see the trail of the rockets coming toward us.

"Those things were huge! They hypnotized me when they left the airplane. They looked like they were moving so slowly, and as they got closer they seemed to pick up speed. I shoved the wheel forward real quick and the rockets went over our head and exploded some distance above the cabin.

"When the German fired the rockets, he made a mistake. He should have broken away, but he came in a little closer. He must have been watching his rockets, and then I heard a .50-caliber open up. The rate of fire was real slow, but the first burst out of our airplane caught him right on the side. I saw pieces about the size of a license plate fly off the left engine cowling, and I saw metal and glass splinters break out of the canopy all around the pilot. I feel certain the pilot was killed instantly.

"From that point the airplane nosed down at about a 30 or 40-degree angle with the left engine smoking a long trail of black smoke. I watched him go all the way down, and it was a long way, till he crashed into the North Sea. After that the second German fighter thought better of having a go at us and he just disappeared.*

"The flight back to England was long and very difficult. It took us about three hours from the time the second fighter crashed. With the No. 3 engine out and No. 4 not developing full power and with about ten feet damaged on the right wingtip, the airplane was trying to fly in circles. We trimmed as much trim into it as the airplane was capable of handling and it still wanted to go in a circle.

"So we had to put some manual pressure on the rudder—and that rudder is as big as a barn door. It was extremely hard to push it in and make the airplane fly straight. I would stand with both feet on the left rudder pedal for a few minutes and try to hold it straight and when my legs would start trembling, Woodie would stand on it for a while until he would start trembling.

"We alternated back and forth like that and finally we both got so tired we called Watts down from the top turret. He sat in my seat for a while and let me rest and then relieved Woodie for a while and between the three of us we finally flew it back to England over a distance of about 500 miles.

"On the way back we experimented with the controls to see if we could land the plane. We found we could maneuver, but the stalling speed was very high—about 150 mph. When we got to Molesworth we waited till everyone else had landed so that if we crashed, we wouldn't block the main runway, which was about 7000 feet long.

"We set the plane down right at the end of the runway going 150 mph and applied the brakes. They got burned out pretty quickly, but they reduced our speed to about 60 mph. We just coasted the rest of the way to a stop almost at the very end of that 7000-foot runway.

---

*It is not possible to confirm this combat from remaining *Luftwaffe* records, since there are no records which list any personnel losses among Ju-88 nightfighter units on Dec. 20, 1943. According to Rudolf Tyrassik, Grover Henderson's identification of these aircraft as nightfighters is also a "fog of war" phenomenon, since no Ju-88 nightfighters were equipped with rockets. "Rocket equipped Ju-88's belonged to either a dayfighter 'Zerstorer' unit or to 'Erprobungskommando 25' (ErprKdo 25), a weapons testing unit specializing in the testing of weapons against four-engined bombers." (Author's translation).

The morning after. The City of Wanette's *officers pose behind her flak-shattered right wingtip. L-R are: Lt. E.L. Matthews, bombardier; Elmer Brown, navigator; Lt. Woddrop, pilot; and Lt. Grover Henderson, copilot. It was of Woddrop that Brown wrote in his diary: "Since we only had one gun working we all have Woodie, the pilot, to thank for bringing us through...it was his beautiful, violent, evasive action that kept us from being blown to Kingdom Come by the many attacks Jerry made with his rockets, cannons, and machine guns." The* City of Wanette *was B-17G 42-31241, GN✪W. The aircraft was later lost to flak on a mission to Berlin on April 29, 1944.* (Photo courtesy Elmer L. Brown, Jr.)

"Woodie got a Silver Star for this mission. We really had no proper business getting back from it at all."

This landing marked the end of the roughest raid Woddrop ever flew. His reaction can be gauged by a passage from the Group's mission report

Bill Watts helped too. Sgt. William A. Watts, engineer and top turret gunner on Widdrop's crew, fingers the 20mm cannon hole punched in the City of Wanette's *side during his furious fight with an Me-210 on the December 20, 1943, Bremen raid. This strike was inflicted on the Me-210's first pass from eight o'clock high. According to Brown's diary, the cannon shell "went through the fuselage heading right for the top turret gunner. It hit his ammunition cans and really scattered his ammo." The German made two more passes before Watts shot him down. Watts also damaged an Me-110, and he topped the day off by downing a Ju-88 on the way home. His copilot, Grover Henderson, considered Watts "one of the two truly great aerial gunners I knew in World War II."* (Photo Courtesy of Elmer L. Brown, Jr.)

to First Bomb Division: "Lt. Woddrop...is very insistent that the A/A defenses at Bremen have been considerably increased and believes definitely that the A/A defenses between Bremen and Wilhelmshaven have definitely been increased." His vehemence is hardly surprising.

The crew's interrogation report underscores how rough the raid was for the enlisted men. The waist gunners both suffered concussions from the flak shell that exploded next to them, and one of them, Valis, came away with 20mm shell fragments. He and Sgt. Hoff went to the base hospital for observation, as did Charlie Baggs for the hit on his head. Sgt. O'Conner was sent there permanently for his "frozen hand," and though the rest of the crew never saw him again the interrogator noted under "crew comments" that "Sgt. O'Conner deserves special mention with possibility of Award."

Nor did Sgt. Bill Watts return unscathed. He was sent to the hospital for a slight 20mm head wound, suffered, no doubt, during his shootout with the Me-210. But he gave far better than he got. Brown had recorded this encounter as a kill in his diary, and there is confirmation in the Group's mission file, where a "combat form" states: "Lt. Brown and Lt. Matthews saw pilot bail out of E/A and chute open."

There is other evidence that Watts made the enemy pay triply for the shell fragments that found him. While there is no mention in the 303rd mission files of the Ju-88 Watts got during the return flight, it appears he also scored against another twin-engine fighter. Both Baggs and Plante saw what happened when, early in the fight, an Me-110 popped out of vapor trails at seven-thirty o'clock, 150 yards out. Watts "fired 200 rounds into E/A cockpit. E/A did a complete loop backwards, then into a spin, and then a pinwheel. Canopy came off. Not possible to come out of spin. Pilot was definitely out." Any way one totals it, Bill Watts had quite a day.

There is nothing in the 303rd's records to show how Grover Henderson felt about this mission, but an incident during debriefing reveals how much of a toll the day's events took.

"At interrogation I didn't have my flying suit on, just the blue bunny suit. There was a little Second Lieutenant in a group of officers who had been there just a few days who wanted to know why I was wearing that blue bunny suit without a flying suit over it. He complained I would get it dirty and went on and on, and somehow it just struck me wrong.

"He was an awful small fellow, so I reached over the table he was sitting behind and snatched him clean over it. I was about to do away with him when everybody else stopped me. It was just that I was keyed up over something he said that made me mad. I was fixing to beat the fire out of him, and for no reason at all. I don't know why I did it, but that was the worst day I had during the war."

The mission had an impact on Elmer Brown that is reflected in his writing. Earlier, he had described some of his roughest missions in a cursory way, providing only a partial picture of what took place. With this mission, his diary begins to show a real passion for detail, a strong desire to pile fact on fact. Looking back today, he explains that "By now I knew I was living through something that was terrible and unbelievable. I felt compelled to write it all down so that people would know that all these things actually happened."

25

Jersey Bounce Jr.

Bremen, December 20, 1943

Up, up the long, delirious, burning blue
I've topped the windswept heights with easy grace
Where never lark, or even eagle flew.
And, while with silent lifting mind I've trod
The high, untrespassed sanctity of space,
Put out my hand, and touched the face of God.

—*from "High Flight"*
John Gillespie Magee, Jr.

With the wounds suffered by Woddrop's men, and the loss of Alex and Leve's Crews—Lt. A. Alex's crew was flying *Santa Anna*, B-17G 42-39764, on their first mission and Lt. F. Leve's men were on their sixth raid in B-17G 42-31233—the 303rd had taken some hard blows during the December 20th attack on Bremen. But the courage and the casualties that day were not confined to the 427th Squadron.

The combination of cold, contrails, and clouds, together with the intense flak that engulfed the formation, allowed the German fighters to sneak in among the bombers even before their escort had left. They made the raid, according to the 303rd's PRO report, "one of the roughest our crews have been on." The 303rd's report to the First Division further stated that up to "125 E/A were seen by this Group. Then were many T/E fighters but all types were included: Me-109, Me-110, Me-210, Ju-88, and FW-190...The attacks started at 1118 hours and continued up to 1216 for a total of 58 minutes."

Like Elmer Brown, Lt. Bill McSween had waited a long time to go on this operation; his last mission had been the one to Bremen on November 26th. He and the rest of Captain Don Gamble's crew were flying as Group lead for the first time with Major Glynn Shumake in *Sky Wolf*. Gamble stated afterwards:

"We were doing fine until we started our bomb run. The formation was perfect. As soon as we got over the target they smashed hell out of us.

That flak was plenty accurate and there was lots of it. Our escort tried to keep the fighters out, but they sneaked through the contrails where we couldn't see them. It was plenty rough all right..."

For Bill McSween, the raid was a navigator's nightmare:

"I never saw ice freeze on the inside windows so bad—almost impossible to see anything outside. That's the hardest job I ever had, trying to scrape ice with one hand and navigate with the other. Luckily, the ground was visible on the Continent, but we drove in straight over the IP, me pulling my hair out every inch of the way.

"In order to make a bomb run, Don had to maneuver into position behind the high and low groups, causing the bomb run to be off the briefed heading. By laying on my belly and looking around the front around the bomb sight, I could see the ground. Sweet [the bombardier] took over after Don straightened out on the run. We flew over the edge of Delmenhorst and I held Sweet back to prevent him bombing it.

"Finally I spotted the aiming point and I talked him into seeing it. He put that load of incendiaries right on the button. Flak was the most accurate I ever saw. Sweet did evasive action with the bombsight. We picked up some holes. Followed course home OK. Don did a good job."

Another who commented on the raid was Lt. John Barker, the pilot who had had such a terrifying ride in the tail of *Mr. Five by Five* during the bomb run on Black Thursday. This mission was his last, and from the No. 6 slot of the high squadron he was well positioned to see the loss of Leve and Alex's ships.

Afterwards, Barker commented: "I have been on lots of rough ones, but that was as rough as any of them. They couldn't stop us, though. I didn't see the bombing results, but the boys say we let them have it."

One of the bombers the Germans couldn't stop was *Jersey Bounce Jr.* flown by Lt. John Henderson's crew in the No. 4 position of the lead squadron. Captain Merle Hungerford was aboard as their Instructor Pilot, together with the regulars on the crew including Sgt. Ed Ruppel, Sgt. Bill Simpkins, and Sgt. Forrest Vosler. That evening Lt. McSween heard a bit of what happened, and wrote in his notebook that "Hungerford and Henderson had to ditch No. 664. They got two men shot up and came to the English coast on two engines. Some luck!"

The full story of what occurred aboard *Jersey Bounce Jr.* that day has never been told, nor can be, this long after the events. But enough remains in the 303rd's records—and in the memories of Ruppel, Simpkins, and Vosler—to present an accurate picture of what took place. The story is one of the most inspiring to emerge from the Eighth's air war.

Lt. John F. Henderson's crew posing before Sky Wolf. *Bottom row, L-R, Sgt. William H. Simpkins, engineer; Sgt. George W. Buske, tail gunner; Sgt. Stanley N. Moody, right waist; Sgt. Forrest L. Vosler, radioman; Sgt. Ralph F. Burkart, left waist; Sgt. Edward Ruppel, ball turret gunner. L-R, top row, Capt. Merle R. Hungerford, Instructor Pilot; Lt. W.J. Ames, copilot (not on 12/20/43 mission); Lt. Henderson, pilot; Lt. Woodrow W. Monkres, bombardier; and Lt. Warren S. Wiggins, navigator.* (Photo courtesy William H. Simpkins.)

Forrest Vosler begins: "We encountered flak as we entered the perimeter of Bremen. It was the usual, terrible heavy flak, and just before we dropped the bombs, we were struck by some antiaircraft. We managed to keep up with the formation long enough to drop the bombs on the target. Then, in our 180-degree turn, we were hit again, and lost a considerable amount of altitude.

"I couldn't see any parts of the aircraft that were damaged. From where I was in the radio room, the only thing I could see, basically, was the vertical and horizontal stabilizer. All I knew was that we had lost an engine—you could tell from the lack of noise in the aircraft—and that we were losing altitude."

From the ball turret, Ed Ruppel was in a position to see more:

"Right after we came up on the bomb run, we got a heavy hit on the left side of the ship by flak. I knew we were hit hard. I could see holes in the wing, but nothing hanging down. I did see that No. 1 engine was on fire, and called on the intercom, 'Pilot, No. 1 is on fire!'

"Henderson said, 'I know it, Sergeant!'

"I moved the turret and checked the bottom of the ship to see the other engines. I called back to Henderson again and said, 'Everything else appears to be okay but No. 1 is still on fire.'

"He said, 'I'll put it out!,' and he rolled over on the left wing and started down. I saw the cowling flaps open on No. 1 as we were going down and the fire blew out, but then we had problems with No. 3 engine. Smoke was coming out, the fan was turning, but she wasn't putting out any power. I knew it was going to be hell from here on in. When you fell out of formation, that's when the German fighters really went after you."

In the top turret, Bill Simpkins also knew it was going to be rough.

"The minute you fell out of formation you were going to have fighters after you, 'cause they liked to prey on the single aircraft, naturally. It was just a matter of time before they jumped us."

It was just a matter of time, and as Henderson's crew waited, Vosler and Ruppel got some glimpses of what lay in store. Vosler next observed "two other aircraft that were damaged off the back of our aircraft. They were at a higher elevation than we were, and were being attacked by fighters. One of them clearly just blew up, disintegrated, and went down. Of course I tried to strain to see if I could spot anybody exit from the aircraft, because you wanted to report this in when you got back on the ground. I didn't see any such thing, but I was limited to a very short period of time before the plane fell below my line of sight."

Ruppel remembers that "As I glanced off to my right, I could see four or five B-17s being attacked by fighters. There was one B-17 that was pretty close to us. They cut one of his wings off and he went into a tight roll. Then they went after the others. They just kept pecking away until they got them all. They chopped up one ship, and another, they hit a third one, and then they went to work on a plane over to our left, and cut him all up.

"I knew people were in there. I hoped like anybody else that chutes would show, but I had no feelings towards them. I knew that when the fighters were finished with them we were next, and I was concerned about myself. When we pulled out of our dive we were down somewhere around 15,000 feet, and it seemed like the entire German Luftwaffe was down there with us. They were single-engine fighters, 109s and 190s."

Things happened with enormous speed after this. The crew later reported that the Germans got Sgt. George Buske, the tail gunner, on their very first pass. Bill Simpkins remembers him saying on the intercom, "I'm hit," and it appears that this first attack was the one that also wounded Forrest Vosler:

"Almost immediately, there was a sound like somebody throwing rice on the aircraft. There was a lot of shrapnel coming through the aircraft. I don't know where it came from, but to the best of my belief it was pieces of our aircraft. A 20mm shell from a fighter apparently exploded near our B-17 and the shrapnel sprayed up from the rear. I was hit in both legs. At this point I hadn't shot at any fighters. I hadn't seen any."

"I stood there for a few moments terribly scared. When I got hit it felt just like 'hot lead,' right out of 'Dick Tracy' and the comics. I could also feel the blood flowing down my legs.

"Several things went through my mind. One of them was that there was no question about my getting the Purple Heart. My next thought was that 'This is a very serious business I'm in, and I've got to do something to protect myself or I'm not going to make it.' Survival is paramount to anybody in combat, so I immediately sat down in my chair to try to avoid being hit again. I figured I'd got an armor-plated chair and it curled up around my back.

"And as I sat there contemplating my next move, I thought how silly my actions were, because I didn't know where the next bullets were going to come from. I had to have the chair facing the right direction, or this wasn't going to work. It wasn't going to stop any bullets. So I figured, 'If this is the way it's going to be, at least I'm going to die standing up. I'll do the job. I might just as well get up because I'm not going to protect myself with this chair. This is stupid.'

"So I got up, grabbed ahold of the machine gun, charged the gun, and got ready to fire."

By this time Ed Ruppel and Bill Simpkins were already engaging the enemy. Ruppel clearly recalls one single-engine fighter that "made a bottom attack from tail to nose. He came closer in than he should have, and he turned over in a half-roll and started to go down. As I brought the ball around on him he was already in the roll, and the gunsights being what they were, I couldn't frame him, and didn't have a chance to fire. He was going too fast. He was the only one I saw coming in from below. All the rest of the attacks were from above."

There were more than enough fighters for Bill Simpkins in the top turret. He remembers them "queueing up eight or nine at a time up at ten o'clock and coming in on us.

"They came in one right after the other. I was lining up and shooting at them. I don't know how many I hit. You didn't have time to count. They were sneaking up from behind, too, and I got a glimpse of Vosler firing. But I couldn't see much more than his head and shoulders out the top of the radio hatch.

"Then I was called back to the waist to help get first aid to Buske. Ralph Burkart, the right waist gunner, called me back and Moody, the other waist gunner, took my place in the top turret. I headed back, and when I got to the tail I saw Buske was slumped over the guns. A 20mm had come right through the armor plate. His flight suit was torn open, and I could see where he got hit, right in the stomach, up front. He was bleeding pretty bad.

"I dragged him back past the wheel well, almost up to the waist where the gunners were. I gave him a double shot of morphine. Burkart was helping me, and he handed me the morphine. It was frozen, so I put it in my mouth—that's how you warmed it up—and I gave it to Buske. Then I put a compress bandage on him. Ralph handed me that, too, and I did the doctoring. Buske was lying there unconscious. I worked on him automatically. It was something you did. If a man was wounded, you helped him. He'd do the same for you."

While these events were occurring, Sgt. Stan Moody was hotly engaged in the top turret. At 1200 hours he spied an Me-110 flying "parallel at nine o'clock, about 800 yards out" and he opened fire with a long burst. He reported that "Black smoke came out and E/A went down end-over-end out of control." Sgt. Ruppel confirmed, and Moody was given a "probable."

Five minutes later Moody scored again. The crew told the interrogator that an "Me-109 came in about two o'clock level, dropped down under the wing, came up at three o'clock, and came in. T.T. [top turret] gunner opened fire first at 800 yards, again at 400 yards. E/A nosed over and went down." Burkart reported a chute coming out, and Moody got credit for a kill.

In the same minute, Sgt. Burkart also got a kill. An Me-210 "came in at five o'clock level" and he "opened fire at 6–700 yards. At 400 yards right wing of E/A came off and [the plane] went down in flames."

Ed Ruppel confirmed, and he recalls the details: "The other boys were calling fighter attacks out. When I heard where they were coming from, I immediately moved the turret over there to try to find them. You didn't look through the scope to see them. You looked through the side shields of the turret in front, the two little windows on the sides, to pick them up.

"Burkart called out an Me-210, and said he was quite a ways off. He was going to let him come in a little bit more. I was waiting and waiting and didn't see nothin', and all of a sudden I saw the aircraft coming down kind of on a tumble, wing over wing. The right wing busted off, and he started to break up. The pilot could never have made it out. I called Burkart and said, 'You got 'em!'"

The fight went on, and five minutes later Bill Simpkins scored from the left waist. He doesn't recall the incident, but the crew reported that an "FW-190 came in to attack about ten-o'clock a little above and then it went in at the waist. E/A came in to 150 yards. L Waist gunner opened fire with long burst at 150 yards. E/A caught fire, smoke came out, and it went down in a tumbling spin, flames all around." Both Moody and Ruppel confirmed, and Simpkins got a "probable."

Bill Simpkins does recall returning to the tail and getting on the intercom to help Henderson evade attacks from the rear, and it was probably while he was here that the climax of the fight occurred. It took place during moments of give-and-take in battle that were the most significant of Vosler's 20-year-old existence. He remembers them this way:

"A twin-engine fighter came up the back. Apparently at this point the tail guns had been shot out, and there was no one at the waist guns, either. So this left the entire rear of the ship vulnerable except for my one single .50 in the radio compartment.

"When this plane came up in back of the B-17, he was so close I could actually see the pilot's face. I would have recognized him again on the ground. He was just off the tail, and had throttled back, and could have knocked us down any moment. He could have easily killed me. I was looking right down the barrel of a 20mm cannon. I could see the lands and grooves of the gun; that's how close he was. If he had had any cannon ammunition, it would have been all over.

"My first burst knocked pieces on the left side of his wing off. I was actually after the engine or the pilot. I moved the gun rapidly over to try to get him. I was firing as I turned, and I went right across the stabilizer and put a hole in it, because this gun had no stops. Our plane seemed to be flying all right, so I didn't bother Henderson with a little thing like my hitting the stabilizer.

"I never saw anybody in my life so scared as that German pilot. He turned white when I was firing, and he dove. Had he stayed there a second longer he'd have been a dead pilot, because I was a pretty good shot. I was after the pilot at this point, not the engines. He was a very lucky young man.

"I radioed the ball turret gunner to confirm if I got the aircraft or not. There was a big plume of smoke as he dove the aircraft; that was normal when the Germans poured the fuel to their engines, so I assumed the pieces I knocked off his wing would not be critical.

"Within seconds after that I got hit again. This time it was much more serious. I got hit in the eyes, the chest, and the hand. The strange part about it is that I had had goggles up on my head. I had pulled them down, and they immediately steamed up. I couldn't see clearly and I pushed them back on my head. I had no sooner put them back, when 'Snap!' I got hit in both eyes. I didn't know what hit me."

Forrest Vosler couldn't know, but Ed Ruppel later saw exactly what had happened to the crew's radioman. One glance told him the whole story:

"Vosler had a flex-held .50-caliber machine gun. A 20mm shell came down the side of his gun till it hit the breech, and that's where she exploded. The cover to the breech was completely blown open, and the gun was black with soft powder from the headspace to the tail of the breech. So I know that's where the 20mm exploded. Vosler must have been bent down over the gun to fire when the shell exploded.

"He was shrapnel from his forehead to his knees, everywhere. There was blood all over him, coming from all those little shrapnel cuts. There was no one place where you could put your hand and stop the blood. I knew he was hit bad in the eyes, too, because I could see the white stuff running down below one eye and onto his cheek."

Vosler recalls his initial reaction to this second, terrible blow:

"It was, 'This is a heck of a place to hit somebody, this is not really playing fair. They're not playing the game right, hitting a guy in the eyes.' I couldn't see well, but when I moved my hand down to my chest where I'd been hit—I was trying to open my jacket to find out how badly—I noticed that my hand was shaking. I couldn't control it. Then I reached up and dragged my hand across my face to see if there was blood, and when I looked at it my whole hand was covered with blood.

"The shell fragments had damaged the retina of my right eye, and I was seeing blood streaming down the retina *inside* my eye, thinking it was on the outside. So my natural feeling was that I had lost the whole side of my face. Having had a lot of first aid experience, I realized you could have this happen, and the shock would be so great that you wouldn't even feel the pain.

"Also, I didn't realize that I had been hit in the hand. It was bleeding profusely, but I didn't feel this injury. The shell fragments had gone through all four pair of gloves I was wearing. But I thought I only had half a face.

"I became extremely concerned. I was out of control, really. Obviously I wasn't going to have a chance to get out of this thing now. I knew I was going to die. I knew my life was coming to an end. The fear was so intense; it's indescribable, the terror you feel when you realize you're going to die and there's nothing you can do about it. So I started to lose control, and I knew then that I was either going to go completely berserk and be lost, or something else would happen.

"And a strange thing *did* happen. I lived every day of my life. I relived my whole life, day by day, for 20 years. It put everything in perspective. For the first time I realized what a wonderful, wonderful life I had had. There were only a few days in my whole life that were bad, and I asked God to forgive me for those bad days, and thanked Him for all the many, many wonderful days He had given me. I said, 'I'm not going to ask you for any more days. It's been too nice.'

"I have never again had that feeling of complete peace that I had at that moment. I only hope that before I die, I might experience that feeling of peace once again. I said to God, 'You've given me 20 years of life. I appreciate it, I thank You for it, and if this is the way it's going to be, I'm happy, let's go.' I even reached out my hand and said, 'Take me, God, I'm ready.'"

The moments passed, and Forrest Vosler began to realize that the Almighty wasn't ready for him.

"I was a little bit surprised and a little bit disappointed, actually, that God didn't take my hand, because I was ready to go at that point. It would have been selfish to ask for even five more minutes of life. But at the same time I became very content, very calm, very collected. I no longer feared death, which is a terrible thing to fear. And I slowly realized that if God didn't want to take me at that particular point, then I had to go on and do the best things I could do."

There was plenty for everyone aboard *Jersey Bounce Jr.* to do. As Ed Ruppel recalls, "In the course of the combat we got two direct hits in the instrument panel. We couldn't tell whether we were flying sideways, upside down, backwards, or what. So we hit the deck, and started to come home that way. We figured we didn't have anything to worry about with enemy aircraft at that low altitude. We were down real low, still over Germany, and we picked up some small arms fire, but by the time they fired we were already gone.

"Nobody was saying anything, and I got out of the turret to see what was going on. Then the pilot asked us to throw everything overboard, to lighten the aircraft up, so we could get as much as we could out of it. People were throwing things out the rear of the ship to lighten the load."

With no fighters around, Bill Simpkins was also on the move, en route to an amazing encounter with his radioman:

"I went back from the tail up to the waist, and Ruppel was getting out of the ball turret about this time. We were unloading everything, trying to make the plane lighter and trying to make it home. We didn't know if we would have to ditch yet. We had two engines out, we were pretty low in altitude, and we were low on fuel too.

"I was on my way up to the cockpit. I helped get Buske to the radio room—we laid him on some life jackets—and I saw that Vosler had been hit. His gun was all shattered. He had his back turned to me, and was standing up working at the radio. I looked him right in the face, and I saw there was stuff dribbling down his right cheek from his eye. He was in a daze, groggy, visibly shook up. He wasn't normal.

"As we were throwing things out, he said, 'You're throwing everything else overboard. Well, why don't you throw me overboard? I'm just so much extra pounds. Throw me out, too.' And he really meant it, because he asked me more than once to throw him out. I didn't say anything, really. I just sloughed it off. I didn't take him real seriously, even though I knew he was getting serious about it."

Vosler offers this explanation for the extraordinary offer that he made: "I was still worried about my face. I saw Simpkins, and I asked him if my face had been shot away. He told me, no, that all I had was a little trickle of blood coming down my cheek, plus what I had smeared on it from my hand. Well, I didn't believe him. So I asked a couple of the others. They all assured me that my face hadn't been shot away, but I didn't really believe them.

"So when we were ordered to throw everything out of the aircraft, I figured if there wasn't enough to lighten the aircraft, why don't you throw me out? I figured I was pretty well shot up anyway, so it didn't make any difference whether they threw me out or not.

"Looking back, I suppose I was still a bit disappointed that God hadn't taken me, but I also think I wouldn't have dared have them throw me out. I don't know. It would have been kind of a traumatic feeling to get kicked out of the airplane without a parachute!"

Bill Simpkins proceeded up forward, and another member of the crew started to focus on *his* wounds:

"I continued up to the cockpit and met Monkres, the bombardier, heading back to the waist. We met in the bomb bay. As we crossed one another he saw that my flight jacket was all tore up from where shrapnel had hit me. I had picked up a piece of shrapnel in the back, and it tore up my heavy sheepskin jacket. He thought I was hit much worse than I was. My hand was stiff, too, and I had it curled up in a fist in my glove. He thought it was frozen.

"When I got to the cockpit, I saw the copilot reading our instruments. They showed we still had fuel, but they couldn't tell how much. So I was doing the fuel transfer, jockeying the fuel in the tanks back and forth from one engine to another as they started to sputter. To do this, I went back towards the radio room. The fuel transfer valve was right behind the firewall of the bomb bay, right near the radio room."

By this time Ed Ruppel had also made it to the radio room, and had seen Vosler first hand. He remembers that "When I came up out of the ball turret, Vosler was lying on the radio room floor. We helped him into his chair. We asked him if he could get a message out, and he said he could if we would put the freq meters in that he told us. These were metal boxes that you plugged into the radio to use different frequencies. There were three or four of us in there trying to help him get the radio so that he could make contact with base. He told us where to put the freq meters, and we were pulling ones out and putting in the ones that he wanted, and he immediately went to the key and went to work. He didn't need to see to use it."

As Vosler recalls these moments, "We struggled along, and by this time I could see some damage on the wings. There were holes in both of them as far as I could tell. I remember Henderson, the pilot, being concerned about keeping the airplane in the air. He told me that he needed to send out a distress signal, because there was no way he was going to make England. Would I send out a distress signal?

"Although my vision was blurred, I could see that we were still over land. We did experience another set of flak over the coast, which I could hear. I'm not aware that it hit the aircraft, but it came awful close. It was pretty dense, but we managed to get through it.

"I told the pilot that I would send out the SOS as soon as we reached some water, out of range of enemy territory. He said, 'I think you better send it now.'

"I said, 'Sir, let me know when we're going down, and I'll send the SOS. When you can't keep the aircraft airborne, let me know. In the meantime, if you keep it up, let's not break radio silence.'

"So I didn't break radio silence. I remember guarding very carefully the critical equipment for that SOS. I had one hand across the table guarding the receiver, and I had the other one guarding the transmitter, so the others wouldn't touch either piece of equipment. Over in another part of the radio compartment there were stacks of frequency modules that would not be needed. I already had the right module in for the SOS. The rest were thrown out by the other crewmembers, and in fact they threw out a lot of stuff. I had to watch them. They threw my shoes out, which I resented. I figured if we landed on land, how was I going to walk? I'm barefoot!"

The men did throw plenty of things out, but luckily there was still one gun aboard that was in the right place at the right time.

As Ruppel recounts: "Coming up across the coast, there was a 109 out in front of us. All our guns in the back had been thrown out, even the tail guns, but Monkres still had his flex gun in the nose. I heard this rapid fire. He fired at the 109. It was making a nose attack, and then broke off. After that, Monkres disassembled his gun and threw it overboard.

"Later, when he came back to where we were, he said, 'I got about three or four good strikes off of that guy, and he pulled off to the right.' That fighter broke off, and never did come back. He didn't want any more, but if he had come in any other way he could have chopped us up."

Jersey Bounce Jr. was now over water, and the stage was set for Forrest Vosler to radio Air Sea Rescue. As Vosler tells it:

"Henderson was an excellent pilot, and managed to keep the aircraft up until we reached the North Sea. At that point I sent an SOS after some of the crewmembers helped me patch up the radio; one of the wires was off the transmission key. I sent the SOS out at different speeds, and I got an immediate response from England. They receipted my message, and asked me to give a holding signal for 20 or 30 seconds while they shot a true bearing on me. I responded and gave them the signal, and they came back, and gave a receipt on that one. They said they had my course, and asked me to transmit every 10 or 15 minutes so that later they could correct their bearing.

"I sent about two more messages after that, and the pilot informed the crew that we going into the North Sea. Apparently, before we had to ditch, we had an Air Sea Rescue plane flying over and around the top of us, because the pilot called me up and notified me that Air Sea Rescue was over the top of us. And he thanked me for the SOS. We must have been in the air some considerable time, at least 20 or 30 minutes for them to dispatch an aircraft out to us. We weren't very far off the coast, probably 60 miles."

British Air Sea Rescue had sent aircraft to intercept the crippled B-17. They sent not one, but *four*. Ed Ruppel remembers them well.

"We were heading for what they called 'The Bulge of England.' When we got fairly close, an Anson came out and a Walrus. The Anson was a land-based plane, a twin-engine job, and the Walrus was a flying boat. And we had two 'Clip Nines' come out. These were marine Spitfire IXs with clipped wings. As we got closer to shore, we finally saw land, and it was a tossup whether we should try to land or set her in the drink. With our instruments all out, we figured the best thing to do was set her in the drink."

What happened next reminds one very much of the Hullar crew's experience. As Simpkins put it, "I remember getting ready to ditch. Everybody got in the radio room except the pilot and copilot. We were bundled all up in there."

Vosler recalls that "By this time everybody was in the radio room, lined up facing the front of the aircraft. It made a nice, compact group. We had eight people in there."

Ruppel remembers: "Everybody got into position for ditching in the radio room. I was the last one to get down into position. Vosler was there at his desk, still banging away on the radio. We didn't talk to him because you leave a man alone when he's doing his job, and I don't remember any conversation in the radio room at this time. All thoughts that the people in there had were kept to themselves, and I imagine a little prayer in there just at that time too."

Vosler, who had already been through so much, believed "The men were all a bit upset, afraid they were going to die. They were really scared. I reassured them that they weren't going to die, that we were going to make it. I'm not putting them down, or anything, but they hadn't been through what I had. I had gone past them. I no longer feared death. All I could do was comfort them, saying, 'Don't worry, you're going to be all right. Don't worry about it. Relax.'"

All are agreed on the details of the ditching. As Bill Simpkins recalls, "We didn't bounce any when we hit the water. We slid right in."

Ed Ruppel remembers "There was a yawl and a trawler out there, and we made a pass in between them and came around and dropped a red flare up to tell our escort we were going in. Henderson did a marvelous job. It happened real quick. He brought the plane right over the water, put the tailwheel down, and fishtailed right in.

"I heard the tailwheel go in. I saw it go down, but I didn't know we were really in the water until it splashed up from the ball turret in my face, and I tasted it and it was salty. I knew we were in the water then.

"It was about the smoothest thing you ever saw in your life. We had training on the ditching procedure, and it was amazing how it all came back. It all fell right into place like a million dollars."

Vosler, too, remembers that "They directed us to ditch ahead of this Norwegian trawler rather than having the Air Sea Rescue drop one of their rowboats with a motor on it; they thought it would be safer for the trawler to pick us up. The pilot was told to ditch ahead of the boat, or as near to the boat as possible, and they would come over and get us.

"The actual ditching procedure we did beautifully. It went without a flaw, with much credit to Henderson, the pilot. He ditched it beautifully. He brought it in on the top of a wave; they had six foot waves there, and he rode the plane in on the top of one just like a surfer. He read it perfectly. The plane stopped quickly, and at about 90 miles an hour it was like hitting a brick wall. But no one was hurt."

Lt. Henderson's men now scrambled out the radio room hatch to their rafts, and like Hullar's men before them, those on the left wing had a relatively easy time while those on the right ran into greater difficulties.

Ed Ruppel was one of those exiting to the left: "I jumped out the hatch and somebody else helped me pick the raft up. We knew enough not to drag it across the wing, because there might be jagged holes there that would put a hole in it. We put it in the water, and decided that we were going to shove off, and then we found out that we were still tied to the aircraft. And the next big joke was, 'Where's a knife?!' We all carried knives strapped to our legs to cut parachute cords. I had mine on me but I couldn't think of that. Finally somebody came up with a knife, and we cut the rope and drifted away. The Anson had marked the spot where we went in with a dye marker, and I saw this trawler coming right toward us. There was a big flare coming out of the bow. When it came alongside we motioned to take the other raft first, which had the wounded in it."

Bill Simpkins was one of those exiting to starboard, and it was here that he witnessed another extraordinary act for a man as badly wounded as Vosler was.

"When we come to a stop we all jumped out the hatch and got on the wings. I got onto the right wing with Vosler. I helped lift Buske out. He was still unconscious. We put him on the wing and went to get the life raft. And while we were doing this Buske started to slide down the wing into the water. Vosler grabbed him and held him till we got hold of him. Then we pulled Buske back and got him into the life raft. By this time the ship was getting ready to go down. The tail was starting to stand up."

Vosler recalls the incident well:

"I remember them all climbing out of the hatch. Some of them went on one side of the airplane, and some went on the other. We got the wounded tail gunner out. Somebody was going to help me out, but I said I thought I could boost myself out all right. I got up on top of the fuselage, looked down, and Buske was slipping into the water.

"I yelled to the pilot, but I could see he wasn't going to respond fast enough. They had pulled the life raft out and it was floating on top of the wing, and Henderson was busy trying to cut the cord on the life raft so it wouldn't go down with the airplane. I knew Buske would be in the water in a fraction of a second. I would have to take action. So I jumped and held out my hand at the same time. I grabbed the antenna wire that runs from the top of the tail to just forward of the starboard radio compartment window. I prayed that it would hold, and I was able to grab Buske around his waist just as he was going into the water, sliding off the trailing edge of the wing. I was bent way over.

"If the wire had broken, both of us would have gone in the drink. It was under an awful strain, and I was yelling all this time. The rest of them responded and got both of us.

"At that point we all got into the rafts. We were pretty close to the B-17 when it went down. I saw the whole front of it under water, and just a few minutes later the tail went. So it was not up very long. And then there was nothing but ocean."

In the meantime, as Bill Simpkins relates, "We got the paddles out, and could see the cargo ship in the distance. We paddled alongside of it, and they grabbed us and got us aboard."

Vosler remembers being rescued, too. "The next thing I can recall is that this ship we had ditched in front of steamed up to us. We promptly unloaded the tail gunner, and when it came my turn I managed to walk up the ladder. This was the first time I experienced any real pain in the legs. I got up on the railing, and jumped down on the deck. A couple of sailors tried to grab me, and although my legs hurt, I thought I was all right, and pushed them aside and told them I could walk all right. I promptly went down in a heap on the deck. They stretched me out there, and I was terribly concerned about my vision, my eyes, and my face. It was very disturbing, because I felt I was really disfigured."

The trawler returned to Ed Ruppel's raft, and "they threw us a rope. It landed across the top of our dinghy and we pulled ourselves over to them. They had a cargo net down over the side with half their crew there to help us up into the boat. I never got my feet wet."

On board the trawler, Bill Simpkins recalls, "They had a doctor who worked on Buske. We were sitting in a cabin drinking some hot tea that they gave us, and the doctor came in and said Buske was in pretty bad shape. They kept us on the trawler for quite a while, and then we transferred to a PT boat. They put Buske on a stretcher, and it was pretty straightforward. The trip back was routine, except that they had a fire in one of their engines."

The fire on the British launch made a greater impression on Ruppel, who recounts that "British Air Sea Rescue, a PT boat, picked us up from the trawler. We were running on it for only three or four minutes when someone yelled, 'Fire!.' I looked back, and one of the engines was fully ablaze. They put that out, worked on the engines a bit, and gave a revised ETA. All day they kept revising the ETA later and later. I thought we would never get in."

For Vosler, the ride was an ordeal. "I was given some morphine against my wishes. The one thing I can recall about the ride was that it was terribly fast. I was up by the bow with my head facing astern and the boat was going at least 45 knots. It was an awful position. When the bow went up, the blood rushed to my head. It was miserable."

But despite his discomfort, Vosler still made a strong impression on Ed Ruppel: "On the boat he was down on the deck. The deck hands laid down some kind of a canvas cover for him to lay on. And I went over and talked to him. Some more of that white stuff was running out of his eye, but when I asked him out how he felt, he said 'All right.'

"The only thing he wanted was a drink. So I went down and found that the boat crew was breaking out tea and rum for us. I told Vosler, and he said, 'I'll take the rum.' That's the kind of guy he was, cocky as hell.

"They took us to a hospital at Great Yarmouth, where we spent the night. The next day we flew back to our own base. I didn't see Vosler until after the war, back in the States, and I never saw Buske again, though I was told he survived his wounds.* For the rest of us, it was more missions."

Vosler's stay at the British hospital was the beginning of a long period of medical treatment in England and America. Both his eyes had been hit by 20mm shrapnel, and for many months he was completely blind, as his left eye reacted "sympathetically" to his totally sightless right one. Only after his right eye was surgically removed did he slowly regain a measure of sight.

There were, however, other compensations for Forrest Vosler. As Ed Ruppel remembers:

"A month or so after this mission, Major Black, our flight surgeon, came to us on the crew and said they took Vosler back to the States, and he thinks they're going to put him up for the Medal of Honor. Then one day Headquarters called us and asked us to come over. We did, and that's when we were told they were putting Vosler in for the Medal.

"They asked us what went on, and we told them what had happened. We were afraid of saying anything except what had happened, because we didn't want to ruin his chances. Nobody can take away from you what happened. That mission has gone through my head many, many times, thousands of times, and I do think the man deserved the Medal for what he did."

The Army Air Forces' leaders did too, and Sgt. Forrest L. Vosler became the second member of the Hell's Angels to win the nation's highest award for military valor.**

Thus ended what is referred to here as "The Battle of Bremen." On the December 20th mission, the Eighth lost 27 bombers, plus three Category Es, and in the five raids flown against the port city from November 26th through December 20, 1943, the Eighth lost a grand total of 77 B-17s and B-24s, wrote off 15 as Category Es, and had 640 bombers damaged.

---

*George Buske did recover from his wounds, and is now retired and living in Rochester, New York.

**The first was a bombardier, 1st Lt. Jack W. Mathis, who despite mortal wounds, laid a highly accurate pattern of bombs on the target during the March 18, 1943, mission to Vegesack.

Twenty-eight crewmen were killed in action, 100 wounded, and 720 were missing. From Charles Spencer and Joseph Sawicki to Forrest Vosler, the Eighth's bomber crews paid a steep price for keeping the pressure up on the enemy.

The heavy bombers were to fly four more raids before the end of 1943, and then, with the help of the new P-51s, press their offensive against the Luftwaffe harder still. As Hullar's crew entered the home stretch of their tour, the conflict was building to its climax.

26

Rocket Area in France

Vacqueriette, December 24, 1943

THE EIGHTH'S AIRMEN took a breather after the December 20th attack on Bremen, as no raids were scheduled for next day. On December 22nd the heavy bombers were off to Germany again on mission against communications centers in Münster and Osnabrück.

Eight PFF ships of the 482nd Group flew with 346 heavies to Osnabrück, while 220 more went to Münster. The raids were disrupted by heavy cloud formations. Twenty-two bombers went missing, and only 434 were able to bomb. The 303rd's target was a railway intersection in Osnabrück, but the Group's bomb run was ruined when another formation of aircraft passed below it. Enemy opposition was light and no Group aircraft were lost.

Hullar's crew did not take part in this operation, but they did fly an air as rescue sortie on December 23rd. As Miller recalls, "We flew for hours and hours low over the water looking for life rafts, but it was all for naught."

Hullar's crew was next slated to lead the Group on Christmas Eve in an attack intended to beat the Germans to a holiday punch.

The surprise had been brewing for some weeks, and George Hoyt recalls the events leading up to it.

"Around the first week of December we were summoned down to Intelligence Headquarters one night to take a close look at some curious objects which showed up on some of the aerial mosaic photos which came from pictures we had taken while leaving France. *Vicious Virgin* had a vertical K-24 'Keystone Camera' under the radio room floor. I turned it on about 10 miles from the French coast each time we returned from a mission. I can particularly recall Dale and me being questioned after we looked over these terrain shots. You could see strange looking shadows shaped like snow skis turned on their sides.

"The most prominent imprints came from shots made in the middle of the afternoon, when the winter sun stood low in the sky. The structures were apparently under camouflage nets, because you couldn't really see them on pictures taken in morning or midday, but the late sun cast shadows under those nets. There were literally scores of these shapes, and we were told that our photos coincided with some that had been taken by RAF reconnaissance.

"The photo interpreters had been doing their homework, because when you lined up the open lengths of these 'skis' on a large map with a ruler, they all converged, amazingly, on the city of London! The French Underground was alerted, and after a few days they came up with the startling intelligence that Jerry was planning a rocket 'buzz bomb' attack on London for Christmas day. So VIII Bomber Command mustered a maximum effort to thwart this insidious strike."

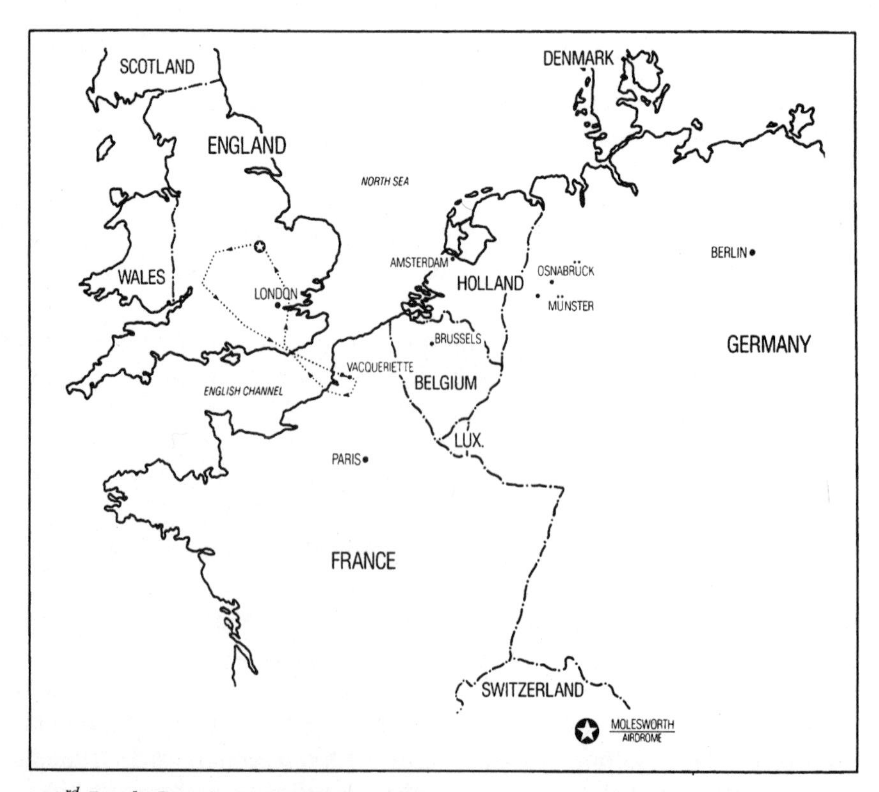

303rd Bomb Group mission Route(s): Vacqueriette, December 24, 1943.
(Map courtesy Waters Design Associates, Inc.)

December 24, 1943. "We were the lead ship leading the wing and the whole division," wrote Elmer Brown of the crew's 18th raid against a hard-to-see rocket launching site near the small French town of Vacqueriette. By this time Hullar's crew was one of the most experienced in the 303rd. Bottom row, L-R, George Hoyt; Merlin Miller; Norman Sampson; "Pete" Fullem; and Dale Rice. Top row, L-R, Lt. E.G. Greenwood, tail observer; Elmer Brown, Asst. Lead Navigator; Maj. Ed Snyder, 427th Squadron CO; Bob Hullar, pilot; Lt. Paul Scoggins, Lead Navigator. Paul Scoggins had a good reason to smile; this was the end of his last mission! (Photo courtesy Elmer L. Brown, Jr.)

The raid certainly was a maximum effort. The Eighth dispatched no less than 722 heavy bombers supported by 541 fighters to attack V-1 sites dispersed in the Pas de Calais area of the French coast. Hullar's crew was leading the entire First Division, 277 bombers strong, to their targets.

George Hoyt recalls, "We flew in old No. 341, *Vicious Virgin*, as 1st Air Division lead plane with Major Ed Snyder in the copilot seat next to Hullar. Good old Brownie was back with us after flying several missions with Woddrop, and I really felt good about this. Marson was not with us, and Merlin took over his waist gun position. We had a third pilot fly in the tail gun position to help coordinate the very large formations that we led behind us."

The tail observer was Lt. E.G. Greenwood. In the nose with Elmer Brown were Mac McCormick and Paul Scoggins, who was on his last mission. Due to the large number of aircraft aloft, the field order stated: "Navigators must realize that this mission is essentially a 'time problem.' Every possible effort will be made to make the briefed time schedule good.

Elmer Brown wrote: "Scoggins and I were navigating. He was doing pilotage and I was doing all of the DR, calculating, and getting Gee fixes."

Scoggins noted, "Another navigator (Brown) and I got the boys there (leading the combat wing)."

The formations used for the raid were not the usual 60-ship combat wings. As Brown described them, "Our group had three nine-plane squadrons flying in trail and bombing by squadron." The 427th Squadron led the first "two-squadron group" with the 427th "high" and the 359th Squadron "low," and a second two-squadron "Composite group" made up of the 360th Squadron and a low squadron from the 384th Group was led by Captains John Casello and "Mel" Schulstad. This composite group flew above and behind Hullar's formation and was to play an important part in the mission.

The 303rd got its ships off between 1100 and 1117. Things went pretty much according to plan until mid-Channel. Here, the 303rd's S-3 report states: "There were two Combat Wings out ahead of us. Crossing the Channel the Groups took interval, and the individual Squadrons got interval by various means, running short dog legs, S-ing, and slowing air-speeds. Our two-squadron Lead Group crossed the French Coast over Burck-sur-Mer, went north of course, and could not pick up the target."

What went wrong was well described by both Brown and Scoggins. Brown recorded the objective as a "Rocket area in France near the very small town of Vacqueriette...We hit the IP on course heading for the target.

"McCormick thought he saw the target and turned off course. We corrected him back but never could make him see the very small, difficult-to-find target. There were almost 3000 [sic] aircraft in the area, therefore we were instructed not to attempt a second run."

Scoggins noted: "It was only a short way in and we had no enemy fighters or flak and had lots of Allied fighter support. Because of a very difficult target to identify, however, we passed the target without dropping bombs, and there were too many planes in the area to make another run, so we brought our bombs back...I'm glad I'm not a bombardier."

Others were far less philosophical about what went on. Lt. Jim Fowler's crew was flying *S for Sugar* on its 48th mission in the No. 8 position of the lead squadron. Fowler attributed the failure to locate the target to "broken clouds beneath us," but he was still "very highly ticked off" when the formation brought the bombs back.

Next to him, in the copilot seat, Lt. Rawlings was also disappointed by the way things turned out:

"We went in at 12,000 feet, which was extremely low, but they wanted accuracy. It seemed to me that we were milling around rather aimlessly and there was another B-17 group that passed over us, maybe 500 feet above us, on an angle with the bomb bay doors open, and the bombs visible. We could look straight up and see this other group, and I felt rather strongly, 'This is *not* the way you fly a bombing mission.' The target was extremely small, but I felt we were our own worst enemy that day. It was frustrating to get all the way out there, and to have it all come to pieces at the end."

McCormick was not the only one unable to spot the target that day. The nine ships of the 359th's low squadron were to bomb separately, but they were unsuccessful for reasons set out in Lt. J.P. Manning's Low Squadron Leader's Report:

"The IP was reached at 1306 hours, where we made a slight left turn to a magnetic heading of 103 degrees. From the IP we took our interval behind the lead squadron and got to the left of our course. We were unable to identify our target, and did not drop our bombs as a result.

"The bombardier was all set up ready to bomb, but even after picking out the checkpoints, the target itself could not be identified. Over what was evidently the target we were flying slightly right of the lead squadron and at that time a high group dropped bombs, which fell very close to our squadron. After our target was passed, we turned right and took up the briefed return course."

The composite group was the formation whose bombs dropped near the 359th. The important thing, however, was that their ordnance scored a bulls-eye. The 360th's lead bombardier was Lt. Jack B. Fawcett.

Elmer Brown wrote that "Only one of our squadrons found and hit the target. He (Fawcett) did a petty good job."

Fawcett commented: "We had a perfect bomb run and I am sure we smashed the target. We had no interference from anything." Photo-reconnaissance showed the bombs well distributed in the target's center.

Thus the 303rd returned home with a measure of success, and no losses. It was a good mission all in all, and gave the men added reason to celebrate the Christmas holiday. As Lt. Vern Moncur, who had just finished his second raid in *Wallaroo*, put it: "There was no injury to crew, or battle damage to the plane. This was a rather novel way of celebrating Christmas Eve."

No one, however, had greater cause for joy than Lt. Paul Scoggins, who jubilantly noted in his diary: "I had one of the nicest Christmas presents! I did my *TWENTY-FIFTH MISSION* and therefore have completed my tour of operations today. I'm one more happy boy, believe me! In one way this was a mission I had been wishing for—another milk run—but I would much rather have put the bombs on the target...I'm still happy for having finished, though, and I guess I have about three months to serve on the Squadron before I can plan on a trip home... Everyone here is so very nice to me and have always been. I think one could never find a better bunch of boys than are here at Molesworth. Of course, most of them are having a big time tonight, but there are several of us sober ones around. My wish is that everyone were as happy as I. I wish a Merry Christmas to all."

The people of London had a Merry Christmas too. Due to the Eighth's efforts, the Germans were in no position to put buzz bombs over the British capital on Christmas Day. Six hundred and seventy heavy bombers managed to lay 1744.6 tons of bombs on the launch sites, and with this mission and many others against these targets in the months ahead, it wasn't until just before D-Day that the first V-1 winged its way to London.

27

"We Had 10/10 Clouds Under Us the Whole Time"

Ludwigshafen, December 30, 1943

THE EIGHTH SCHEDULED two missions between Christmas and December 29th, but both were scrubbed due to bad weather. On the 30th the Eighth's leaders got off a raid against the city of Ludwigshafen, described by VIII Bomber Command as a "key inland port and home of one of Germany's two largest chemical plants."

The plan called for PFF bombing, with five wings of the First Division in the van, five from the Third Division following a half hour behind, and two wings of B-24s from the Second Division trailing closely on their heels. A total of 710 bombers were sent, escorted by 583 of the Eighth's fighters with RAF Spitfires providing additional support.

The 303rd and the 384th Group were to make up the second combat wing going into the target. The 384th furnished the wing's lead group and the high squadron of the high group. The 303rd filled out the balance of the formation.

Hullar's crew stayed home, with the main group formation being led by Major Walter Shayler and Lt. W.C. Bergeron's crew of the 360th Squadron. However, the mission roster included two well-known navigators: Elmer Brown, flying once again with Woddrop's crew as lead of the low group's low squadron, and Lt. Bill McSween with Captain Don Gamble flying *Vicious Virgin* at the head of the low group's high squadron.

Gamble's diary provides a good introduction to the raid:

"Up at 0400—Lay in sack and cussed for 15 minutes. Got one egg and an orange for breakfast. Briefed for a lengthy little flight to Mannheim Ludwigshaven to drop eggs on rubber works and a chemical factory. The lights went out about halfway through the briefing, that is the light in the projector. We later found out that it wasn't important to

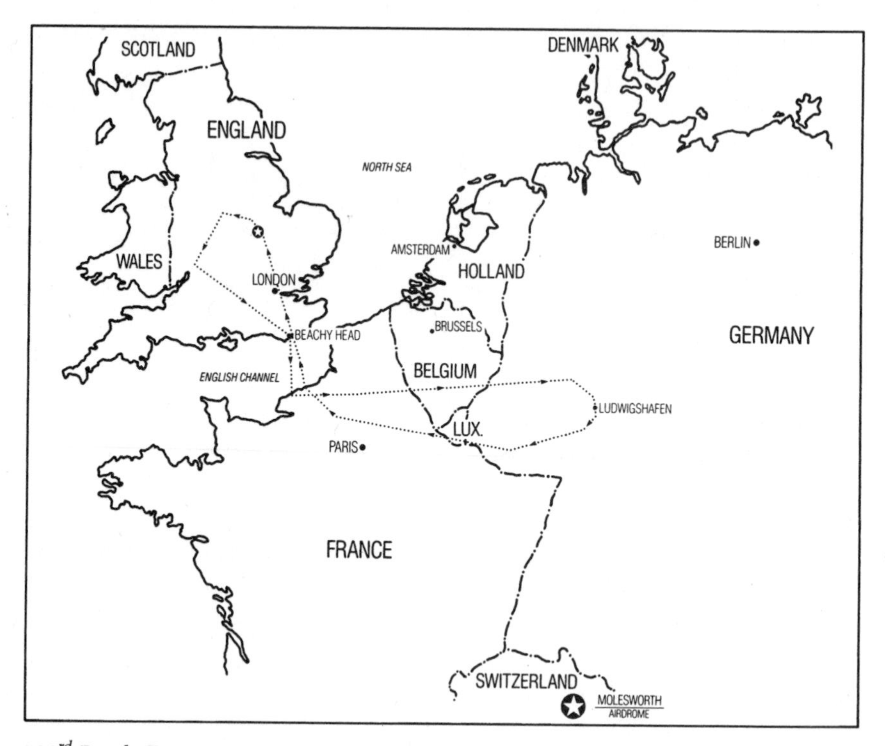

303rd Bomb Group mission Route(s): Ludwigshafen, December 24, 1943. (Map courtesy Waters Design Associates, Inc.)

see pictures of the target. Col. Kermit Stevens gave his usual lengthy oration and I wished I was back in the sack.

"Just time to get my stuff and out to the ship—*Sky Wolf*—had low oil pressure on No. 1 engine. Loaded on a truck and went over to take Q-341, a 427th ship. Took off last and found the group at about 2000 feet. Relieved Lake of the squadron lead."

McSween's notebook describes the outbound trip:

"We flew 427th ship 341-Q, a good warm plane, temp. −25, altitude 22,000 feet. Flew south to Beachy Head to FeCamp, from there due east to IP. Solid undercast except right at coast. Computed wind crossing channel. Was stronger by 15–20 knots than briefed."

Lt. Fowler's crew was also flying and his notebook offers a lively account of the mission:

"We were briefed to fly a spare aircraft on this mission, falling out at mid-Channel and returning to base only if no one dropped out of the formation, and continuing on with the formation if a plane dropped out and left a spot open. We were assigned the *Knock Out Dropper*, a 359th B-17F. This was its 57th mission. We took off at 8:05 and fell in behind the formation as they formed over the field, and trailed along behind as they rendezvoused with the rest of the combat wing and began the climb to altitude. Our base was furnishing the low group and two squadrons for the high group this day, and our instructions were to fly in either of these groups.

"No one aborted up to the time we reached the English coast and started out over the sea, but the leader of the second element of the low squadron in the high group fell back a ways and the leader of our three spare aircraft took off to fill the spot. Before he got in place, the lagging plane pulled back in formation. Our chances of going on the mission looked bad. We were at mid-Channel, but we pulled up and looked over the high group, thinking there was a hole. But they were all filled in. Then the tail gunner called out that tail-end Charlie of the low squadron low group was aborting, so we dove down, came up under, and beat the other spares to the hole."

Barney Rawlings remembers this as "a tension-filled time for everyone aboard. I can recall very vividly the feeling of despair that we had after we had gone through this entire rigamarole of getting briefed, taking off, and chasing the formation, hoping for a hole to fill in, and it looked like we would have to go home. Then there was a feeling of delight which was shared by the entire crew, I think, when we finally got a chance to go. I think we all wanted desperately to go on this mission."

Fowler went on to write: "We picked up a little flak at the Coast and some thereafter," and the Fortresses met enemy fighters going in as well. Brown wrote, "We had 10/10 clouds under us the whole time while over enemy territory except for scattered clouds over 20 to 30 miles of the French coastal area. The only attack we had was about 30 minutes before the target, when we didn't have any of our escort with us. The attack was made by a squadron of nine FW-190s flying in close formation. It was a head-on attack at our squadron, which we were leading. They came close enough that they turned over on their backs as they went under our ship, but didn't do any damage."

Gamble likewise noted: "Had one fighter attack on low squadron, but they didn't lose any ships. P-47s covered us all the way in."

The 190s had a definite impact on Lt. Rawlings, who was seeing enemy aircraft in action for the first time.

"I felt a fascinated, horrified admiration for the guys flying the fighters. They barreled right through at a fantastic closing speed, and seemed to be quite skillful at what they were doing. I had a feeling of real terror about them—not that I became a gibbering idiot, but I was frightened. I felt that if these guys wanted to get you, they probably would. I also had a terrible feeling that a pilot could be dead and still kill you from a head-on collision. They came awfully close to the formation and our airplane. I saw the turrets on about seven or eight of the planes in our formation turning to follow a fighter that was coming through above us, to my right. All these top turrets were shooting and gradually turning so that they appeared to be shooting right at us. I realized that you could quite easily get shot down by your own people. I formed the definite opinion that while flak might be spectacular, the fighters were *really* dangerous."

The flak on this mission would also prove dangerous. The field order stressed that "Care will be taken by lead navigation not to overshoot IP due to concentration of flak installations east of this point," but that is exactly what occurred.

McSween wrote that "IP reached nine minutes early, flew slightly past, turned back on target. PFF hit it on the nose. Flak was damn accurate through 10/10 clouds. Violent evasive action was made."

Gamble wrote, "Dodged a rough bit of flak over city," and Elmer Brown recorded that "The flak over the target was moderate and very accurate."

Lt. Fowler also observed: "There was heavy flak, very accurately aimed, but our group leader took very effective evasive action, and we picked up only one small hole in the right horizontal stabilizer."

The Group ran into enemy fighters on the return trip, too. Fowler commented that "Soon after bombing the target a few Me-109s came through firing 20mm. They hit and ran without effect.

"More 109s came across our nose and rolled through and scattered the group at our right soon after the target. They went through once and started pestering a Fort with one engine feathered back behind us. After a couple of passes, some P-47s came down and chased them off."

The attacks did cause a loss in the 303rd. In the No. 5 slot of the high group's low squadron was *Woman's Home Companion*, B-17G 42-39795, flown by Lt. W.C. Osborn's crew of the 360th Squadron. They were reported "shot down by enemy fighters," with "conflicting reports on what happened to this plane."*

---

*Much more was learned of the fate of Lt. Osborn's men when the crew's engineer, Sgt. W.E. Wolff, successfully evaded capture and returned to Molesworth on April 30, 1944 after crossing the Pyrenees Mountains into Spain and reaching Gibraltar. He reported that the crew's B-17 was hit by flak that severed the rudder cable shaft causing the aircraft to leave the formation. Lt. Osborn turned back at 11:25 hours at 15,000 ft., under control, with three FW-190s heading to attack. The German fighters made three attacks during which a number of crewmen were wounded and Sgt. L.B. Evans, the tail gunner, was shot out of the aircraft and killed. The attacks ceased when Lt. Osborn managed to hide in a cloud bank at 12,000–15,000 ft.

After flying a short distance, Lt. Osborn crash-landed the B-17 in a farmer's field at Froid Chappel, near the Luxembourg-French-Belgian border. Sgt. L.W. Fitzgerald, the ball turret gunner, died of his wounds shortly after the crash-landing. Belgian civilians arrived at the crash site and directed those among the crew who could to scatter. A Belgian doctor treated two of the wounded, Sgt. G.L. Daniel, the radioman, and Sgt. E.D. Wolfe, the left waist gunner, who were later captured by the Germans and became POWs. Lt. E.L. Cobb, the navigator, was also captured by the Germans and became a POW.

The return trip over France held a tough problem for Lt. McSween: "The flight back was slow, bucking a headwind. I learned some DR. My dead reckoning wasn't working out and I didn't find out what was wrong until I reached the French coast. Lesson: You don't double the drift angle when the wind is cut in two for computer purposes."

Lt. Fowler also noted the long time it was taking to reach the coast, and other things as well. "We were two hours and ten minutes in getting back to the enemy coast going out. Two B-24s flew with us for almost an hour until they located their group. Spits met us then. Crossed the coast at 14:20 and began the descent out over the channel. It felt very good to get off constant-flow oxygen; had been on it about five hours."

The Group let down over the Channel to the top of a cloud layer, then let down in England over a haze layer. Lt. McSween wrote: "Visibility was poor over the base on the return. Don nearly landed an top of a B-17. I was worn to a frazzle. Flying time: 7:30 hours."

Gamble noted: "Nearly landed on a plane. Went around in pattern and made sloppy landing. Very good ship."

Lt. Fowler "Sweated out gas coming home, but made it OK. Landed with about 25 gallons left in each tank. No losses for our squadron."

Total losses for the mission were 23 B-17s and B-24s, plus 13 missing fighters. The Eighth claimed 20 German fighters. Bombing results were a disappointment. Although 658 of the heavies dropped their bombs through the 10/10 clouds, not one touched the chemical plant.

December 31st saw one final strike into occupied France. The Eighth sent 572 bombers and 548 escorts against a variety of objectives. The 303rd's target was the blockade runner *Orsono* located near Bordeaux with a cargo of crude rubber, but 10/10 clouds area again made "daylight precision bombing" impossible. The 303rd lost no B-17s, but the Eighth

---

Lt. Osborn and Lt. J. Jernigan, Jr., the copilot, evaded capture for six months with the help of the underground. However, they were later betrayed while leaving Brussels for Paris in an automobile when they were driven to a Gestapo prison. They remained in the hands of the Gestapo for two months before being transferred to a regular *Luftwaffe* prison camp.

The bombardier, Lt. N. Campbell, also evaded capture, successfully making it to Gibraltar by way of Belgium, France, and Spain. He returned to Molesworth a month and a half after Sgt. Wolff, on 17 June 1944.

Most remarkable and heroic of all was the fate of Sgt. Vincent J. Reese, the right waist gunner, who made contact with the underground and opted to stay and fight with them rather than evade and try to return to Allied lines. He participated in underground actions involving the destruction of German installations, but was captured along with 22 other members of the underground and executed on April 22, 1944 in a woods near Chimay, Belgium after the group was discovered hiding in a shack. Sgt. Reese is buried in the Netherlands American Cemetery, Martgratten, Netherlands.

returned with 25 missing heavy bombers, 15 Category Es, four missing fighters, and claims of 33 German aircraft shot down.

With the end of 1943 Hullar's crew had completed 21 of their 25 operations. Slowly but surely, the stress of the many missions was having its effect, and as the old year died and the new one began, it wasn't at all clear that the crew would keep together.

28

Combat Stress

The Tour Takes Its Toll

OVER TIME, THE flight surgeons and psychiatrists charged with evaluating the Eighth's combat crews believed they had identified three phases of a crew's response to combat. The first usually coincided with the first five missions a crew had to fly, when the men were either overly self-assured or "mouse-quiet," asked too many questions or none at all, and were susceptible to jibes from veteran crews who had already "seen it all." For Hullar's crew, this phase never took root. With First Schweinfurt as their second operation, they were instant combat veterans who had won the immediate respect of their peers.

Merlin Miller remembers that "Shortly after we arrived at Molesworth, a couple of fellows from another crew told us we weren't going to make it through our tour. When we came back from Schweinfurt that day, those same fellows came up to us and said, 'If anyone makes it, you guys will.'"

The second period, lasting through the 10th raid, was the one in which fear was openly acknowledged and accepted. Here Hullar's crew conformed to the norm. The Münster raid was the crew's 10th, of which Elmer Brown wrote "We were all so scared."

Paradoxically, it was on the home stretch, when the medical profile indicated that the crew should all be "effective, careful fighting men, quiet and cool on the ground and in the air," that the stress of the missions came to a head. The subject is a sensitive one even now, but two men on the crew have decided to share the way in which the critical question "Can I continue?" was answered for them.

For George Hoyt the question had been brewing a long time. It came into the open after the Christmas Eve attack on Vacqueriette. As Miller explains:

"We all knew he was having trouble. You could tell it from the way he called out fighters on the intercom. We tried to help him, especially Rice, but I never realized how serious his situation was."

As Hoyt tells it, the problem was reaching its height during the crew's last mission to Bremen, on December 16, 1943: "I remember climbing aboard the airplane with a high fever from a bad case of the flu. It was a real double whammy. I hung on somehow and performed my duties, including a series of 'on target' fixes, and headings to get us back in the thick weather. But I was losing weight and felt nervous and weak.

"Dale Rice and Charlie Baggs became aware of my failing condition, and 'Dr.' Baggs prescribed two aspirin at bedtime and a double shot of brandy. I lived with this prescription—and that bottle of brandy under my bed—until our next raid against the buzz bomb site in France on Christmas Eve.

"I was still feeling bad, and as we got into the plane, I remember Dale saying, encouragingly, 'The oxygen will help to knock it out.' I made it, but I was afraid of having to go on sick call, for I knew it would mean two or three weeks in the hospital, and that I would miss some of the missions with my crew. So I decided to try to hang in there.

"I felt all of us were keyed up and under a lot of stress during this period, and to let off steam one night, when the weather had everything socked in and we were sure we wouldn't be flying next day, we hung one on of historic proportions.

"We—meaning Dale, Charlie Baggs, Bill Watts, and me (I don't think Merlin was along, he was always quite cool, quiet, and well behaved)—sneaked off the base without passes to one of the pubs in the local villages. Here we embarked on a rip-roaring drunk that would have absolutely appalled Bob Hullar. We got back on the base with the help of an RAF recon car driver who literally charged the gate in his vehicle, and to top the evening off, Bill Watts and I engaged in a hog-calling contest in the middle of the squadron area at about 1:00 A.M. We could have raised the dead with the noise we were making.

"It all caught up with me just days before our next mission—to Kiel—on January 4, 1944. Bob Hullar came to me and asked me to visit our squadron flight surgeon. So I went to his office for a 10:00 A.M. appointment, and he gave me a full '64' flight physical. When he was finished I walked back to his office in my olive drab GI undershorts and stood in front of his desk, nervously waiting to hear what his diagnosis was. My heart was pounding and I was very much on edge.

"He stared a long time at my physical exam form, and then he looked at me and spoke very slowly.

"He said, 'You are badly exhausted physically, and your weight has taken a nose dive. I am going to recommend that you be sent back to the States. You can be part of a War Bond Drive crew. You have done your part here. You have 21 missions completed, and that's an outstanding achievement in this theater of operations. What do you say to that?'

"I was horrified. His words hit me right in the pit of my stomach. The whole idea was incredible! Never in my life had I ever given up on anything I had attempted. This went against everything I felt was important—the sense of duty which was so forcefully instilled in me by my parents, the discipline I had at the academy where I attended high school, the superlative training that I received in the Air Force—everything.

"I was desperate. It was a desperation as great as any I have ever felt. I knew in my heart that I could never accept this and feel the same about myself again. It was one of the great turning points of my life. So I let it all hang out. I strode up and down the room in front of his desk telling the doctor again and again that I would never consent to his suggestion. I can recall saying that I would complete my tour of missions even if I had to stow away on the plane. I even banged my fist on his desk to emphasize my point. I went on and on until I had exhausted myself, and the doctor just sat there and stared at me wide-eyed. It was the most impassioned speech I ever gave.

"After I was all run down, the doctor rose slowly from his desk, and I began to wonder if he was going to have me court-martialed for gross insubordination. But he put his hand out to me across his desk as I stood there in my GI undershorts.

"He said, 'Son, go back to your crew and complete your missions. I had no idea what this all meant to you. Good luck, and God bless you.' And he strode out of the room with tears in his eyes."

Elmer Brown had a very different encounter with the Squadron flight surgeon, which he recalls this way:

"After I had flown about 20 missions, right about the time of that mission I went on to Bremen on December 20, 1943, I experienced two incidences of memory failure. In the first one, I couldn't find my bicycle. I reported that it was missing to the Military Police, and they located the bike in an area of the Base that I seldom went to. Forty years later I can't recall the details, but when I learned all the facts I was convinced that I should have known where the bike was parked. But I could not remember anything about it.

"The second incident occurred when we were swinging a compass on an airplane. I looked at my wrist to see what time it was, and my watch was missing. A watch is a navigator's most important instrument, so I was very much disturbed when I couldn't find it. I expressed my distress to the other members of the crew that were present, and Dale Rice said, 'I have your watch.' He had wanted to time something, so I had loaned him the watch. It had only been a short time, an hour, that he had borrowed it. But again, I could not remember having loaned it to him.

"It was expected that I would be a lead navigator in the rest of the missions I had to fly, and this could mean a Group lead, a Wing lead, a Division lead, or even the first plane of the Eighth Air Force to go on a mission. In any of these lead positions the safety of hundreds of airmen rested on the lead navigator's performance. I remembered the mission to Münster, when we were supposed to fly north of and parallel to the Ruhr, 'Happy Valley,' but because of the lead navigator's error, we flew directly over Happy Valley for several miles, and the flak was the thickest I had ever seen.

"The success of hitting the target with hundreds of tons of bombs was also dependent on the lead navigator's ability to direct the formation to the IP and the target area on the correct heading. The lead navigator had to be very accurate in all his readings of instruments and in making his calculations and data interpretations. He had to plot all the courses, headings, winds, and other information on his maps. He had to exercise good judgment in every decision that he made. It followed that a lead navigator had to have a good memory.

"Not saying anything to my crew, when the opportunity presented itself, I slipped away and went to see the Squadron flight surgeon. I told him that on two occasions my memory had failed me. I told him exactly what had happened to me on these two occasions. I assured him that I was not one of these guys that was trying to get out of flying combat. I enjoyed the idea of flying lead navigator, because he was the one who was telling everyone where to go instead of being the guy in the back of the formation who told everyone where the lead navigator was taking them.

"I felt it was much more fun and challenging being lead navigator, and that the lead ship had better protection, surrounded by the firepower of all the other ships in the formation. But I also felt that I would be remiss if I didn't tell someone about my problem, in case they wanted to relieve me of the duties of flying lead navigator.

"The doctor said that there was nothing he could do for me. He said that the two experiences I told him about did not warrant any change in my duties. And tears came to his eyes when he said this. He also said that I should come back if I had any more bad experiences, but I left it at that."

The strain also affected other members of Hullar's crew. Miller explains:

"Rice and I went to Liverpool one time to visit his brother up there. His brother, Charlie, was with some Army transportation unit. We stayed in a hotel that had been turned into a Red Cross Club. The three of us were in one room, which had one single and a double bunk, over and under. Charlie had an overnight pass, and the three of us went to a pub someplace and had a few beers. We went back to the room and Charlie slept in the upper bunk. When we woke up the next morning, he looked a little bit bleary-eyed.

The two of us said, 'What's the matter, didn't you sleep good last night?'

"He said, 'Hell, no. You guys flew a combat mission all night.'

"I asked him, 'What are you talking about?' I didn't remember dreaming that night at all, and Dale said he didn't either. Both of us said we slept pretty good.

"Charlie said, 'Maybe you did, but you guys were calling out fighters and talking to each other all night. You'd call out a fighter, and Dale would yell, "I see him. I see him. Get on him, mother," and pretty soon he would call out one and you'd talk back. That went on almost all night, and I never did get any sleep.'"

It was an evening Charles Rice still remembers: "They were literally reliving their missions. When you're asleep you can hear things going on that may not wake you up, but they are nevertheless things you can react to. And this was exactly what was happening to them. When they first woke me up I thought immediately, 'They're flying one of their missions.' This is the sort of thing you read about in books, the terror of it all. But in my opinion it wasn't so much a question of terror as alertness to the danger.

"I remember another incident where the three of us were riding on a train in England somewhere. The English trains had these little coaches in the cars where six people could sit, three on one side and three on the other, facing each other between a little aisle, with baggage racks right over their heads. We were all sitting together, and another train shot past us on the track, just like that. Both of them jumped up and hit their heads on the baggage rack. That's how tense they still were from what they were going through.

"This was what was keeping them alive, being able to react instantly, trying to do something to protect themselves and their crew, because things happen so fast in combat."

Charles Rice has real authority to speak of such things; he lived them himself. After Hullar's crew finished their 25 missions he followed in his brother's footsteps, wrangling a transfer to the 303rd and flying a 35-mission tour in 1944.

Merlin Miller remembers another evening when "Sammy went to bed early while the rest of us were playing cards. All of a sudden he sat up in his bed, yelled 'Fighters!', made a complete turn, and fell off the bed on his head."

As Norman Sampson recollects, however, "The stress of combat did not seem to affect me at the time of the missions. But on the way home, on the boat, I was playing cards and my hands began to shake. I had a hard time hanging onto a glass of water. My hair began to come out by the combfuls, also. My hands still tremble sometimes, but I have all my hair back now."

Rough as it was, all the men of Hullar's crew flew their missions without a single "personnel failure," the cryptic reference in the Group's records to a bomber turning back due to a physical or other handicap in one of its crew.

But as Bob Hullar indicated to Lt. Rawlings that morning in the mess hall, the missions didn't get any easier the more you flew. And while the crew now had only four more to go, their next two were to show how long the odds still were against completing their tour.

29

Duel with a Smokescreen

Kiel, January 4, 1944

THE FIRST DAYS of 1944 were quiet ones at the Eighth's airfields. There was no mission scheduled New Year's Day, nor the day after. At Molesworth there was aerial activity, but it was mostly at the hands of the RAF. On January 1st British pilots buzzed the base with a trio of captured German aircraft—a Ju-88, an Me-110, and an FW-190—landing after their demonstration and allowing the Group's personnel to get a hands-on look at their enemy's best aircraft. The next day another British pilot put on a show with a captured Me-109, and the field also hosted two four-engine Lancasters of RAF Bomber Command homeward bound from a raid against Berlin.

A mission laid on for January 3rd was scrubbed, but it was a different story on January 4th. An ambitious attack against the U-boat yards at Kiel was in the works, a round trip of 825 statute miles from the English coast. The escort was to be continuous with the drop tanks then available, but the effort would put the B-17s and B-24s near the outer limit of friendly fighter range.

The plan called for six First Division wings accompanied by seven PFF ships to lead. Three more PFF ships were to follow with three B-24 wings of the Second Division and four wings of the Third Division, making up a total force of 569 heavies. The bombers would be shepherded by 70 Lightnings from the Eighth's P-38 groups, the 20th and 55th, together with 41 Mustangs from the 354th Group of IXth Fighter Command. A sizable diversion was also planned. Seventy-five B-17s in two wings from the Third Division were to attack Münster. Ten Thunderbolt groups, 430 P-47s in all, would serve as their escorts.

Members of the 303rd watch as a British pilot warms up a captured FW-190 for take-off at Molesworth on New Year's Day 1944. Note the base's huge "J" hanger in the background and the Fortress at the extreme left with the "triangle C" on its tail. (Photo courtesy Wilbur Klint.)

For the Eighth to dispatch 15 combat wings on a single day's operations was no mean accomplishment; to achieve it, corners had to be cut. One place where the knife was wielded was the fourth CBW of the First Division force. It was a Hell's Angels show all the way, Elmer Brown noting that "our 303rd Group put up two complete groups of approximately 20 planes each, and called it a wing."

He recorded something else of no small significance for a man who had recently questioned his own ability to fly lead: "I was the lead navigator for the wing and Kaliher was the other navigator in our ship who worked with me. I was flying with Hullar's crew. Col. Stevens, our CO, was our copilot and the wing leader. Orvis was the bombardier." They were in *Vicious Virgin*, with Lt. E.G. Greenwood as tail observer and Merlin Miller taking over Marson's position at the right waist.

Chuck Marson did not stay home, however. This was Bud Klint's first mission with his own crew and Marson manned his right waist gun.

"This happened," Klint explains, "because of the ruptured eardrum which Charlie got when we ditched on September 6, 1943. He also was hospitalized by a trick knee that continued to give him trouble. When he was unable to fly with Bob's crew they got a replacement gunner, and when my crew was assigned, he was available in the gunner pool. Charlie was assigned to me and I was certainly proud and happy to have him on my crew. He was an excellent gunner."

Klint's crew was assigned the No. 5 position of the lead squadron in *The Flying Bitch*, the chin-turreted B-17F that Hullar's crew had taken to Norway on November 16, 1943.

Lt. Jim Fowler's crew was assigned the No. 3 slot of the lead squadron in *G.I. Sheets*, the B-17G they had flown on their first mission. At the briefing, Lt. Barney Rawlings was drawn less to this fact than to Colonel Stevens, the "Old Man" who was leading the wing today.

"He struck me as a big, rough bear of a man, with an admirable quality of courage. In the briefing he always made the same remark: 'Well, it may be rough, fellows, but you got to bow your neck. You got to bow your neck.'"

Today Colonel Stevens would be risking his neck along with the rest of them.

Vicious Virgin took to the air right on time at 0800, and soon 37 Forts were winging their way to Kiel. The task of guiding them there now rested with Elmer Brown. He wrote: "According to the metro winds we got in our briefing flight plan, the winds were supposed to be 320 degrees at 110 knots at bombing altitude of 25,000 feet. I could tell from the way the winds were drifting us on the way to the English coast that they were not as metro had forecasted and the metro winds for the balance of the trip wouldn't hold true either.

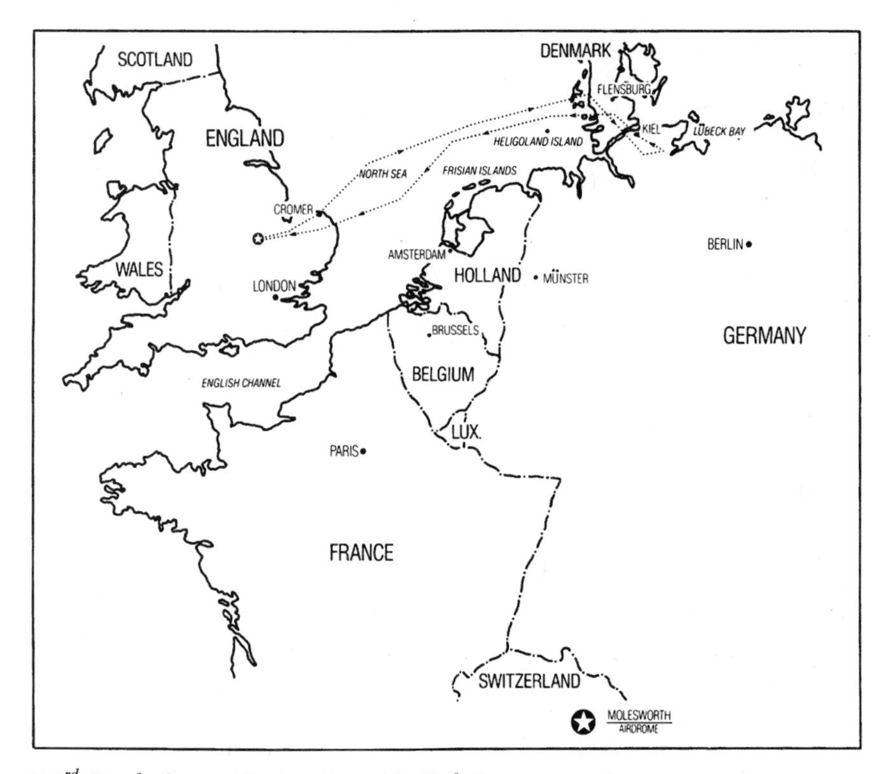

303rd Bomb Group Mission Route(s): Kiel, January 4, 1944. (Map courtesy Waters Design Associates, Inc.)

"We left the coast at Cromer at 14,000 feet climbing on course over the North Sea. By use of Gee fixes and air plot I determined an average wind in the climb from 14 to 20,000 feet. The wind was 07 degrees/60 knots. I couldn't get any more Gee fixes, so I had to estimate what the wind would be for the balance of the trip. I figured the wind from 25,000 feet would be from the same direction, 07 degrees, but I increased the velocity to 80 knots.

"This change in wind meant that we would reach the target about 30 minutes later than scheduled, but we would have a tailwind coming home and should make it back to the base at approximately the expected time. We were to make a rendezvous with P-38s and P-51s in the target area, so I had the radio operator advise the base that we would be late at the target."

George Hoyt remembers getting the word out: "Our briefing told us that P-38s would cover us over the target area, but as we were running late due to the winds, Elmer calculated we would miss our escort without word getting back to England. So he asked me to break strict radio silence and call back to HQ that we were going to be 30 minutes late getting to the IP. I tapped out the message in Morse code, and Jerry immediately chimed in with jamming. My response was with the appropriate 'Q' signals for 'Weak Signal,' and 'Jamming.' In seconds a signal of vastly increased power came from England acknowledging our position. It nearly blasted me out of the radio room."

The 303rd needed all the help it could get, for its ranks were seriously diminished—there were six aborts in its low group. Lt. Jim Fowler pressed on, despite worries about the gaps in the Group's ranks:

"We flew the group as a combat wing composed of a high [sic] and lead group—a flimsy formation, made more flimsy by numerous aborts...Felt very vulnerable with our small formation and with no friendly aircraft preceding us to the target. Sighted fighters just before passing to the north of Heligoland but they stayed off to the south. Sighted fighters, probably P-38s, as we neared the Frisians. Missed our landfall and came in too far south, and increased error by turning south."

Brown was actually bringing the Group in far closer to its planned landfall than Lt. Fowler realized, and his diary contains a suspenseful account of his struggle to find the target:

"[My estimated] wind worked very well as I DRed about 350 miles over the North Sea, and I was only a few miles north of the intended course upon reaching the German coast. As we approached the German coast we had a 10/10 undercast, so we decided the bombing would have to be by Pathfinder and flashed a green Aldis lamp at the PFF ship that was flying the No. 2 position to take over.

"The pilot of the PFF ship was Brim, a fellow who was in 42-1, my flying class with me. That morning was the first time I had seen him since I washed out of flying in July 1942. We were about 10 miles from the coast and we wanted him to take over so he could make a good landfall.

"PFF gave us a red light indicating his equipment was not working. The Colonel said to me, 'Take us in and we will have to do visual bombing.' The clouds thinned out a little bit at the coast, and through a thin cloud I picked up the island where we crossed the coast.

"From there on I was strictly on DR. I turned on what I thought was the IP. We went down the bomb run but couldn't see anything but 10/10 solid clouds below us. The ETA ran out but we continued on the same heading for three minutes more.

"The Colonel said, 'We had better do something.' As the flight plan called for a 90-degree turn off the bomb run, I had them turn 90 degrees to the left.

"The Colonel said, 'Let's try the secondary target' (Flensburg). So I gave him a heading for Flensburg, which took us east of Kiel.

"Soon after making the turn the bombardier saw a smokescreen on the ground through a thin cloud and haze. I told the Colonel that was the Kiel smokescreen and that we ought to drop the bombs in this area. He said, 'OK, if you are sure we are over Germany.' The bombardier took over, made a sharp turn to the left, and tried to synchronize on a couple of high smokestacks. Clouds crossed over before we could synchronize, but we dropped the bombs anyway.

"Our other group couldn't stay in formation in the turn and their lead navigator was afraid we were over Denmark, so they didn't drop their bombs. We were the only wing over Germany at that time. The flak over Kiel was moderate, but very accurate. We got a small flak hole in the fuselage right below where I was sitting. The Colonel and the crew were getting restless when we were flying around and couldn't find the target."

There was plenty to be concerned about. Lt. Fowler wrote: "Wandered around Danish peninsula. Passed over Tubick Bay [Lübeck Bay] and followed Baltic Coast up for possible visual run. Caught heavy and very accurate flak at point assumed to be Kiel and dropped bombs. Got out of there."

George Hoyt remembers how close the flak came, too: "As we came over the Kiel area there were 10/10 clouds obscuring the ground completely. I heard all the talk over the intercom about what to do, as the flak was increasing in intensity. During the run over the target, when the bombardier spotted those smokestacks on the ground, I was looking back over the vertical stabilizer from the radio room hatch. I saw four black puffs burst in a row about 150 feet behind us, right at our altitude. Then four more puffs burst only 50 feet behind us. My experience told me that this was radar-directed fire, and I figured the next volley would be right on us.

"I called the pilot and said, 'This flak back here has got us zeroed in exactly, you had better kick it around.'

"But Colonel Stevens said, 'We're on the bomb run and we're just going to have to sit here and take it.'

"That third salvo of four bursts went off, and I realized that, while they had our forward motion tracked right in, this volley was off on altitude. It burst right below us, even though the other two volleys were right on in altitude. I thanked the Lord at that moment. It was really close."

Colonel Stevens's only comment afterwards was: "There was quite a bit of flak around that area...We flew a pretty good formation over there and should have dropped our bombs in a good pattern. We saw a lot of heavy black smoke coming up through the clouds as we left."

Bob Hullar also observed that "We had good navigation and I think we hit the target. Our bombardier aimed at some tall chimneys sticking up through the smoke."

Klint wrote: "'Brownie' was our lead navigator, and took the Group in by a combination of dead reckoning and visual navigation. He was further hindered by a beautiful smokescreen which had been laid over the city, but in spite of that, we got some bombs in the general area. The center of our bomb pattern was about six miles off the aiming point, but considering the way we went in that wasn't too bad. Flak over the target was as accurate as any we had yet seen, but only moderate in intensity."

Elmer Brown and Lt. Orvis had done a good job in the circumstances. Brown learned later that "most of the bombs hit in the Kiel Fjord (water) about 100 feet short of a torpedo factory. We did get a couple of hits on some buildings. This was at the eastern edge of the town of Kiel about five miles from our aim point, which was the center of town."

On the way out, Brown also observed a thrilling sight: "As we left Germany, we met about seven or eight wings of Forts and Libs going in." But it was at this point that the Luftwaffe arrived, too.

As George Hoyt recalls, "After we turned off the bomb run, the bombardier called to say he could see many German fighters climbing to meet us from below."

Hullar's crew reported about six Me-110s and 210s trying to sneak up the *Virgin*'s vapor trails to make tail attacks, one of which made a strong impression on Hoyt.

"A couple of calls of fighters came over the interphone, and I saw a blue-gray Me-210 swing wide around our tail. He took up a position off at three o'clock and somewhat higher than us, about 400 yards out. I called him out, and swung my gun, but it wouldn't move over that far.

"Suddenly four single-engine fighters zoomed over my top hatch from eleven o'clock to five o'clock. Silver-looking objects dropped from them, falling behind our formation. I called on the intercom, 'It looks like they are dropping bombs on us back here.'

"Then, seconds later, four more fighters flashed over us from five o'clock high, dropping what I now saw were wing gasoline tanks. I called, 'Those are P-51s, they are '51s dropping their wing tanks!' Up above us a P-51 came by chasing a German fighter, and then several more came into view on Jerry's tail.

"All of this flipped by in about 10 seconds, and as I shot my vision back to see where that Me-210 at three o'clock went, I couldn't believe my eyes. He was still flying along in that position, only this time a P-51 was pulling right up behind him. The P-51 opened up with his wing guns and I could see strikes hitting all over the fuselage of that 210 with pieces flying off. He caught the Jerry flatfooted. Smoke belched out of one of the 210's engines, and he dove down out of my field of vision with the '51 right on his tail."

Lt. Fowler also noted: "Me-109s and Me-110s came in from below but didn't offer much trouble. About this time P-51s showed up and with P-38s took care of most enemy aircraft. Our gunners got a few shots but no claims...Not much trouble going out. Enemy fighters were kept down by P-51s."

Sitting next to Fowler, Lt. Barney Rawlings had a feeling of near euphoria: "Everything that bomber pilots were supposed to encounter, we did. We had a satisfying dose of flak and a satisfying dose of fighters. I felt, finally, for the first time, that I was able to call myself a combat pilot. I was at war."

Another side of being at war would soon show itself to Lt. Rawlings. As the Group passed over light flak at the enemy coast the 427th Squadron was about to lose one of its most experienced crews.

Lt. F.C. Humphreys and his men had been through 15 missions, including Second Schweinfurt. This day they were aboard B-17G 42-31526, *Sweet Anna*, a brand new bomber on its first combat flight. What happened to the crew caused many to take note.

Lt. Fowler wrote: "Humphreys, leading second element, evidently got hit by the flak, and started dropping back. McGarry, on his left wing, moved up on ours..."

Bud Klint, flying off of Humphreys", right wing, recorded: "After we left the target, he began to fall behind the rest of the formation. We stayed

Klint's crew. Bottom row, L-R, Sgt. McGrew; Sgt. C.E. Walsh, engineer; Chuck Marson, right waist; Sgt. F.B. Knight, ball turret; Sgt. L. Ratliff, radioman. Top row, L-R, Sgt. N.F. Smith; Lt. E.L. Jenkins, copilot; Lt. E.P. Eccleston, navigator; Lt. R.W. Meagher, bombardier; Bud Klint, pilot. (Photo courtesy Wilbur Klint.)

with him awhile because we could see no apparent signs of trouble. However, when he began to drop farther and farther behind and below the rest of the Group, we and the other wing ship had to leave him and return to the rest of the formation."

Elmer Brown wrote: "Soon after leaving Germany, Humphreys dropped out of formation and started lagging behind. Klint being his wingman went with him until he realized Hump was in trouble, so Klint caught up with the formation. Hump was losing altitude fast and was last seen by Woddrop's crew, who had to lose altitude and left the formation, as they were low on oxygen. Hump was last seen flying west about eight miles off to Frisian Islands about 06 degrees 30 minutes E."

After the Group got back to base, the crews learned that Humphreys had radioed for help. Lt. Fowler "heard Humphreys had sent an SOS and asked for QDM [a request for course heading]. We hope he may have ditched or headed for Sweden."

Klint wrote, "'Hump' later sent an SOS to the base, but nothing further was heard from him. He was reported MIA and presumed to have ditched in the North Sea or to have gone to Sweden."

Brown recorded that "His regular bombardier was Orvis, who was with us, and his regular navigator was Culpin, who was with Woddrop."

The hopes for Sweden were not borne out. As with *Mr. Five by Five* and Captain Cote's crew, no one knows the exact fate of *Sweet Anna* or Lt. Humphreys's men.*

Elmer Brown ended his diary entry by writing: "The rest of us got back OK," but two men were wounded by flak and there were five cases of frostbite. Nineteen bombers were missing and two fighters failed to return; in return, a total of 11 enemy aircraft were claimed. The Münster bomb wings reported good PFF runs, but bombing results over Kiel were inconclusive because of heavy clouds and the smokescreen.

VIII Bomber Command was unhappy enough with the outcome to order a major effort against Kiel the next day. On this operation, however, the bombers would get an especially early start. Takeoff at Molesworth was scheduled for 0715, a full 45 minutes before the launch on January 4th. It was a change in plan that would have fatal consequences.

*The body of the crew's right waist gunner, Sgt. E.P. Madak, washed up on Ameland Island, Holland on Oct. 4, 1944. All other members of the crew were listed as KIA.

30

"The Biggest Ball of Flame I Have Ever Seen"

Kiel, January 5, 1944

THE OPERATIONS PLANNED by Pinetree for January 5th again had Kiel as their prime objective. A total of 247 bombers in three B-17 wings from the First Division, followed by three B-24 wings from the Second, were to hit the U-boat shipbuilding center accompanied by six PFF ships in case blind bombing was required. They would again be escorted by the Eighth's two P-38 groups and the single group of P-51s available.

Meanwhile, the Eighth's P47s, helped by RAF Spitfires and Typhoons, were to escort 274 heavies in attacks against other targets in France and Germany: an antishipping bomber base and repair depot at Bordeaux-Merignac, the "important advanced bomber training base" at Tours, and a ball bearing plant at Elberfeld, Germany.

The 303rd fit into the plan at two points. The 358th Squadron was to contribute six ships as high squadron of the lead group in the second wing going to Kiel. This force would be led by the 379th Group, which was supplying 20 ships to fill the two other squadron positions in the formation. The main 303rd effort was a group of 19 ships forming the low group of the next wing. The 359th Squadron was Group lead under Major Richard Cole and Lt. J.P. Manning's crew, with Mac McCormick as Group Bombardier. The 360th Squadron was high and the 427th was low.

Hullar's crew was leading the 427th in *Pougue Ma Hone*, B-17G 42-31060. The ship, whose name means "Kiss My Ass" in Gaelic, may have belonged to Lt. F.X. Sullivan, a new pilot who was filling the copilot seat next to Bob Hullar. All of Hullar's enlisted regulars were aboard, including Marson, who wanted to make up one of the missions he missed while Klint's crew was sitting this one out. Occupying the nose with Elmer Brown was Lt. M.L. Cornish, an experienced bombardier.

Getting the Group off the ground and properly assembled would be touchy. At the 0415 briefing, the crews were told they would be taking off into darkness. The first bomber was to roll down the runway at 0715, and from there, Hoyt recalls, "the tail gunners were briefed to flash their hand-held Aldis lamps for the plane behind to see." *Pougue Ma Hone* was the 14th Fortress to go, getting off at 0740, and things started to go radically wrong almost as soon as she was airborne.

Elmer Brown wrote: "We took off and circled the field about once when a B-17 went whizzing by us under our left wing going in the opposite direction. Then the left waist gunner called a plane at eight o'clock level. We were all very much on the alert for other aircraft. I looked out at about eight-thirty and bang—there was the biggest ball of flame I have ever seen and the concussion of the explosion tossed our plane around like a cork on water."

To George Hoyt, it seemed that "no sooner had Pete Fullem uttered his warning on the intercom than a huge and awesome white, yellow, and orange fireball exploded outside my left window. It lit up the whole quadrant of sky around eight-thirty and revealed the terrain all around just several hundred feet below us. Out of the fireball I saw debris arching skyward, including a human body with arms and legs thrown out as it hurtled upward. It was utterly grotesque.

"I heard Hullar yell, 'Let's get the hell out of here!' as the concussion slapped our plane upwards as though it were a leaf in a storm. Then our plane began to drop like a brick. I thought we were fatally damaged and I grabbed the radio table to brace myself for the crash as we neared the ground. But Bob Hullar saved us. He revved up the engines and we began gaining altitude again."

Merlin Miller remembers "Pete letting out a yell about a plane whizzing by us in the dark going the wrong way, which of course attracted our attention considerably. We started looking, and it seemed like just a few seconds after that the explosion occurred. It was pretty much under us, off to one side, and it was quite a shock. The plane started to rise, and it felt to me almost like we were riding a ball of fire straight up, just like we were in an elevator or something. It was a big black ball with red tongues of flame sticking out that seemed to be right under us, and it stayed with us as we went up. We didn't know what caused it, but it sure scared the hell out of us."

Norman Sampson was in the ball turret at the time, and remembers "The explosion and ball of fire were just terrible. I didn't know what it was. I thought maybe the Germans were bombing our field."

Almost the same thought occurred to Major Ed Snyder. "I was in the control tower on the field. There was a hell of a big boom, and then I saw a big fireball and burning debris in the sky for a moment or two. When I saw that I thought, 'Jesus, do we have an enemy fighter in the air?' I thought maybe a night fighter had shot one of our planes down. But it just wasn't right for that.

"We got on the horn to Kimbolton tower right away and within seconds the collaboration between us made it apparent that one of their ships had run into one of ours. There was very little separation between the two bases and somebody obviously had strayed, cutting a corner or something in the pattern. It was tough, but the mission had to go on."

The unlucky aircraft were No. 747 from the 379th Group and B-17G 42-31441 of the 360th Squadron, flown by Lt. B.J. Burkitt's crew. They had taken off as No. 5 ship of the high squadron, just three ahead of *Pougue Ma Hone*.

There was still more tragedy as the 379th and 303rd tried to complete assembly in the dark.

"There were four big fires on the ground," Brown wrote, "and we learned after we returned from the mission that apparently two other planes just got screwed up on their instruments and flew their ships into the ground. I believe a waist gunner is the only one that lived in the four planes and he had some broken ribs. The bombs in these ships continued to explode throughout the day. They say at noon there was a big explosion."

The 303rd never recovered from this terrible beginning. The 358th Squadron failed to make its contact with the main 379th formation and returned to base early.

Of the main 303rd formation, Brown wrote: "The Group assembled and left the base as scheduled," but the balance of his entry makes one wonder if the effort was worth it:

"We had a layer of stratus clouds about 10,000 feet out over about 100 miles of the North Sea so we could not start our climb when we left the English coast (Cromer). Soon after leaving the coast we started running collision courses with other wings of B-17s and B-24s. We were a little late and never did get in our correct slot. We circled to avoid these other formations and ended up trailing the whole Eighth Air Force.

"There we were, leading the low squadron of the low group, the tail end of the whole show. We only had four planes in our squadron and the same was true of the other two squadrons in our group—a group of 12 airplanes. Pretty sad. After we made the circle, our squadron was lagging and Hullar never did catch the group, although we were indicating 170 most of the time."

Everybody aboard *Pougue Ma Hone* was agitated about the state of things at this point. Merlin Miller recalls, "We had trouble with one of our

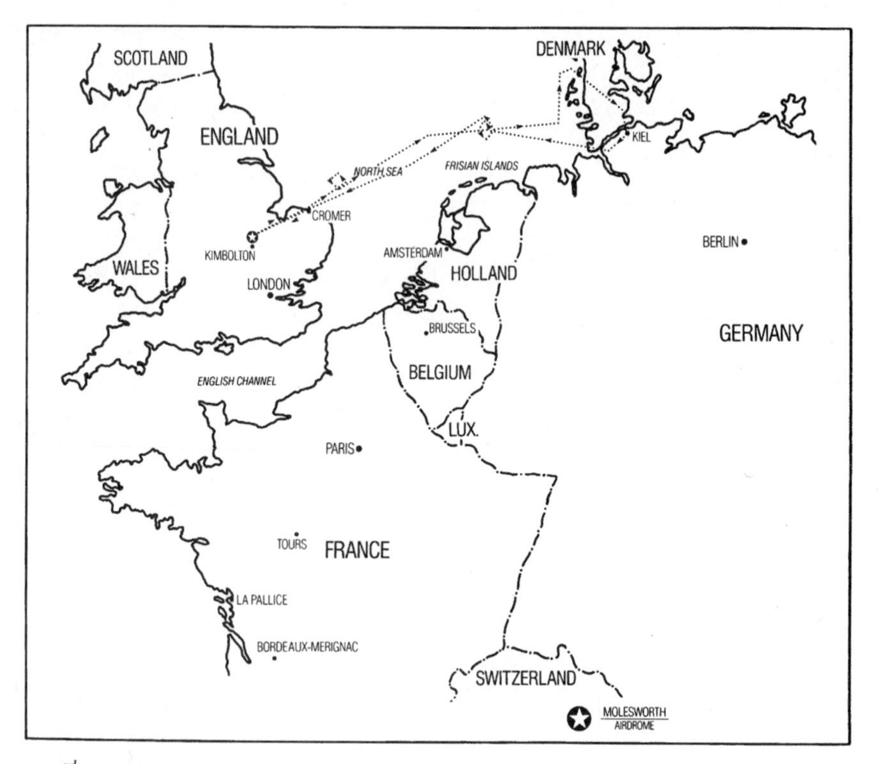

303rd Bomb Group Mission Route(s): Kiel, January 5, 1944. (Map courtesy Waters Design Associates, Inc.)

superchargers and we had a lot of difficulty. We couldn't hook up with our group and we seemed to hook in with somebody and then somebody else and it all got to be rather confusing."

George Hoyt remembers: "Out over the coast of England and the North Sea we circled so much that several times we were flying with the sun at six o'clock heading back towards England. Each time I figured that the mission was scrubbed and that we were going home."

It was not to be. As Miller recounts, "We decided to go on with the mission anyway," and Norman Sampson remembers having "a very hollow feeling about completing this one."

Elmer Brown's diary picks up the narrative:

"About halfway cross the North Sea, clouds started breaking up and we started a gradual climb. The climb was too slow as we reached sight of the German coast about 20 miles south of course at about 18,000 feet and we were supposed to be at bombing altitude of 25,000 feet. We went north along the coast trying to gain altitude and get on the intended course.

"I had Hullar cutting corners, trying to catch the group because we were supposed to drop our bombs on the group leader. I did pilotage up to within a few miles of the IP, then we started getting attacks from enemy fighters and there we were, still about a mile and a half behind."

The crew now faced a running battle with fighters while trying to set up on the bomb run. At interrogation the men reported encounters with 12 to 15 Ju-88s, Me-109s, 110s, and 210s.

Elmer Brown wrote: "I got busy watching and shooting at enemy fighters and trying to figure out how to catch the group. I was depending on DR for navigation, but it didn't work out as we had a strong, unexpected tailwind and reached the target much sooner than we were supposed to. Our group released the bombs but we were so far behind we could not see them go and we didn't think they bombed."

The main Group had bombed "with results reported as fair to good" and then, Brown recorded, "they made a gradual turn to the west and we made a sharp turn to catch up with them. In doing so we cut across a corner of Kiel, the target, but I was so busy I didn't realize it. I figured they were going to circle around and take another run. When we realized they were going home, we dropped our bombs in the sea hoping that we might hit one of the ships down there."

These were "12 naval vessels...on a heading of about 280 degrees true" which Hullar's crew had spied, but Brown wrote: "no luck."

Another reason for Brown's preoccupation over Kiel was the flak. It was "moderate" but "fairly accurate" from the IP into the city and the crew reported that "we had to take pretty violent evasive action."

Miller remembers "The young lieutenant who was flying with us said later he thought he knew what evasive action was until Hullar started bouncing us around. Then he learned what it was *really* like."

All the while, George Hoyt recalls, "German fighters were around, and they were not at all bashful about making attacks. B-24s were flying on both sides of us, and they drew off a lot of the fighters. We considered them our best 'fighter escort.'"

Brown "saw several planes, B-17s, B-24s, and enemy go down in flames," and he was drawn to the B-24s, which "were flying very erratic. Those guys would abort just before the bomb run."

One B-24 made a vivid impression on him: "A vertical fin and rudder came off a B-24 and sailed through the air like a paper plate, but the B-24 went flying along with one rudder like nothing happened."

Merlin Miller remembers that "We did get a fair number of attacks coming out. There were a good number of German fighters that jumped us the first time, and then some of them went after the B-24s. It did seem to me that they had some kind of problem with their formations. But we were still getting fighter attacks, and Hullar was still taking violent evasive action. Some P-51s finally took the German fighters off of us about the time that things started to get real serious."

At 1158 Hullar's crew reported that a "P-51 got a T/E E/A" that "was smoking on the way down," and during the fight other Group members saw the Mustangs at work.

Sgt. Edward Carter, top turret gunner of *Baltimore Bounce* in the No. 3 slot of the lead squadron, later said: "Two Ju-88s dove down on us and then two P-51s dove on them. In a few seconds all that was left in the air was the two P-51s and just scattered bits of the two Ju-88s."

What the men were witnessing was a P-51 field day. By the time they were through, VIII Bomber Command reported that the Mustangs had decimated "a large group of rocket firing Me-110s approaching our bombers" and had "destroyed 16 E/A in all in the target area, with no casualties to themselves."

The overwater flight back was uneventful, but Hoyt remembers that "shortly inside the coast the bombardier reported two B-17s wrecked on a hillside. The cards were stacked against us today."

Elmer Brown wrote: "We had about a 500-foot ceiling and less than a mile visibility for landing at the base, but it wasn't as bad as I have seen in fog when we were coming in for a landing." *Pougue Ma Hone* touched down at 1510, after exactly seven and a half hours in the air.

There was much anger among the 41st CBW's bomber crews at debriefing. As Elmer Brown recalls, "I was furious about them having sent us up in the dark that way, so that those midair collisions could happen. This was one the high command really screwed up."

Many other crews showed their displeasure at both Molesworth and Kimbolton, where the 379th was also licking its wounds after a flight over the Continent just as chaotic as the 303rd's.

Two pilots of the 379th summed up the lesson best. Lt. Kenneth Davis commented: "*No more night takeoff*. Four fires seen in vicinity of field. Just missed collisions. Nothing gained—never got into formation till daylight anyway."

Lt. Edward R. Watson, Jr. said, "Dispense with night assembly—too dangerous."

The Eighth's planners took the comments to heart. As Ed Snyder recollects, "After this one there weren't any other predawn assemblies I can recall during the time I was there."

The Eighth's operations concluded with mixed results. Strike photos over Kiel showed "extensive damage at various localities" in the city, but a total of 10 bombers were lost. The three Third Division wings bombed the Bordeaux-Merignac Airdrome with good results, but strong enemy fighter attacks were made on them after two P-47 groups providing penetration escort left in the vicinity of La Pallice. Seven B-17s fell. The two First Division wings dropped on Tours at a cost of one Fortress, and the last two wings of the Third Division missed the Elberfeld ball bearing plant altogether because of PFF equipment failure and bad visibility. They lost one B-17.

With the collisions and crashes during morning assembly, the total number of bombers lost was 25, with 12 missing fighters. The price was high, but it would pale in comparison with the casualties from a raid soon to be set in motion by an entirely new command team at Eighth Air Force Headquarters.

31

Laurels and Disaster

Oschersleben, January 11, 1944

THE JANUARY 5th attack against Kiel marked the last mission mounted by the man who had nurtured the Eighth and watched it grow from humble beginnings in the early months of 1942 to the powerful Air Force it had now become. Effective January 6th, General Ira C. Eaker was transferred to command of all American and British air forces in the Mediterranean. Eaker bitterly regretted the change, but the shuffle was part of Allied planning for D-Day, and all but inevitable in light of General Eisenhower's preference for his own man to direct the Eighth's fortunes.

Ike's choice was Lt. Gen. James H. "Jimmy" Doolittle, the aggressive combat leader famous for his prewar aviation exploits and his daring Tokyo raid with a tiny force of twin-engined B-25s launched from the Navy carrier *Hornet* (CV-8) on April 18, 1942. Doolittle went on to serve Eisenhower well as air commander during Operation Torch, the November 1942 invasion of North Africa, and General Arnold was hard put to deny Eisenhower his choice as the Supreme Commander of SHAEF, the Supreme Headquarters Allied Expeditionary Force, now building in England for the invasion of Europe.

Doolittle reported to an intermediary, General Carl A. "Tooey" Spaatz, as did Eaker in his capacity as operational commander of the second U.S. strategic air force in Europe, the Fifteenth, based initially in North Africa and later out of Foggia, Italy. Spaatz was responsible for coordination of the two U.S. heavy bomber forces as commander of the United States Strategic Air Forces in Europe (USSTAF). His chief operational deputy in USSTAF Headquarters at Bushy Park, near London, was General Fred Anderson, who had worked so closely with Eaker in planning the raids of the previous autumn.

With these organizational changes, VIII Bomber Command was officially disbanded, and the name "Pinetree" passed into history. Direct command of the Eighth now lay in Doolittle's hands at a new Headquarters in High Wycombe.

He would not wait long before putting his new weapon to a supreme test.

The first Doolittle mission was set for January 6th; it was scrubbed after briefing due to bad weather. Next day the bomber crews were up again for a return to Ludwigshafen and another crack at the chemical works the Eighth had sought to eliminate on December 30, 1943.

Due to heavy clouds this raid was a PFF strike, like its predecessor. Five hundred and two bombers were dispatched, supported by no less than 571 fighters. The 303rd put up 36 aircraft: 18 and a 482nd PFF ship as lead group of the 41st CBW under Major Kirk Mitchell, 358th Squadron CO, plus 18 more under Lt. Woddrop in the low group. The Group's war diary noted that "32 A/C dropped their bombs on the target" but as with most PFF missions, "Results of the bombing were unobserved due to the heavy cloud cover." A total of 420 bombers dropped 1001 tons of bombs for a loss of five B-17s, seven B-24s, and six lost fighters. Total claims against the enemy were seven aircraft by the fighter escort and 30 from the aerial gunners. It was a typical mission, very unlike the one that would follow.

The next three days at Molesworth gave no hint of what was to come. The 303rd's war diary noted: "Normal routine duties throughout the period. No missions were scheduled and local transition and formation flying were held."

Hullar's crew was given the go-ahead for a few days leave and while Elmer Brown and Mac McCormick stayed on the base, Klint and Hullar set out for London. The enlisted men went to one of the Eighth's "rest homes," a place George Hoyt remembers as Mouldsford Manor:

"It was a very large country estate on the Thames River, with a marvelously beautiful 18th Century house surrounded by formal gardens. The Squire had his own nine-hole golf course, a private fishing dock on the river, and a staff of servants you wouldn't believe. There were several butlers, and one of them kept the fireplaces stoked at all times. There were upstairs and downstairs maids, nightly entertainment, and the company of the pretty Red Cross girls. There was a barrel of beer in the Great Hall at all times, and the meals were just sumptuous. We loved it."

While Hullar's crew relaxed, their comrades back at Molesworth were getting ready for General Doolittle's first "maximum effort." The strategic objective was the German aircraft industry, in particular a complex of plants bunched together in the cities and environs of Brunswick, Halberstadt, and Oschersleben. It was the deepest penetration since Second Schweinfurt.

The field order stipulated that the First Division would lead, followed by the B-17s of the Third Division and the B-24s of the Second, making up a force of more than 650 bombers. The 303rd's objective was the A.G.O. Flugzeugwerk A.G. at Oschersleben, responsible for approximately 48 percent of FW-190 output and last visited by the Eighth during Blitz Week on July 28, 1943. The Hell's Angels were to put up two groups—the lead and low in the 41st CBW—with the 379th Group taking the high group position.

This time, long-range fighter escorts were supposed to make the route less dangerous than it was for the bombers in 1943. There were now 12 P-47 and two P-38 groups available to take the heavies to and from the target area. But only the Mustangs of the 354th Fighter Group had sufficient

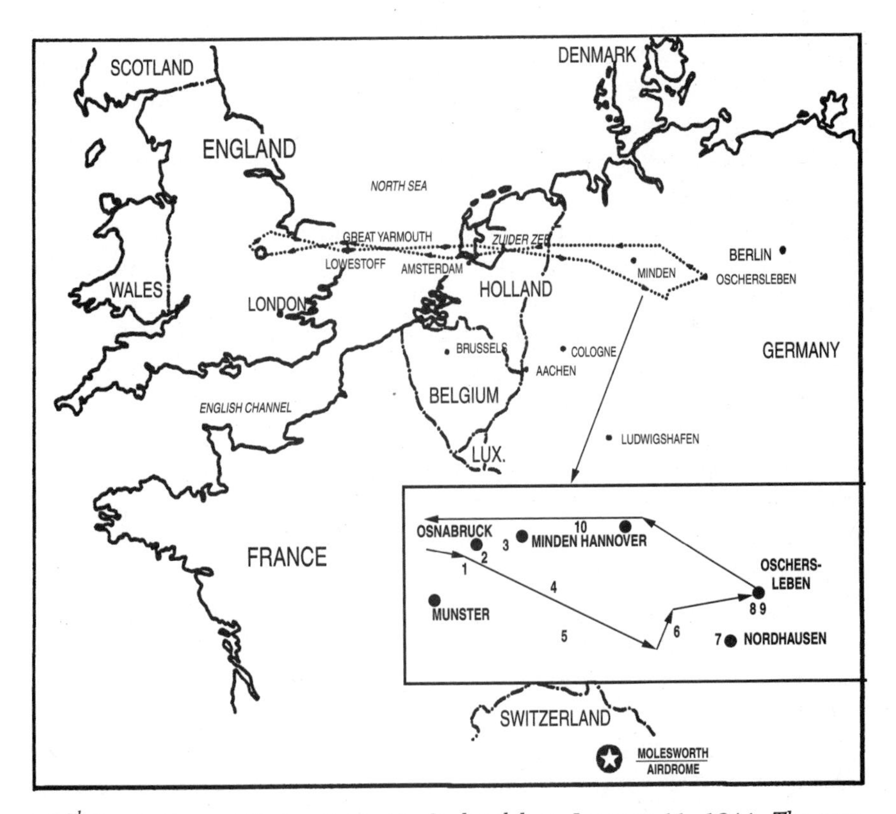

303rd Bomb Group Mission Route(s): Oschersleben, January 11, 1944. The map inset shows 303rd aircraft losses en route to and from the target. Losses are: (1) Bad Check, *Lt. G.S. McClellan; (2)* Flak Wolf, *Lt. J.W. Carothers; (3)* No. 896, *Lt. R.H. Hallden; (4)* War Bride, *Lt. H.A. Schwabe; (5)* Baltimore Bounce, *Lt. W.A. Purcell; (6)* S for Sugar, *Lt. T.L. Simmons; (7)* No. 865, *Lt. P.W. Campbell; (8)* Sky Wolf, *Lt. A.L. Emerson; (9)* No. 794, *Lt. W.C. Dashiell; (10)* No. 448, *Lt. H.J. Eich, Jr. Positions are approximate.* Source Harry D. Gobrecht, *Might in Flight (2d ed.* 1997). *(Map Courtesy Waters Design Associates, Inc.; inset added by author.)*

range to protect the "Big Friends" over the target itself, and for the plan to work against the heavy German fighter opposition that was expected, the Eighth needed clear skies and flawless coordination between bombers and fighters. In the event adherence to the plan proved impossible.

A good gauge of the calamity that befell the 303rd is provided by Brown's diary, in the only entry he wrote of a raid he did not fly:

"January 11, 1944 (Monday)—I didn't get to go on the raid again today. I am grounded by order of the Eighth Air Force except when they can use me for lead. The raid was quite rough. They hit an FW factory about 70 miles SE of Berlin. They really flattened the target. We lost 10 of the 40 planes our Group sent, and our Squadron lost three planes and crews. They were McClellan, Carothers, and Simmons—all a nice bunch of fellows. Kaliher, Cornish, and Fisher, a former B-26 pilot on his first mission as copilot, were all with McClellan. The other three (Kaliher, Cornish, and McClellan) had 17 raids in. Really tough."

Elmer Brown wasn't alone in compiling the casualties. That evening in the 358th Squadron area Lt. Bill McSween made his own grim accounting:

"Not my mission; 303rd put up two groups; 358th put up two squadrons, high squadron, low group and low squadron, lead group. Emerson led low squadron, lead group in No. 562, Skywolf. Campbell led high squadron, low group...Group lost 10 ships; 358th lost four ships and five crews. Campbell, Emerson, Schwaebe, and Dashiell went down. Watson bailed his crew out over the Zuider Zee—plane on fire—and landed back in England. Lost 865 and 562, all lead ships, all DR compasses and GEE equipment."

Completing the catalog of losses were three other planes and crews. The 359th Squadron lost B-17F 42-3448, flown by Lt. H.J. Eich, Jr.'s crew, and *Baltimore Bounce*, carrying Lt. W.A. Purcell's crew. In the 360th Squadron, Lt. R.H. Hallden's crew failed to return in B-17G 42-37896.

Neither Brown nor McSween got to go on the raid, but Lt. Darrell D. Gust did. January 11, 1944, was an especially important day for him. Like Elmer Brown, Gust had married before going into combat and this day marked both his first wedding anniversary and the last trip of his tour. Few Eighth Air Force airmen ended their combat careers in a more terrifying way, but Darrell Gust is quite capable of telling the tale himself. His recollections are confirmed by the navigator's log that he has kept all these years.

"In January 1944 I was the 359th Squadron Navigator, having transferred from the 358th Squadron to this slot in December 1943 after the 359th Squadron had lost its Squadron Navigator on a mission. At the time the Group Navigator was Captain Norman Jacobsen, a fellow classmate of mine from Mather Field Navigation School, Class 42-15.

"Jake called me early in the evening of January 10 and said, 'Come on down to headquarters and inspect what the teletype is bringing in.' When I arrived, the message was that the 303rd would not only be leading the Wing, but also the 1st Division and the entire Eighth Air Force! And the 359th was to be the lead Squadron in the Group, which meant that Jake and I would fly together with Brig. General Robert Travis, Air Commander of the First Division, and Lt. Col. William Calhoun, our Squadron CO and Group lead, in the first aircraft of the whole show! Jake and I completed all the pre-mission navigational details and hit the sack around midnight."

Also slated to ride with General Travis was Capt. Jack B. Fawcett, the same bombardier whom Elmer Brown felt "did a pretty good job" during the December 24, 1943 attack on the V-1 site at Vacqueriette. After the mission Fawcett, writing as "an exuberant 22 year old," prepared an outstanding narrative of his experiences. His writings, together with Darrell Gust's account, provide a fascinating dual look at the events that were to unfold in the lead ship's nose.

Fawcett explains his presence on the raid this way: "My 24th mission (Vacqueriette) and 25th (a ship in the Bay of Biscay) were so easy that I persuaded myself to go on an extra five. Nothing to it, so I thought. Colonel Calhoun was going, Jake was going, so I figured I'd go too. It was to be a deep penetration and our Group would be leading the Eighth Air Force, and General Travis accompanied us. It was a meaty target."

Gust continues: "It seemed I had just gotten to bed when the orderly awakened me for briefing. The emphasis at briefing was on the importance of the FW-190 assembly plant at Oschersleben as a target, plus the number of enemy fighters we could expect to encounter—around 300 single and twin-engine fighters, as I recall. But then we were briefed that we would have P-47 and P-51 escort, which made it a little more palatable.

"It had been several months since my last mission (the October 8, 1943, raid on Bremen), and at the aircraft after briefing my old stomach was really doing flip-flops. I went behind our plane, *The Eight Ball*, and tossed my cookies—'Nervous in the service,' we used to say.

"We took off per my log at 08:10 1/2 Z time and climbed through the overcast on what was known as 'B-Plan.' We got the Group assembled and left Molesworth at 09:18 Z time. We then assembled the Wing, and the Division, and left the English coast at Lowestoff, climbing over the Channel and crossing the enemy coast at 10:33 plus, right on course. All this time Jake and I worked as a team, ascertaining our position, getting winds, making out ETAs, etc."

At this point Jack Fawcett was busier than he thought he would be: "As we assembled over the field I wanted to steal a few winks, but in the dawn's gray brown I had to keep alert for wandering aircraft from other squadrons and other neighboring air fields. Planes were all over. I could see the winking Aldis Lamps and the pyrotechnic flares, their colors denoting the different groups. It was an early, busy sky. The weather across the coast didn't look too promising. Maybe we'd have to use our Pathfinder. Well, that wouldn't be so bad. I've hollered long for permission to accompany a pathfinder mission. But of course as bombardier, I naturally prefer visual sighting."

The initial engagements began as soon as the Group crossed the enemy coast. General Travis reported that "The fighters started their attacks at the Zuider Zee despite our fighter escort and came at us in bunches—Our first attacks were four FW-190s, the next was 30 FW-190s, the next was 12, and they just kept coming. They attacked straight through the formation from all angles without even rolling over. They came in from all sides and it was quite apparent that they were out to stop the formation from ever reaching the target."

Fawcett felt the same way: "We were hardly across the Zuider Zee, when I looked up to discover what seemed like hundreds of planes milling around. Friendly and enemy. A formation of enemy fighters pulled up at nine o'clock level, ten o'clock, then at eleven o'clock they peeled off and came at us in threes and fours—in rapid succession. This wave barely engulfed us before another was positioning itself for attack. Some squadrons had 12 planes, others had 30. It had been so long since I had seen the type of ferociousness now attacking us that I was momentarily spellbound. One o'clock and eleven o'clock; wave after wave—they certainly were determined. Most of their attacks seemed to begin on us as the lead ship, but then were diverted to lower squadrons or groups."

One of the enemy tried to ram the lead ship, causing Lt. Colonel Calhoun to comment: "The fighters were desperate today. A group of 30 FW-190s came at us head-on and I had to lift a wing once to keep one from hitting us." Jack Fawcett relates what the experience was like:

"Oh, oh, here comes a fellow—after us—Good Lord, I fired as well as I could, but the gun position was awkward and the Plexiglas was a bit dirty at that position. He kept boring in at us, but I could no longer bear on him—I could only stand there with my mouth hanging open, watching and trying to convince myself that this fellow couldn't hit us. Hit us, hell! He wasn't concerned with fire power, he was going to ram us! My aching back! Cal lifted our right wing and just then the FW passed right through where we had been. Whew they shouldn't do that. One of our men called out to say he thought the German was wearing a new type of oxygen mask, another said that only 15 rivets were used to hold the FW tank brace on. The FW was close! Nice going, Cal!"

*The Eight Ball** did not go down but Darrel Gust recalls that "On their first pass they got the 'Mickey' aircraft on our left wing." This aircraft was B-17F 42-3486, an H2X "Mickey" aircraft from the 482nd Group, based at Alconbury, ten miles from Molesworth, which had joined the formation over England. It crashed in the Zuider Zee, where its remains were discovered by the Dutch military in March 1968, together with the bodies of all ten men aboard. A second H2X ship from the 482nd Group, B-17F 42-3491, piloted by Capt. Lecates and crew, took off from Molesworth just after General Travis, and took station off of *The Eight Ball*'s right wing in the No. 2 position. This PFF ship made it on into the target and back home.**

Fawcett continues: "I don't know how long these attacks continued. The general was calling them fast and furious until one gunner, not knowing who was calling fighters, said in exasperation, 'Yes, yes, but don't call them so fast, I can't shoot at 'em all anyway.'"

The Group's report to First Division provides further details: "The number of E/A seen range from 30 to 300 with the predominant figure being about 150. At least half of these were FW-190s and Me-109s. The others were Me-110s, 210s, and Ju-88s. There are a few reports of Ju-87s: one report of FW-190s with long noses and inline engines [FW-190Ds],

*The full name of the aircraft was *The '8' Ball Mk. II* but it was widely known by the name in the text which is used for ease of reference.

**Readers of the first edition of this book will note significant differences in the accounting and chronology of the losses suffered by the Group on this mission, for which some explanation is in order. The account in the original version of this chapter was based on a chronology derived from crew observations in the Group's mission file, together with certain inferences drawn by the writer regarding which PFF aircraft was lost, and when, from Roger Freeman's original 8th Air Force history, *The Mighty Eighth*. The new chronology is based not only on the Group's mission file, but also on the additional personal accounts of Jack Fawcett and Vern Moncur contained in this chapter, the extensive MACR (Missing Air Crew Report) data summarized by Harry D. Gobrecht in his 303rd history, *Might in Flight*, and much more specific aircraft loss information contained in Roger Freeman's latest B-17 book, *The B-17 Flying Fortress Story: Design, Production, History*, which establishes that Capt. Lecates' PFF ship, B-17F 42-3491, was not lost in action until March 3, 1944. Even with this new information, however, it is impossible to reconcile all contradictions in points of view and time in a battle this intense and chaotic. It goes without saying that the chronology presented here represents the writer's best analysis of what happened when, and that I bear full responsibility for any errors that remain.

and one report of a single-engine ship which resembled a Miles Master [a British advanced trainer]. This E/A was firing rockets."

"The S/E attacks by FW-190s and Me-109s were made with outstanding vigor and pressed home to minimum range. Formations of 10 and 15 E/A came right through our formation in line and abreast, making their attacks from ten to two o'clock and level to low...There were some tail attacks and a few reports of E/A coming almost straight up from below to make belly attacks...Rockets were fired by T/E ships from ahead into our approaching formation."

The battle reached its peak around 1105, near the town of Minden, continuing along the inbound route south of the line Hannover-Brunswick. As Gust recalls:

"We were somewhere over Germany near Dümmer Lake when General Travis called up and said, 'The weather is closing down over England, so the 2nd and 3rd Divisions are being recalled, but because we are this far into Germany and weather is constantly improving, I'm electing to continue on and bomb the primary target.'

"At the time we had P-47s all around, but the General no more than told us this than the P-47s did 180s and headed for home. I've always thought there must have been some mix-up in signals by the fighters at this time. No sooner had the P-47s left us than we were hit hard by about 50 to 60 190s making frontal passes at us, wingtip to wingtip in groups of six to twelve. This occurred at 11:06 Z per my log, and their attacks were vicious; they were certainly the cream of the Luftwaffe! They always tried to down the lead aircraft and for the next half hour I went from navigator to gunner, manning the two cheek guns in the nose...On their second pass they got our second element leader and one of his wingmen."

Gust's memories are partly confirmed by Lt. Vern Moncur, whose crew was flying *Wallaroo* in the No. 6 slot of the lead squadron. Moncur's journal provides a gripping account of these fierce German attacks and their effects on both the lead 303rd group and the low Hell's Angels group echeloned below them to the left.

"The first pass made at our group included 30 to 35 ME-109s and FW-190s. The low group, to our left, had three Forts go down from this first pass. We also saw three German fighters shot down by this group during this time. The No. 4 ship, lead ship of our element and on whose wing we were flying formation, had its No. 1 engine hit. It immediately burst into flames and dropped out of formation. A few minutes later, this plane exploded."

It is impossible to say who was in the "No. 4 ship" that Vern Moncur saw go down. The fight was far too concentrated and violent for an accurate accounting of all the casualties, and the Group's records are unclear.*

It is easier to account for the first three losses in the Hell's Angels low group that Moncur recorded. *Bad Check*, a Fortress Hullar's crew had flown early in their tour, and one of the Group's original aircraft, was one of the first to go. Lt. G.S. McClellan's crew was aboard her in the No. 7 slot of the low group's low squadron. Lt. Robert Sheets's crew aboard the *City of Wanette* in the squadron lead saw *Bad Check* at 12,000 feet circling in a tight turn. Other crews reported her going down with the wheels down, and Lt. James Fowler learned that night that 10 chutes were seen to come from her. Lt. McClellan was on his 18th trip and the rest of his crew was not far behind except for Lt. W.A. Fisher, the copilot from B-26s who was on his first B-17 raid. *Bad Check* was on her 45th mission. She reportedly went down some time between 1055 and 1105 near the town of Lienen, 20 miles Southwest of Osnabrück.

Next to die was probably *Flak Wolf*, Woddrop's favorite and the Queen that had taken Hullar's crew on their first mission. She was flown by Lt. J.W. Carothers's crew, most of whom were on their fourth mission. They were in the No. 6 slot of the low group's low squadron, and from the nose of *The Flying Bitch*, at the head of the low squadron's second element, Lt. E.L. Cronin, bombardier of Lt. K.A. Hoeg's crew, "saw Carothers pull off to the left and explode. Had time to get men out. Saw three chutes plus some objects, perhaps men."

Flak Wolf's end was also observed by Lt. T. Lamarr Simmons aboard *S for Sugar* in the No. 5 low squadron slot: "I saw several bombers explode before this, maybe three or four, but this one made a really vivid impression on me. I saw a wing fall off the plane, and then the whole fuselage just came apart with a whole bunch of pieces in the air and fire all over the place. I didn't see anybody get out." *Flak Wolf* was on her 40th raid. She crashed at Kloster Oesede, just South of Osnabrück.

---

*There are two lead group formation diagrams in the 303rd's mission file for January 11, 1944 and they are in conflict. One shows *The Eight Ball* with PFF ships on each wing, and Lt. H.J. Eich, Jr.'s No. 448 in the No. 4. slot. Another shows Lt. W.A. Purcell's *Baltimore Bounce* in the No. 3 position with No. 448 in the No. 4 slot. But Purcell's loss is well documented shortly after this initial attack, and Eich's aircraft apparently fell much later in the battle. Appropriately, one of the lead group formation diagrams contains this caution: "N.B. The formation varied and A/C frequently changed position as a result of our losses." The writer's present conclusion is the "No. 4" ship was actually the lost PFF ship, B-17F 42-3486, and that only two other B-17s were lost from the lead group's lead squadron—Lt. Purcell's *Baltimore Bounce* inbound to the target and Lt. Eich's No. 448 after the target. Anyone having additional information on this subject is encouraged to contact the author.

Lt. Hallden's No. 896 in the No. 6 position of the low group's lead squadron was another early loss. From the No. 4 squadron position, Lt. F.F. Wilson's crew in B-17G 42-31471 saw Hallden's bomber at 1055, just as an FW-190 was attacking from seven o'clock low. They reported his ship "in distress at 19,500 feet on a heading of 120 degrees...The aircraft was on fire and went out of formation into a spin. The tail section came off. Three men but no parachutes were seen." Most of Hallden's crew were on their fourth raid. No. 896 crashed near Kirchlengren, due East of Osnabrück and North of the Group's inbound track to the IP.

The 190 that got them did not get away. Lt. Wilson's tail gunner, Sgt. W.G. Hubley, opened fire and "the fighter blew up and pilot bailed out." Hubley got credit for a kill.

At 1110 Lt. H.A. Schwaebe's crew, flying the *War Bride* in the No. 2 slot of the low group's high squadron, peeled out of formation. This Fortress, which had taken Lt. Jack Hendry home from so many missions, was last seen at 17,000 feet by Lt. E.S. Harrison's crew from B 17G 42-39885 in the No. 6 position of the lead group's low squadron. They reported her going down "under control" but no chutes were noted. It was the *War Bride*'s 35th mission, and the seventh for half of Lt. Schwaebe's crew. The ship crashed near Detmold, about 30 miles Southeast of Osnabrück.

These observations are consistent with what Lt. N.E. Shoup's crew, flying B-17F 41-24605 in the No. 5 slot of the low group's high squadron, reported. At 1113 they saw a "B-17 out of control, eight chutes," together with another B-17 that exploded with no chutes. Two minutes later they saw a third Fortress with the tail shot off and no chutes.

At 1117 another Fortress fell. This bomber was Lt. W.A. Purcell's *Baltimore Bounce*. Lt. Purcell's crew had gone on the raid as a spare, taking a position in the lead squadron for one of four 303rd ships that aborted from the two group formations. Vern Moncur had the best view of her end:

"Soon after [the loss of the No. 4 ship], the No. 3 ship ahead of us also caught on fire in the No. 1 engine and peeled out of formation. This ship exploded, also. Lt. Purcell was the pilot, and he and his crew didn't have a chance. (Purcell and I had been together through all of our training.) I then moved my ship up into the No. 3 position, flying on the left wing of the Wing Leader, General Travis."

The Eight Ball's crew saw *Baltimore Bounce* blow up, as did Lt. H.S. Dahleen's crew from B-17G 42-31183 in the No. 5 slot of the lead squadron formation. Other crews saw the ship "leaving formation and later rolling over on back and going down," while from the No. 2 position of the lead group's high squadron, Lt. G.N. Bech's crew in B-17G

42-31843 observed Purcell's plane catching on fire, "with the left wing coming off." Only two chutes were seen before the explosion, but all aboard were killed. The remnants of the aircraft came down near Laubke/Lippe, Germany.

Moncur continues: "Several fierce attacks were made on our squadron—the other groups were getting worked over by the Krauts, also. We were all really catching hell. We made several evasive maneuvers to get away from the fighters during this time. It looked like the Germans thought we were headed for Berlin on this mission, and were making an all out effort to stop us."

The low Hell's Angels group was being especially hard hit with Lt. Dahleen's crew reporting "seven B-17s leaving their formation in various forms of distress from 1117 until 1215," while Lt. Darrell Gust in *The Eight Ball*'s nose observed that "the Germans were exacting an even heavier toll down there."

Oschersleben Casualty. S for Sugar, *B-17F 41-24619, GN✪S, was one of three original 303rd B-17s lost on the January 11, 1944 Oschersleben mission while being flown by Lt. T. Lamarr Simmons's crew. She belonged to the 427th Sqdn. and was considered a prime Squadron symbol because of the "Bugs Bunny" nose art, which was very similar to that on the 427th Squadron's flight jacket patch. This publicity photo of* S for Sugar *was taken during "Black Week," on October 10, 1943, and it is believed to be the first time it has ever been published. Note the immaculate condition of the aircraft and the words "Hi Doc!" barely visible to the left of the rabbit's left ear.* [(Photo courtesy National Archives (USAF Photo A-60921 AC).]

One of the Forts in distress during this time was the 427th Sqdn.'s *S for Sugar,* now one of only two ships left in the second element of the low group's low squadron. From the pilot seat Lamarr Simmons recalls her last moments:

"It was a great shock getting shot down, because of course you never think it's going to happen to you. But on this mission I had my doubts. I saw those other planes go down, I didn't see any of our own fighters, and the German fighters just came in and kept battering us. They were coming in from all sides and my gunners were screaming at me to shake the plane around. We had already taken some hits. Part of the left wing was shot up and the left waist gunner told me that the rudder was shot up, too.

"Then we got another attack from the rear and these two 190s also came in from about one o'clock. They fired at us, rolled over, and went down underneath us. It was real pretty flying on their part. The ball turret gunner reported to me that No. 1 engine was on fire right after this, and I started to have trouble holding the plane. We fell off to the left, and I had the copilot help me try to hold it.

"We flew on as long as we could, maybe three minutes, but the fire kept getting worse and I realized we couldn't stay aboard. I expected that the plane would explode. So I hit the alarm bell and everybody started to jump.

"I was the last to leave. I went back to the bomb bay, and saw the engineer there hung up on something, and I helped him get loose. It didn't take long and then I jumped. I delayed opening the chute because I didn't want to get shot at, and I landed okay."

Most of Lt. Simmons's men were on their fourth raid, and all bailed out safely. They spent the rest of the war in German POW camps, where one died of pneumonia.

Pilotless *S for Sugar* flew on to the northeast, finally crashing in a dense forest near the town of Oberhaus. The Germans salvaged the wreck, using its scrap metal for their own fighters. *S for Sugar,* long-time symbol of the 427th Squadron and the second of the Group's original ships lost this day, was on her 52nd trip.

Another of the Forts "down there" was the 358th Squadron's *Yankee Doodle Dandy,* which had been second element lead of the low group's high squadron. This B-17F was manned by a familiar crew: Lt. John Henderson's. On board were 11 men: Henderson in the pilot's seat; Lt. W.J. Ames, the copilot; Lt. Warren Wiggins, the navigator; Lt. Woodrow Monkres, the bombardier; Sgt. Ralph Burkart at the tail guns; Sgt. Stan Moody in the top turret; Sgt. Ed Ruppel in the ball turret; Sgt. Bill Simpkins at the left waist; Sgt. Robert Jeffrey at the right waist; Sgt. Robert King at the radio gear; and Sgt. Cyril Dockendorf in the radio room with his aerial cameras. During the half hour that the 303rd battled its way from Dümmer Lake to the IP, Henderson's crew put up a fight unparalleled in Eighth Air Force history. The skeleton of their story lies in the Group's records, but Ed Ruppel and Bill Simpkins bring it to life again.

Bill Simpkins's account begins when the Group received word of the recall:

"There was a recall out and everybody else went back. I don't know why we went in, but we did. We wondered why. The clouds opened up. We had very little fighter escort, if any, and there were very few bombers going in that we could see. Then the Germans came."

The first attack against *Yankee Doodle Dandy* took place at 1110 hours at 21,000 feet near Minden. Fending it off from the right waist was Sgt. Jeffrey. There was no airman in the Eighth that day who had a stronger desire to fight. Ed Ruppel remembers Jeffrey as "an overweight boy, on the ground, who wanted to fly. He went on a very severe diet, ran every day, and finally got down to flying weight and they let him fly. This was his first trip up."

Simpkins explains that "He came out of the mess hall, and he was going to be a regular member of our crew. He had an English girlfriend, and she had been killed in a bombing raid, and he was really eager for combat. This was his first mission. He wanted to kill Germans.

"I was coaching him. An FW-190 came over the top on my side, and around the back. I knew it was coming around and in the waist you could see out both windows. So I watched on the other side for it, and it came in around three o'clock, right in toward Jeffrey. I was looking over him, and I saw him fire on it. I saw it catch on fire then."

The combat interrogation form for this attack adds that "Pieces started to fly off the FW-190 from all around it. It caught on fire and went into a spin with flames shooting back well beyond the cockpit." Jeffrey was awarded a "probable."

It might have been this same German fighter that caused *Yankee Doodle Dandy* to fall out of formation. As Simpkins recollects:

"One of the enemy aircraft raked us up top. It hit the vertical stabilizer where it comes down over the waist. All the control wires above the waist came flying down on the first rake. The cables were hanging down, and they were in my way, in the window. That's why we dropped back out of formation, and from there on we were under attack from all sides. I don't know how Henderson kept the plane up. I think he used automatic pilot a lot."

Henderson's problems were worsened by further damage to the bomber's tail. As Ed Ruppel remembers:

"We had a nice big hole in the tail. It was from a rocket. It came from a 110 laying out there, tail down and nose up. He was hanging right off the wing at nine o'clock. There were three of them out there, but he was the only one that launched. And he lobbed it right in. I came around to him, but I didn't have a chance to shoot. As soon as he fired, the three dropped off. They were gone in a fraction of a second.

"It startled me to see this rocket coming at me. It was like a football spinning. It was a swirling mass, twisting and coming right towards me. I swore to God it was coming right for my lap.

"All of a sudden I didn't see it anymore. I kept looking and looking for it and it was nowhere around. I didn't even know it hit. The plane didn't vibrate at all."

The rocket nonetheless found its mark, knocking a hole in the vertical stabilizer the size of the bomber's "triangle C" marking. As Ruppel explains, "We talked about it later on, and figured the rocket didn't explode because it had a predetermined fuse, and he was too close. The tail section was all eaten out. There was nothing from the forward rib back, just a big hole. The rudder was all torn up, and Henderson had to be flying on trim tabs."

Yankee Doodle Dandy was now all alone, trailing the group formation, surrounded by a swarm of German fighters. Bill Simpkins saw them "queuing up about 15 at a time on us, peeling off, and coming in one after the other at ten o'clock."

They were attacking from all sides, however, and Lt. Warren Wiggins, up in the nose, was the crewman who scored next. At 1113 an FW-190 approached at eleven o'clock, coming in "100 feet above to attack. Gunner fired at 500 yards, long bursts. FW came in to 200 yards. When hit FW smoked and flame came from under cowling. Then pilot bailed out." Henderson and Moody confirmed, and Lt. Wiggins got credit for a kill.

Back in the tail, Sgt. Ralph Burkart was firing as well, but the details of his last moments will never be known. All that can be said with certainty is that he died at his guns, fighting for himself, his crew, and his country. Ed Ruppel will never forget seeing him after the battle:

"I went back to the tail. Burkart was slumped over his guns. I reached up, and pulled him back by his helmet. His helmet came kind of loose and I pulled it a little more, and I saw that the whole back of his head was laying in his flying helmet. I looked at his face and he had a small hole in the right side of his nose between the nose and the eye. It was from what we called a 7.9 'liner,' a smaller caliber bullet that the Germans used to line their shots up before pumping the 20mms in. I knew Burkart was finished then."

Sometime during the fight Bill Simpkins also went to the tail to find out why the guns were still, and "I saw they were all shattered, and I can remember seeing that Burkart was dead and telling Henderson to move the plane around just like I did on the mission when Buske got hit back there."

At 1118, however, Simpkins was at the left waist gun, and it was then that he scored a kill. The interrogator wrote that "A/C was lagging behind because of damage and was by itself about 2000 yards behind formation. FW-190 was skidding in from eight o'clock and slightly high. Tail guns were out. Left waist started firing at E/A when 300-400 yards out and held it under fire when cowling came off and E/A started to slide off and then blew up."

Bill Simpkins provides further details: "I remember it coming in, and I had to wait for it to get past the end of the wing so I could shoot at it. You could shoot your tail off if you didn't watch it. You had to wait till they got past that so you could fire at them. I just did it automatically. You didn't think about it. No time."

Sgt. Moody confirmed Simpkins' kill, and moments later the man Simpkins called "the gun expert from Maine" claimed the first of three fighters he would score against this day. With the tail guns out of action, many of the enemy were now attacking from the rear, and at 1120 an FW-190 "came in just over [the] vertical stabilizer—Top [turret] opened up at about 600 yards—came into 100 yards. Gunner held long burst and plane exploded at 100 yards. FW fired at 100 and 150 yards. This E/A could not be corroborated due to [the] position that [it] exploded," the interrogator noted, but Moody got credit for a kill anyway.

Three minutes later two more FW-190s fell to Sgt. Ed Ruppel's guns. He remembers neither, but the interrogator wrote of one that: "was climbing straight up to make a belly attack. It was slightly down and ahead of this A/C. As E/A came into range, ball turret opened fire and as E/A slowed up in its climb, it exploded." Henderson confirmed, and Ruppel got credit for a kill.

Ruppel told the interrogator that the second FW-190 "was about 200 to 300 yards below our A/C when he spotted us. He immediately pulled his nose up for an attack. I first fired when E/A was 150 yards away. Hit him at 100 yards away. He rolled away and I shot another burst and he burst into flames. Saw cowling and small pieces fly off. Spun down and lost sight of him at cloud base, which was about 15,000 to 17,000 feet below our A/C." Henderson saw this one too, but Ruppel only got a "damaged" out of the encounter.

Meanwhile, other enemy fighters were continuing to attack from the tail. At 1124, Moody encountered another FW-190 that "was following [the] one that attacked three minutes before and came in at six o'clock in the same fashion. Top turret opened up at 800 yards with long burst. FW came into 400 yards, where he exploded." The kill was confirmed by the radioman, Sgt. Robert King, a man who Simpkins remembers "was so tall he could stick his head out the radio room and look out the side."

Sgt. King himself was to score sometime between 1125 and 1130 against an FW-190 that "attacked from seven o'clock and quite high. E/A was first seen at 300–400 yards and 10 rounds were fired. E/A leveled off and something was seen to fly off his plane. E/A caught on fire and was seen by top turret to dive down and to explode a short distance below."

Unhappily, Sgt. King was also on the receiving end of enemy fire. Ed Ruppel recalls that "He got hit in the side, shattered his bone between the knee and the hip."

Without question, the cruelest blow against Henderson's crew came at 1127. It occurred during a combat encounter so close and intense that Ed Ruppel retains photographic recall of it even now.*

"I was facing forward, and the battle seemed to be on line and above me. The fighters were making head-on attacks coming over the nose.

"Then something told me to turn around, and I did and saw this German fighter. It was a 190 laying underneath our wing. He was directly in line with our tail, tail low, and nose high. He was about a hundred yards back, but he seemed so close that I could reach right out and touch him. Everything he had he was firing at the time. He was ablaze from wing to wing, and it seemed like he was firing right at me.

"It was an awesome sight, believe me, but I had no sense of what the German was doing, I didn't think about it at all. All that was going through my mind was to line him up, set him up, and pump it to him. I dumped about three nice bursts into him, and he staggered a bit and swung over to my left, and silhouetted real nice. He was vertical to the ground. I framed him, pumped two more nice big bursts right into him, and he lit up like a match. Then he disappeared. He blew all to pieces."

When this happened, Simpkins remembers Ruppel yelling on the intercom, "I got one! I got one!" But Ruppel recalls it differently.

"It scared me at first. To this day I have a vision of something coming out of the cockpit area. I think it was the pilot. I looked to see if a chute would open up, but I never saw anything. I set back for a second and said to myself, 'What did you do?!'

"Then I heard the crackle of gunfire on my right. I swung around, saw more Germans, and got back into action. It was over in seconds. Combat is very fast, terrifically fast. Two minutes of combat is a lifetime.

*When the author interviewed Ed Ruppel in the 1980s for the first edition of this book Ruppel described this incident to me without the benefit of any records to refresh his recollection. When I later obtained the combat report for this incident and compared it with the description which Ruppel had given me on tape, there was virtually no difference between what was on the tape and what Ruppel had told the combat interrogator over four decades earlier.

"I think this plane was the one that got Jeffrey, from the angle it was firing, and what happened to him. Jeffrey got hit in the lower groin with 20mm fragments."

Bill Simpkins was standing next to Jeffrey at the time, and he agrees:

"A 20mm shell came up between us. I'll tell you why Jeffrey got killed and I didn't. He was wearing his flak suit and I wasn't wearing mine. I was standing on it. And my flak suit took all the brunt for me, but the fragments came right up between his legs and killed him. All I got was hit in the back—pieces of twisted metal. I had several pieces in me. It stung, and my heavy fur-lined jacket had fur all sticking out where the shells had hit it and my flak suit on the floor got all tore up.

"I had my back to Jeffrey and he was down. I turned around to help him but there was nothing I could do. He was out cold, and I couldn't really help him. It wasn't like Buske, where I could patch him up a bit. His flak suit was all tore up too, and he was hit pretty bad—all through the stomach. I moved him forward a little bit but we were still being attacked.

"When I was in the hospital recovering from my wounds, they came in and they asked me why I was alive, 'cause it looked like I was wearing the suit too. I said I wasn't wearing it. So from there on they made what they called 'flak mats' that they put in the plane so you could stand on 'em. Jeffrey got hit with a 20mm with phosphorous. The shell penetrated from underneath."

When the interrogator recording Jeffrey's "probable" learned of the gunner's wounds while interviewing Simpkins, he wrote down something not strictly required by the combat form: "This gunner was on his first mission and was mortally wounded in combat." One gets the feeling that he wanted this death to mean something.

The battle continued. In the same minute that Ruppel blew up the fighter that got Jeffrey, Moody scored against yet another German. The "Top turret was facing backward looking for [A/C] supposed to be B-17 on wing when he saw Me-110. Gunner called to pilot to take evasive action. Pilot dropped A/C abruptly, bringing Me-110 into range of turret at 400 yards. Gunner fired long burst; Me-110 started smoking one engine. Pilot bailed out." Moody was awarded a "probable." Nor was this the only time that Henderson's quick action facilitated a kill. Though what follows is not mentioned in the Group's mission file, Ed Ruppel is certain it took place on this raid.

"There was an Me-109 coming in, sliding in over the nose at a 45 degree angle around ten o'clock between No. 1 and No. 2 engines. I don't know why he was sliding that way. It seemed like he was drifting in, having trouble of some kind or another, just laying out there asking for it. But I couldn't get a shot at him because when I tried to fire, the interrupter cam stopped me. It kept me from shooting our props up.

"So I yelled to Henderson to raise the left wing a little bit. Henderson was aces, and he did so right away. He could see the German, too. I popped the German with a couple of good bursts and he blew, went all to pieces; there must have been a million and one of them, and we flew right through the debris. There wasn't even enough left to disturb us in flight."

In the nose, Lt. Woodrow Monkres may have been the next to score. His victory occurred between 1120 and 1130 hours, against an "FW-190 with belly tank, one of several which had come through the formation 2000 yards ahead of this A/C. The particular E/A came in on a head-on level attack against this straggling A/C and was fired on by Bombardier with a single nose gun at about 500 yards, rolling up on one wing at about 100 yards, then leveling out. At that time pilot of E/A bailed out being seen by Top Turret Gunner and then E/A burst into flames and blew up."

At 1134 Ruppel fielded yet another belly attack, by an Me-110 coming in very low at six-thirty o'clock. He informed the interrogator that the "E/A flew below us milling around for a position to attack. He pulled up very sharply, hitting all along our belly. He was about 400–500 yards away when I hit him. He rolled over and flames and grey smoke were seen coming from the engine. He continued on down, but couldn't tell whether he was under control or not." Ruppel was given credit for a "damaged" fighter.

The Group was now five minutes from the IP and *Yankee Doodle Dandy* fought on. Despite everything the enemy threw at her, Lt. Henderson's crew was one of 20 Hell's Angels that bombed the target that day.

But other Group ships were being lost heartbreakingly close to the objective.

The lead ship of the low group's high squadron, B-17F 42-30865, was the first to fall during this period. One of the survivors was Lt. John Nothstein, her navigator.

He recalls today that "We were flying a brand new B-17 with a chin turret, and it hadn't been named, only designated No. 865. This was a maximum effort, and every ship and man that was available flew. Consequently, our crew was somewhat of a mixture or pickup, and I was not too familiar with the others. However, I did know Paul Campbell, the pilot, Lt. Doty, the copilot and Lt. Millner, the bombardier, from contact in the Officers' Mess and Club. I had seen the enlisted men, too, but I was not familiar with them. Campbell was the only one I had flown with before.

"Our ship was a squadron lead in the low group, which was a rather vulnerable position, but I gave no thought to that. I felt I was soon on my way home, having to complete only nine more missions to end my tour in England. But the other two divisions got a weather recall, and our route to Oschersleben was the same one we would have had to take to Berlin, and I later understood their beloved Berlin is what the Germans thought was our objective. They threw everything up at us, determined that we would not get there.

"We were hit by enemy aircraft of all sorts and from all angles. Our biggest worry was mass frontal attacks of mostly Me-109s. We also experienced mass attacks from the rear by massed rocket ships. Campbell took violent evasive action, and I was standing with my head on the ceiling most of the time. I navigated and made entries in my log book as best I could. I estimate we had to abandon ship about 12 minutes from the target.

"We got repeated word from the men in the rear that they had been hit, but we could not find out the extent of their injuries because of a fire in the bomb bay. Finally Millner, the bombardier, was hit and knocked back on my Gee box. I was able to check him out, and made a report to Campbell on his condition. He caught a direct blast to the midsection and was definitely gone.

"Shortly thereafter I got a blast myself. I was hit in the leg, breaking my femur, and also received hits in my right arm and the right side of my chest. I unhooked my intercom and made my way as best I could to the bombardier's guns, which were better than mine. About that time I looked over my shoulder and saw the pilot go out the escape hatch. That brought me to realize I had better find a way out, too, as there was no one flying the plane!

"I crawled back to the hatch and made my way out, which is something I never thought I would have to do, but there was no one pushing me. The plane must have been going straight up, since I barely cleared the bomb bay doors. My right shoulder actually touched one of the doors.

"When free of the plane I pulled the ripcord. The chute had barely opened when I saw the plane blow up. I received a blow to the side of my head. It was probably a piece of the plane debris, and it knocked me out. I later learned I had a fractured skull.

"I did not regain my senses until I was underwater. I thought I was in Hell. I managed to struggle to the surface and saw that I was in the middle of a lake. My adrenaline must have taken over at that point. I unhooked my chute and swam to shore. The Germans were waiting for me and I was immediately taken prisoner."

Lt. Nothstein spent the next 13 months in POW hospital camps, returning to the United States in February 1945 as an exchange prisoner.

The crew's two pilots also survived, as did the engineer, Sgt. Stan Backiel. He was still in No. 865 when she exploded, and he came to in

Oschersleben Casualty. Sky Wolf, *B-17F 41-24562, VK✪A, was a second original 303rd B-17 lost on the January 11, 1944 Oschersleben mission while being flown by Lt. A.L. Emerson's crew. The main photo provides an overall view of the aircraft (with Capt. Don Gamble's crew) and the inset is a much earlier view (March 1943) showing the distinctive "Sky Wolf" artwork that graced her fuselage behind the radio room window. The "Sky Wolf" artwork shows a winged wolf with a "bow and arrow" (the arrow is actually a bomb) and the long-time service of the aircraft is indicated by the fact that the 358th VK squadron code was painted* over *the artwork. The third original 303rd B-17 lost on this mission was* Bad Check, *pictured on p. 69.* (Main photo courtesy Charles S. Schmeltzer, the right waist gunner on Gamble's crew. Photo inset courtesy Frank Hinds, the early 303rd veteran who took the picture).

the air, not knowing how his chute opened. The pieces of No. 865 came down near the town of Nordhausen. The other members of the crew found final resting places in France, where they were buried in a military cemetery.

Also lost in the target area was *Sky Wolf*, the favorite ship of Captain Don Gamble's crew, and yet another of the original B-17s the 303rd brought with them when they first arrived in England in October 1942. She was flown by Lt. A.L. Emerson on his 20th mission, together with the remnants of Lt. Bill Fort's old crew, who were about halfway through their tour. One of those aboard was Fort's old engineer, Sgt. Grover C. Mullins, who had won the

Silver Star on the November 26, 1943 Bremen raid. Mullins recalls this remarkable heat-of-battle exchange with Lt. Emerson, his new pilot:

"I remember we lost an engine and started to fall out of formation. I tried to get the pilot to salvo the bombs and get us back into formation, but he wouldn't do it. I remember getting out of my position [in the top turret] and shoving the throttles forward to get us back in formation, but the pilot told me it would burn up the engines and he was going to court-martial me when we got back, and I told him we weren't going to get back if we didn't get back into formation, and sure enough we got shot down."

Lt. M.L. Smith's crew, flying B-17G 42-31239 in the No. 3 position of the lead group's low squadron, reported *Sky Wolf* "in distress at 20,000 feet in target area...peeled off although all four engines seemed OK. A/C was apparently hit on bomb run. No chutes were seen." *Sky Wolf* was on her 60th raid. She crashed in the town of Wolsdorf.

As the Group neared the IP, in *The Eight Ball* Lt. Darrell Gust had switched from the role of aerial gunner back to his principal job of navigator. He picks up the narrative from here.

"According to our flight plan, we should have been at the IP at 1131. Since I was manning the guns, Jake was really doing all the navigating and it took me a few moments to realize that we had not yet reached the IP. The fighters let up a bit and I went back to the job of assisting Jake in navigating to the target.

"At 1139 we were at the IP. I could see the primary target plainly, the bomb bay doors were open, and for lack of anything else to write in my log, I recorded: 'God, this run seems slow.'"

For many below Darrell Gust there was no opportunity to think of time. At 1147 Lt. H.L. Glass in B-17G 42-37841 was filling what had been the No. 6 position of the low group's high squadron. His tail gunner, Sgt. James Roberts, reported an "FW-190 [which] attacked a 'stray' B-17 that had taken position behind our ship. B-17 exploded (one chute came out). Attack was from six o'clock and FW-190 held its course and attacked us from six o'clock. I fired quite a number of bursts from both guns. FW-190 exploded in midair and was blown to bits."

This "stray" Fortress was B-17G 42-39794, flown by Lt. William Dashiell's crew. Most of them were on their eighth raid and they had taken the No. 3 slot in the squadron formation, but they were all alone after the loss of their wingmen in the first element, Lts. Campbell and Schwaebe. Since Lt. John Henderson's bomber had long since been knocked out of second element lead, Dashiell would naturally have sought cover from the two planes left in the squadron formation: those of Lt. Glass and Lt. N.E. Shoup to his right.

The pilot of one of these ships reported Lt. Dashiell taking a direct hit just as the Group was going over the target, and he also believed they fell out of formation and started down in a long glide. He thought they could have gotten out, but he was unaware of Sgt. Roberts' report. The pilot also had no way of knowing what Sgt. Bill Simpkins observed from *Yankee Doodle Dandy*'s left waist.

"I saw a ship fall out of formation on my lower left. It was below us, and it started to spin, and all of a sudden it blew up. It was a big explosion, with parts just falling through the air, and it got my attention because of that. I remember seeing control cables from the plane just falling through the sky."

Sgt. Roberts reported one chute emerging from the explosion, but no one in Lt. Dashiell's crew survived.

Just one minute later, at 1108, the 303rd's bombers dropped their loads. Jack Fawcett's account describes both this moment and the nine minute run up to the target which Darrell Gust had found so slow.

"We came in south of the IP, but Jake spotted it and we headed straight for it. I was able to confirm it by a nearby stream. Then we were off to the target. Surprising view . . . thirty miles away was the forest near which my factory target was located. The woods showed up clearly, but the little town was lost in a gray haze. So I put the sight on it and just waited. In fact I had time to set up my camera so I could possibly get some target pictures. As we approached. I had time to check my preset drift, etc. It was all good. Soon, I could discern the runway, the town, and then the target. I had plenty of time and good visibility, so my synchronization was good. Because of the time we had, everything was quite deliberate; I would have no excuse for missing. I had one eye on the indices, and one on the bomb rack indictor. The indices met, the lights disappeared. No, two lights remained, so I jumped my salvo lever to make sure all the bombs dropped. With the plane again in Cal's hands, I grabbed my camera and crawled under the bombsight, camera poised for my bomb-fall. Oh, boy, there they were, right in the middle of the assembly hanger I had aimed for. The nose glass was smeared, so I imagined the picture would be no good. But I watched the bomb pattern blossom, covering the target completely. That, then, was my justification for number 26. That FW shop would be closed for a long time."

According to Darrell Gust "Bombs were away at 1148 and our tail gunner-observer reported an excellent clustering of bombs right on the target." The Group's photo interpretation report provides further confirmation of just how good Fawcett's aim had been:

"The pattern of bomb bursts is seen centered squarely on the target with a heavy concentration of both high explosives and incendiaries scattered on and among the buildings of the plant. Three hits are seen on a storage area in which aircraft are stored under a camouflage netting. An undetermined number of hits are seen on the Main Machine Shop, the Final Assembly Shop, and a probable Components Erecting Shop. Direct hits or near misses are seen on another Components Erecting Shop, a possible repair shop, and seven other smaller unidentified buildings...In addition, high explosive bursts are seen scattered over approximately one-third of the factory airfield and on an adjacent road and railway. Incendiaries fell in the target area and across the railway sidings and the freight depot immediately south of the target...The high explosive bombs on the target were dropped by the 303rd lead Group and apparently by the 303rd low Group. Incendiaries dropped by the 379th Group flying high fell on the target and also immediately south of it...Fires appear to have been started in the plant as a result of the attack."

The Hell's Angels had succeeded in their mission, but the enemy continued to make the Americans pay. As the formation pulled out to the north, homeward bound on a westward track running from Brunswick to Hannover, the German fighters were making attacks only slightly less intense than those going in. At 1152 Lt. Shoup's crew saw a Fortress going down. It carried L-K markings on the tail and was, in all likelihood, the sole bomber from the high 379th Group that was lost. The 379th's ships carried a "triangle K" marking on their tails, and the L would have been an individual aircraft identifier. Ten chutes were seen.

One minute later Sgt. Bill Simpkins got his second kill of the day. The combat form stated that "Soon after turning off target, FW-190 came in at ten o'clock level and was in to 200 yards when left waist gunner opened fire. Part of FW-190's right wing came off and other pieces started to come off. FW passed under wing of this A/C and ball turret gunner saw pilot bail out of E/A." The interrogator added, "At this time, the left waist gunner had been wounded in the left arm, leg, and back."

Ed Ruppel remembers "the aircraft tumbling underneath and breaking up. I didn't know who fired at it. When I called up, Simpkins said he had fired at it and I said the pilot had bailed out. I saw him go out and a few moments later the plane 'puffed,' blew up."

Lt. Darrell Gust saw much of the action during this period, and his narrative continues:

"After dropping our bomb load we ran into some fairly accurate flak. At 1204 we were still under fighter attack. The FWs were red and silver with belly tanks. At 1209 one of the few P-51s which managed to stay with us got an enemy fighter, and at 1230 a twin-engine Me-110 started to make a frontal attack on us. Out of nowhere a P-51 came blazing down and the 110 blew up in a burst of smoke about 600 yards in front of our aircraft. A fraction of a second later we flew through the black puff. I remember thinking, 'Christ, a couple of seconds ago that was an airplane with two Germans aboard!'"

The German attacks lessened as the bomber formation limped westward, but the enemy continued to cause casualties. At 1253 Lt. W.R. Kyse's crew, flying B-17G 42-37893 off *The Eight Ball*'s left wing, "reported an unidentified B-17 on fire at 20,000 feet...near Steinhuder Lake. Ten parachutes were seen." This was probably Lt. H.J. Eich, Jr.'s No. 448 flying the No. 4 position in the lead group's lead squadron. His aircraft crashed in Steinhuder Lake, and the body of the only member of his crew who was killed, Sgt. D.S. Harvey, washed up ashore near the town of Nienberg on March 6, 1944. The rest of the crew became POWs. Eich's men was on their eighth mission and their ship was on its 13th trip.

Lt. Vern Moncur recorded other casualties as the Group approached the German border.

"I...saw another Fort (ahead and to our left) do a very steep wingover, nearly going over on its back, and then go down in flames. About this time I saw a German fighter get hit by a flak burst and explode. This made us all chuckle! High above and ahead of us, a P-47 hit a German fighter, and the Jerry's plane exploded. And to our left, a P-47 knocked down a JU-88 at about the same time. (We had a few P-47s and P-51s come out to help us on our withdrawal as soon as Bomber Headquarters found out that there were two wings of bombers which had gone on to their targets.) As an added feature during all of this time, we were continually being shot at—and far too accurately, too—by some very good Kraut flak gunners."

There were other, more startling sights in store for the 303rd's crews as the homeward trip continued. Darrell Gust noted that:

"At 1310 we began a slow letdown. Somewhere near Amsterdam, Holland, I witnessed an event that left me dumbfounded. Down below us and slightly ahead were a couple of crippled B-17s streaking like hell for home. The one at the higher altitude suddenly began jettisoning everything they could get their hands on to lighten the load. Someone threw out a full box of .50-caliber ammunition. It went plunging down and hit the lower B-17 right in the wing between No. 1 and No. 2 engines, leaving a gaping hole. At 1331 1/2 I noted in my log that five men bailed out of a B-17 of the 'C-K' Squadron. I've often wondered what happened to the other five—maybe they made it back, but I doubt it. What a hell of a way to go down!"

Since there was never a 'C-K' Squadron on the Eighth's rosters, this ship had to be *Meat Hound*, B-17F 42-29524, carrying the 358th Squadron's 'V-K' code and the individual aircraft letter K. Many crews reported the strange sight of men bailing out of a crippled B-17 that later rejoined the formation, and Lt. Shoup's crew positively identified this aircraft as No. 524. Shoup's crew saw the plane with two feathered props, and three parachutes. Other crews saw nine men jumping from the bomber over Holland "making mostly delayed jumps."

Meat Hound was being flown by Lt. Jack W. Watson and crew, most of whom were on their eighth raid. He ordered them to hit the silk after the bomber had lost two engines and had caught on fire, and was about to jump himself when the flames abated enough to risk staying with the airplane. Watson managed to land *Meat Hound* at Metfield, but "didn't have much to say" about the incident during interrogation.

Meanwhile, Lt. Gust was helping to guide the Group's remnants home:

"I got Gee fixes going across the Channel and we crossed the English coast at 17,000 feet at Great Yarmouth. The weather was really starting to sock in and B-17s were scattering like quail and heading for the first field they could find because of fuel shortages. Molesworth put out magnesium flares to help us find the runway. We landed at 1505, home from a mission that lasted just under seven hours for our crew."

For Jack Fawcett, the trip home from mid Channel to Molesworth was actually the most frightening part of the whole raid:

"We penetrated the overcast in midchannel and came through at 3500 feet. All too soon there was nothing to see but fog. (Jolly Old England.) We were 700' above ground but couldn't see it. Nothing seemed visible! Whooee! Jake was pinpointing like mad. Just a little patch of ground was all that was visible.

"Obviously, too soon I had thought ourselves safe. Zoom-zoom, an element of B-17's drifted by. We saw them when they were half way past. Ulp! Now I was really sweating! Harder than ever before. This was sudden death staring us in the face. Plane after plane loomed, then disappeared. Yi! That was close, really close. Much too close! Ahh there was the 360th area. Good God, I'll bet there are thirty unseen planes circling the field. Many at our level! For the first time I began to resign myself to fate. This was a horrible mess—far worse than being fired upon and being able to fire back. At this point I can honestly say I was afraid. I'm not exactly sure of what I was afraid of, but I was shaken. It seemed such a senseless way to end up.

"Cal was flying at close to stall speed and only 300' off the ground. He spotted a runway, flew up one side, and turned sharply around for position to land. As we came in, we found a ship just ahead, and planes were appearing from every which way. But we settled to the runway behind three other ships. Good piloting and safe at last! As we rolled down the runway, we could see that landed ships were sitting everywhere on the field. Some wheel-deep in mud. Hmmm—we still risked having a desperate ship settling on top of us. But of course we still had our marvelous luck, and finally ended up in *The Eight Ball*'s dispersal area."

Fawcett adds today: "When the flak and fighters of the Oschersleben experience appeared, you can be sure I questioned the sanity of my decision to do five more [missions]."

Of the 41 ships that had taken off from Molesworth that morning, only 20 came home from the mission, the last setting down at 1523. Four had aborted, turning home early; nine landed away from the base, including *Yankee Doodle Dandy*, *Meat Hound* minus all but her pilot, and Captain Lecates's PFF ship; and 10 would never come home at all. Of those that returned, 19 had battle damage. Despite the great bombing—and claims that resulted in credit for 29 fighters destroyed, five probables, and nine damaged—the day had been a disaster for the 303rd.

Darrell Gust sums it up this way:

"We attended the postmission briefing as a crew, gave the facts as we saw them, and called it a day. Captain John Lemmon, the pilot of my original crew, was at briefing to meet and to greet me. He also gave me hell for picking this type of a mission as my 25th and final one.

"I told him, 'Vince, more than once today I realized I had somewhat volunteered for this mission and thought: What have I gotten myself into?! I somehow had to get myself out. I prayed a lot."

"Looking back, I know there was an Almighty looking out for our crew. As the lead aircraft we should have been one of the ones that were shot down, but by some miracle we were alive to tell about it. The 303rd's losses were 10 out of 42 lost by the First Division on this mission, an almost 24 percent loss rate, and 25 percent of all aircraft dispatched by our Group. I believe it was the worst loss we ever experienced, and as compensation we were awarded the Presidential Unit Citation."

The story was much the same at General Doolittle's new Eighth Air Force Headquarters. No fewer than 60 bombers were missing, together with five Category Es and five missing fighters, making raw losses equivalent to Second Schweinfurt.

The whole First Division, and the Third Division's 94th Bomb Group, were given Presidential Unit Citations for going on to their targets, and bombing heavily and accurately against the ferocious opposition of over 300 German fighters.

The 303rd lead crew on the infamous January 11, 1944 mission to Oschersleben pose beneath their ship, The Eight Ball, *the morning after the mission. Pictured L-R, Top Row are: Lt. R.H. Halpen, tail gunner/observer; Capt. Jack B. Fawcett, lead bombardier; Capt. Norman N. Jacobsen, lead navigator; Brig. Gen. Robert F. Travis, copilot/mission leader; Lt. Col. William R. Calhoun, pilot; and Lt. Darrell D. Gust, assistant navigator. The pictures of the enlisted men in the front row cannot be matched with their names but they are: Sgt. K.P. Fitzsimmons, radioman; Sgt. G.R. Keesling, engineer and top turret gunner; Sgt. L.L. Mace, ball turret gunner; Sgt. A.C. Santella, right waist gunner; and Sgt. H.F. Jennings, left waist gunner.* (Photo courtesy Darrell D. Gust).

Nor did the efforts of the "Little Friends" go without notice. The P-51s of the 354th Fighter Group claimed 15 of the 31 German aircraft shot down by the escorts that day. And one of their number—Major James H. Howard, a veteran ace of Claire Lee Chennault's famous Flying Tigers in China—won the Medal of Honor for single-handedly taking on 30 German interceptors, shooting down six, and foiling their attack against the First Division's 401st Bomb Group.

These future honors afforded no comfort to the men of the 303rd in the immediate aftermath of the mission. Though Lt. Gust's combat days were through, the rest of the Group had to carry on somehow. There was some consolation from the outstanding job of bombing the Group did, but the mood at Molesworth was black that night and was best summed up by a single, bitter line from Lt. Bill McSween's notebook: "We took an all-time whipping."

32

Aftermaths of Oschersleben

January 11–13, 1944

FOR THE MEN of the 303rd there were many aftermaths to the terrible Oschersleben mission of January 11, 1944. The Group as a whole faced a tremendous organizational challenge, for there are few tests of a military unit's staying power more telling than its ability to carry on after a 25 percent loss rate in a single battle. The way the Group's personnel reacted to the event says much about the 303rd's leaders, about the *esprit de corps* of its combat men, and, on another plane, about the horrors of war.

Many saw the price the Group had paid and looked angrily at General Travis. Lt. Bill McSween was one of them: "We felt this disaster should never have happened. A bad high-level command decision was made, and I will not say more."

Others did say more, however, especially at the Headquarters of the First Bomb Division.

Captain John J. Casello of the 360th Squadron had led the 303rd's hard-hit low group and, a day or so later, sat in on a First Division debriefing where all participating group leaders were present. As he recalls:

"General Travis was asked directly by one of the more junior officers there why he didn't return on the recall instructions. He was a Brigadier General, and the highest ranking officer there, and without a moment's hesitation he said, 'I received no recall.' That stopped the questioning dead in its tracks."

Travis's response was true and it raised a provocative point. If the decision was an error, did the blame actually lie elsewhere? For General Travis was not alone in "pressing on regardless." The leader of the Third Division's 94th Group was only 25 miles from his target when that Division's recall order was received and he went on as well—at a cost of eight B-17s. Looking back, Ed Snyder is the one who puts the command decisions into their proper perspective.

"I do not fault Bob Travis, even though there is a good possibility that a mistake was made on his part. He was the kind of guy who, if the scales hung in the balance, would say, 'Go for it.' But he wasn't blind dumb, and it wasn't a screwup in my opinion. The weather over Europe was always bad, and Travis had the decision to make. Nobody recalled him. And I'm sure he felt the thing to do was to go ahead and try to bomb the target. I can't believe that anybody in Travis's position would have turned back unless he had been ordered to, or unless the weather was obviously impossible.

"You have to understand that the Eighth Air Force was *never* turned back by a threat of the consequences. Nobody ever said, 'We've got too many fighters ahead of us, so we better turn around and go home.' If we had a mission, we knew that the fighters and the flak were going to be there, and we hoped for the best. That was the creed of the people who were leading these missions."

It was very much Ed Snyder's creed, too. He was one of the crew interrogators that day, and he well recalls his own reaction.

"It was a terrific loss, and my feeling was, 'This is terrible.' But the questions I asked myself were, 'How in the hell do we get out of this one? What do we do? How are we going to get enough planes and crews together for the next one? We've got to go on.' I never had the feeling, 'Gosh, we're dead.' We got together and did something. It might not have been much, because of all the crews and planes we lost, but we did whatever we could."

The 427th Squadron's personnel reacted to the losses with a mixture of emotions. There was grief over the death of close friends and concern about the fates of the missing. Those who had not been on the mission also experienced a strange kind of guilt. And for almost everyone there was a strong desire for revenge.

Hullar's crew had an equal share of all these feelings. Sampson grieved for McClellan's men—"They had a lot of missions in. It was a sad day for us"—and George Hoyt still recalls "the sad news that greeted us on our return to base, from the Rest Home.

"One of the crews that was missing was McClellan's. They had given us great cover fire on that trip back from Bremen when our wing nearly ripped off, and it still upsets me to think of them going down. I remember those boys well."

Merlin Miller adds: "How did we feel? We were all mad. We wanted to get even with the Germans for what they had done to our Group. I suppose

I've always had a guilty conscience concerning that raid. I can't say why. I don't know that we would have made a difference. We might have got our fannies shot off. But we were an experienced lead crew. We should have been out there. That's how I felt then, and to a certain extent that's how I still feel today."

The combination of grief, guilt, and rage was also felt in Lt. Jim Fowler's crew. Fowler had been quite friendly with Lt. Simmons, and he wrote in his notebook: "Simmons, Carothers, and McClellan shot down. Carothers blew up. Don't know about Simmons. Nobody saw him. McClellan crew all bailed out; 10 chutes were seen to come from it. Sure hope Simmons did the same." He also recorded that his crew was "Briefed to go as spare on mission to Halberstadt," but that they "Could not get No. 2 engine started."

Fowler's notation about the engine was brief but the underlying facts provoked an extreme reaction in Lt. Barney Rawlings, and today he still has strong feelings about the subject:

"We couldn't get the damn No. 2 engine started. We couldn't get the damn airplane in the air. It wasn't our fault that the engine hadn't got started, but when I thought about it, I wondered if we could have done more. Maybe we could have walked the prop through one more time and got the engine to turn over. Maybe we could have done something crazy and heroic, like taking off on three engines and windmilling the fourth prop to get the engine started.

"I had a real feeling of guilt about it. I had not been in the Group that long, and I wasn't that close to the people that we lost. But I still had this intense feeling of guilt about not having gone, and about having been spared this exposure. I also had a feeling of absolute, savage rage that this had happened to my outfit, that it had all happened without me, and that if we had gotten into the air, we could have done *something* that would have materially changed the outcome. I still get emotional thinking about it."

With these kinds of feelings current, it comes as no great surprise to hear Bill McSween say, "Morale did not disintegrate in my Squadron even though we lost five crews."

His assessment is borne out by the 303rd's wartime history, *The First 300*, which declares: "Even after the Oschersleben mission when the 358th took the brunt of the losses, the remaining crews were ready to go the next day. Sixty empty beds might have wrecked the morale of a lesser outfit. It just made the crews of the 358th a little madder. As a matter of fact, more than a score of ground men have left the comparative safety of line jobs to volunteer for service as gunners."

The fighting spirit behind these words is reflected by an incident involving Sgt. Bill Simpkins. After he returned to Molesworth, Simpkins recalls, "General Travis asked me if I wanted to fly again. He asked me when he gave me my Oak Leaf Cluster for my second Purple Heart. He had given me my first one, and this was the second time I was wounded. It bothered me to lose people—Campbell's crew had shared our barracks, and it really got me to see all them empty bunks—but I told him, 'As long as my crew flies, I'll fly.' That's how I felt, even though I didn't have much of a crew left."

There was, however, another aftermath of Oschersleben for a certain number of the Group's men. The original tail gunner on Lt. William Dashiell's crew had been left behind with a case of the flu, only to learn that his crew was lost when the rest of the Group got back. A few months afterwards, he wrote the mother of one of the missing, saying:

"That was the end of my world—the war was over for me. From then on I lost interest in planes and bombing.

"Our barracks housed 12 men, the noncommissioned members of two crews. Neither of those two crews returned that day and I alone was left in the barracks that night—a night that was the longest and the loneliest of any I hope I ever must have. The next day I packed their belongings and saw to it that they were properly taken care of. Later that day I was moved to the hospital. I was finally moved to an evacuation hospital in the north of England and later embarked for the States in a hospital ship, arriving here on March 27, 1944...

"I am well again, but shall never forget the grandest guys it was ever my great fortune to meet."

There is another Group member who is brave enough to say honestly and openly what happened to him. The path Ed Ruppel took after Oschersleben differed from Simpkins's, but it is one no less deserving of our attention and respect. To know how, and why, Ruppel parted from the rest of his crew after this raid, it is necessary to travel with him on part of his journey. It begins in the air over England, as *Yankee Doodle Dandy*'s pilot searched for a place to set her down.

The place Henderson picked was Watton, a former RAF base turned over to the Eighth as a strategic air depot for the overhaul and repair of B-24s. Henderson's crew was about to discover that the station was ill-equipped to handle wounded men. It is at this point that Ruppel starts his story, with some help from Simpkins here and there:

"Henderson said we were going to land at this base. 'It don't look like much,' he said, 'but we're going to land here anyhow.'

"Before we landed Henderson dropped the gear, and he asked me to check it over and see how it looked. I checked it and couldn't find anything wrong. Everything looked good, and I told him so.

"He said, 'No ripped tires or anything?' and I said 'No, none whatsoever.' We fired off four red flares to tell 'em we had wounded aboard, and we reversed our normal landing pattern to land right-to-left. There were no other aircraft around.

"As soon as we sat down we pulled off to the left. That left front tire was flat. When I looked at it on the ground there was a hole in it about six inches long, like somebody cut it with a knife. It had looked fine from the air.

"There was a 'Follow Me' jeep that came up to the plane. The driver says, 'Can I do something for ya?' And I says, 'Yeah, you can get an ambulance, and help us take care of some of these wounded we got aboard.' And he went off at the same speed he came in on! He came back about 10 minutes later with a couple of ambulances.

"The people that were on the ambulances were as ignorant as the day is long as far as handling wounded in combat is concerned. They had no 'wickers' that you could lay the wounded on and carry them straight out the plane. We had to do it the best way we could.

"All wounded moan. We immobilized the radio operator in the plane, put him in a basket, but every time we went to move him, he would holler out in pain. So we hit him with a couple of shots of morphine, and slowed him up a lot, and then we brought him out through the window."

Simpkins rode in the first ambulance with King, the radioman, and he adds his recollections:

"I came out the nose by the navigator's compartment and some Major came out in a jeep. I hollered at him too to get an ambulance, and he said 'Why?' He was confused. They didn't have combat planes land there, and he didn't know what had happened to us.

"So I said, 'Well, we got wounded aboard.' It took them a long time to get an ambulance out there. There were two of them, and they took me and the radioman to the hospital in one. They brought us bottles of whiskey in the hospital, and they said, 'Here you are. Drink it.' But I didn't drink that much. They operated on me that night."

Meanwhile, as Ruppel recalls, "There was a spark of life in Jeffrey. He was semiconscious. He didn't say anything, but he moved, and he had feelings. I helped bring him out. I got his shoulders over by the door, and somebody else got hold of them, and I tried to grab a hold of him by his legs and I couldn't because they were all bloody and slimy. So I had to hold onto him by grabbing my hands on the inside of his boots, and squeezing. He was moaning, but I had to do it because he was so slippery with blood, and I couldn't hold on to him any other way.

"When I squeezed, a lot of blood squirted out, and I heard four or five ground guys behind me start to throw their cookies. And three or four more ran away from the plane. We lost about half the crowd that was around the plane. We put Jeffrey on a stretcher that was on the ground, and then they slipped him into an ambulance.

"Burkart come out last, cause he was DOA. We took him out the tail, and they put him in the same ambulance with Jeffrey. They took Jeffrey to the hospital and Burkart to a little morgue they had there."

The surgeons now had their own battles: to save Jeffrey, and to care for Simpkins and King. While they worked, Ruppel was preoccupied with thoughts of the mission and Burkart.

"I had known Burkart real well. He and I had got into a couple of fist-fights back in the States. There was one evening when we were going back to the base in a cab, and we started to fight in the back seat, and the cab driver threw us out to finish the fight. But that didn't mean nothin'. We were both drunk at the time.

"I volunteered to bring Burkart's body back. I said to the rest of the fellows, 'You go on back to the base, I'm messed up enough as it is. I'll stay here and see what the hell happens.' And I *was* messed up, from the mission. I wasn't in the mood for anything, but I was concerned that the rest of the boys could go back and relax a little bit, so I said, 'I'll take it from here.'

"I went to the mess hall, and I ate a little bit. I didn't want much, and everybody was lookin' at me. And I thought, 'What the hell is the matter with me? Is somethin' showin' or what?' Of course my clothes, my pants and shirt, still had blood on 'em, but it didn't really bother me, not at all. Combat men are kind of callous towards blood.

"There was one sergeant that was nice to me, and they wanted me to go into town that night. And I went into town, but not with them. I walked around that town, got on a bus, and came back to the base. I wanted to be alone. I didn't want to talk to anybody or say anything.

"I went in the next morning to have breakfast, but I didn't go into the mess hall. I went back behind the regular mess hall, and there was that sergeant that knew who I was from when he saw me come in the previous day, and he brought me in the back and made up some eggs and bacon for me—not powdered eggs, real ones. But I didn't even want him around, and he was so nice to me.

"The back of the door of the kitchen was right near the door to the morgue. But I didn't know it at the time, and I came walking out the back door of the kitchen to see what the agenda was for the day, and somebody opened the door to the morgue, and I just had to look in that door.

"There were two bodies in there. They were the members of my crew. It wasn't right that I should see them propped up the way they were in there. It was horrible, and I dumped my cookies right there. And this sergeant, being the beautiful fellow that he was, grabbed a hold of my arm and he took me around to the back of his kitchen, and he brought out a bottle of whiskey and he poured me a water glass full, and he gave it to me and said, 'Get this into ya.' And he said, 'How do ya feel now?' And by this time I was shaking, a bundle of nerves.

"So I had a black coffee and another half a shot—I didn't want to take more 'cause that stuff was precious over there—and he kept talking to me.

"He asked, 'How did you get the hole in the tail?' and I said, 'A rocket went through.'

"And he said, 'A rocket!' He acted like it was Buck Rogers. All that he said, I don't remember to this day, but he was talking to me like a Dutch Uncle.

"I asked him if I could see the head honcho and he said, 'Sure, come on, I'll take you in.' And he did. So I went in to see the head honcho, the chief doctor.

"Those doctors over there were miracle workers. They fought to save a man. If there was an ounce of life left, they'd fight to try to save a man. There was no time schedule on that. They worked 24 hours a day. But the head honcho took a pan of water and a pair of forceps, and he picked up this piece of shrapnel from a 20mm and he dipped it in the water, and it fizzed. And he told me, 'That's what Jeffrey had inside of him. There's nothing we can do to clean it out, there's no method that we know to clean it out. We tried as much as we could, but he passed away. We couldn't save him.'

"We had a nice big hole in the tail. It was from a rocket." The 358th squadron's Yankee Doodle Dandy, *B-17F 42-5264, VK✪J, photographed shortly after Lt. John F. Henderson made an emergency landing at Watton, England on the way home from the Oschersleben mission of January 11, 1944. Close scrutiny of the photo suggests that the aircraft still carries the obsolete red outline around the national insignia on the fuselage.* [Photo courtesy National Archives (USAF Photo)].

"You know, when I got back to the States I had this R&R in Atlantic City at one of the hotels there for servicemen. Jeffrey's family was from New Jersey, and his parents found me at the hotel. It wasn't right that they should find me, a combat man, like that. They found me in the lobby and they started asking me questions. And I had to excuse myself real quick by saying something stupid like I had to go upstairs and put on a tie because I was out of uniform. I went up and asked some guy who worked there to send them away. What else could I do, I ask you? How could I have told them what happened to their son?

"So now I had *two* bodies to take back instead of one. The head honcho told me what the arrangements were, and I said, 'When can I get the bodies and go?'

"And he said, 'They should be ready in an hour or so.' So I wanted to go back to the barracks and lay down, and that sergeant wouldn't let me.

"He said, 'No you don't, you're going to stay here and I'm going to walk around with ya.' So I stayed there in the mess hall, and every once in a while he would come back and talk to me."

At some point Ruppel went to the hospital, since Bill Simpkins remembers him being there: "The next day I woke up from my operation,

A real combat man. Sgt. Ed Ruppel in happier times, posing at the tail guns of a B-17 during Stateside training before deployment of his crew to the ETO. (Photo courtesy William H. Simpkins.)

and Ruppel was standing there, and he had in his hand a little cigarette box with pieces of shell fragments in it. They had taken them out of me."

Ruppel doesn't remember this visit, but he does recall the next stage of his journey:

"They finally had the bodies ready, and I had to sign for them. They were mine, I was totally responsible. They loaded them up in the ambulance, and I was introduced to the driver. And I sat up in the front, and we started to go, and he didn't know where he was going. We went here and there, he had to stop and ask people, and about eight o'clock that evening the damn ambulance quits.

"We were just coming into this little town, people walking all around. So he calls up this base and they send a new ambulance.

"And he says to me, 'What are we going to do? We got to take them bodies out, turn them around, and put them into the new ambulance.' 'Cause per military regulations the head must go forward in an ambulance at all times. I didn't know this, but these were things I was learning real quick. We were supposed to turn them around, but because we were right in the middle of town, people were trying to gawk into the ambulance and these bodies were naked laying there. The only thing we did was take a sheet and threw it over them.

"So I says, 'No. You'll cause more disturbance over here than you can shake a stick at. Run 'em in the other way.'

"We argued, and I finally said, 'Run 'em in. I'll take care of it. I'll sign for it.' So we put the ambulances back-to-back, run the bodies in from one to the other that way, and we finally got to this morgue.

"When we got there, a Second Lieutenant proceeded to chew me out because the bodies were the wrong position in the ambulance. I didn't say much to him, but I looked at him, and I think he kind of got the hint that he better get the hell out of there, 'cause I was about to explode.

"I walked in and the guy there had to sign to receive the bodies, to make sure that the right ones go in the right places. So he opens up a drawer with Burkart's name on it and checks the deceased, Sgt. Burkart, to make sure, and he puts him in there. He did the same for Jeffrey.

"And I look and see that there's a drawer with a name on it for every member of our crew, including *me!* Why was it that way? Because we were listed as MIA. We didn't return to our base, and after a period of time we were automatically listed that way until actual facts were known. They didn't know how many were killed. They didn't know how many weren't killed, or who was wounded, or what. They set that up just in case. And when the man took Jeffrey where he belonged, he signed for the two bodies. They weren't my responsibility anymore.

"Now I'm anywhere from about 60 to a 100 miles from my base, and the ambulance driver, he's going back home. So I said to this one guy who looked like he knew what he was doing, 'Would you call my base, so I can get transportation to get up there?' But he didn't know where I was from, and he could've cared less.

"So I told him, 'I don't know what the story is, but the rest of my crew is back at the base, and I would appreciate it if I could get back to the base.'

"He said, 'Where is your base?' But we were always taught that you don't tell them that, they ought to know by your unit. So I didn't tell him, and he was reluctant to give me any kind of service.

"So I said, 'Is there a CO I can see?' He said there was a second in command, and I asked to speak to him. I called our base and told them the situation, and they called him and those people had a real change of heart from whoever talked to them. Before you could say Jack Robinson there was a covered jeep available to take me back. They got an enclosed jeep that picked me up and took me to my base.

"And when I got there, my crew wasn't there. They were out at Blackpool, R&R.

"Major Black, our flight surgeon, took care of me. Whenever anything happened, he was the one who looked after us. He made the arrangements. They cut separate papers for me, gave me money and a whole set of new clothes, and they drove me to London and put me on the right train, and I got to Blackpool. That was the end of the mission for me.

"When they said you're going to Blackpool, that was the greatest thing in the world. It was the Atlantic City of England, great big hotels right on the water. I didn't need a reservation, and they did everything for me. Breakfast in bed, great service, and nobody bothered ya."

Despite his rest at Blackpool, it wasn't too long after these events that Ed Ruppel took a deep look inside, only to discover that he had given the Eighth Air Force all that he had.

"Ten men will depend upon one another up there," is how he states it. "One link spoils it and your whole chain is broken."

He never flew another combat mission, but no one knowing his story can ever say that he didn't give the Eighth enough.

33

Hell's Angels

January 14–20, 1944

THREE DAYS AFTER Oschersleben, the Eighth's leaders sent 552 bombers plus 645 fighters to strike 21 V-1 "buzz bomb" sites in the Pas de Calais area of France. The 303rd's target was a site at Le Meillard, and Bud Klint's crew was one of those tapped for the raid. The effort wasn't much, Klint felt.

"On January 14th we had another mission to fly. A few weeks ago we could put up 40 airplanes for a maximum effort, but on this day we did good to get 18 off the ground. We bombed from 12,000 feet. The Kraut gunners liked that—they knocked one ship out of our Group and put a wide assortment of holes in most of the others. We bombed in formations of nine ships with excellent results—all our bombs were in the target area."

The missing ship was another veteran that Hullar's crew had flown: *Wallaroo*, on her 35th raid. She was said to have "gone down before the target after being hit by A/A gunfire," and "from three to ten parachutes were reported." Her crew was a green one from the 358th Squadron; it was only the second mission for Lt. A.R. Arimdale and most of his men, but their Instructor Pilot was Captain Merle Hungerford, who had flown with Lt. John Henderson's crew on the day Forrest Vosler won the Medal of Honor. Hungerford was, Bill McSween recalls, "among our best pilots."

Overall losses for the Eighth were light: two B-17s, one B-24, and three missing fighters.

Several more missions were scheduled over the next five days but all were scrubbed. "Normal routine duties" were the rule, while at Molesworth an event of no small significance was brewing. But Brown, Hullar, Rice, and Hoyt were unaware of it as they took off on an errand with Major Ed Snyder and another officer, Lt. Knutson.

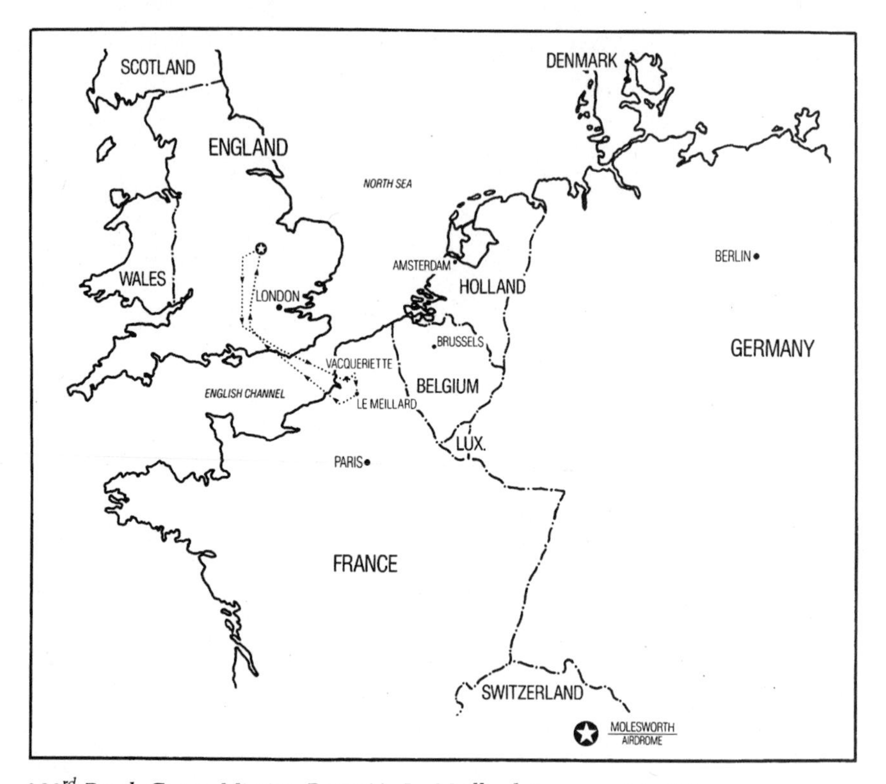

303rd Bomb Group Mission Route(s): Le Meillard, January 14, 1944. (Map courtesy Waters Design Associates, Inc.)

Brown wrote: "January 19, 1944, Wednesday. We went to Langford Lodge, which is just east of a big lake that is about 18 miles west of Belfast, Northern Ireland. At takeoff the ceiling was about 800 feet and visibility less than a mile. We flew in soup (clouds, light rain) with barely wingtip visibility all the way to the Irish Sea. It started breaking. Over the Isle of Man and Ireland there were just a few scattered clouds."

George Hoyt remembers that "when we got over Ireland the visibility was good, and Bob flew in at low altitude, giving us a 'Cook's tour' of the magnificent Irish countryside. It was overpoweringly green—even the ocean rocks at the coast were green, covered with algae—and as we went inland we saw the thatched huts that brought alive the geography lessons we had all had in school as kids. When you see Ireland this way for the first time, you know why they call it 'The Emerald Isle.' It was all unbelievably quaint. I felt like I was flying over a movie set for a fairy tale."

The clear sky over The Emerald Isle was fleeting. Foul weather rolled in off the ocean. Brown wrote that "They would not clear us back to our base because of the weather, we had to spend the night in Belfast. It was after dark that we got to Belfast and we got up before daybreak, and as the town was blacked out I didn't get to do any sightseeing. From what I could observe, it is similar to any English city. They do have a lot of streetcars here; some of them are double deckers. While in London, busses are about all that is available. We slept in a Red Cross Hotel. Major Snyder and Hullar had to sleep in one double bed, Knutson and I in another, all in one tiny room." Hoyt and Dale Rice shared similar accommodations.

The next morning the airmen got down to the business at hand. "The reason for going to Ireland," Brown recorded, "was to bring a plane of ours back that Coto [an officer in the Group] had to leave up there." Hullar, Brown, and Hoyt flew her back while Snyder, Knutson, and Rice returned in the B-17 that had brought them there.

While both bombers were returning to Molesworth, the men discovered a crowd at the field. The scene was one worth recording and Elmer Brown wrote:

"We got to our home base about 1400 Thursday, January 20. We were surprised to see practically all personnel of the field out by the runway. We soon learned that *Hell's Angels* was about to take off on the first leg of its trip to the States. Everyone that could painted their name on the ship, for it was covered with names. It was a thrilling sight to see everyone's hand raised practically in unison waving goodbye to them as they took off."

Hell's Angels *on her way home. "It was a thrilling sight to see everyone's hand raised practically in unison waving goodbye to them as they took off."* (Photo courtesy National Archives (USAF Photo)

Hell's Angels' reputation had been firmly established in the States long before her return home. She was one of the original bombers the 303rd's crews had taken to England in late October 1942, and on May 14, 1943, she was the first Fortress in the Eighth to complete 25 missions against the Reich. Though this fact was obscured somewhat by the earlier return of the 91st Group's *Memphis Belle* (the *Belle* was first Fortress to fly home with a *crew* that had completed a 25-mission tour), *Hell's Angels* soldiered on, establishing a record of reliability unequaled during the early, grim months of the air war by any other bomber in the ETO: 40 straight missions without an abort.

In 1943, wartime security measures still precluded use of the numerical designations the Eighth's units carried, and the public came to identify with the 303rd through the many news items that featured the Group's most publicized ship. The name stuck when, after what *The First 300* calls "several weeks of suggestions, debate, and argument," the Group decided to make the bomber's name its own.

Hell's Angels flew her 48th and final combat flight on December 13, 1943, when she was withdrawn from frontline service. She had, in all her raids, established another remarkable record: Not one of her crewmen was ever killed or wounded in action. Now, while the 303rd replenished its ranks with brand-new bombers and crews, *Hell's Angels* would show the American public that the Group bearing her name was far from finished. Individual ships and men might fall, but the Eighth would carry on until the job was done.

The raids ahead would show the tide of war beginning a slow turn against the enemy. With each day the fighter pilots and flak crews of the Luftwaffe would find themselves increasingly unable to counter the hard-won experience of the Eighth's veteran crews when combined with the ever-growing weight of American manpower and *matériel*.

The next two raids were cases in point. Elmer Brown never flew a better sortie than the one he was slated for the day after *Hell's Angels* departed. And the mission after that, led by all of Hullar's crew, would show off the Eighth in all its waxing strength.

34

"A 'No Ball' Target in France"

Beaulieu, January 21, 1944

ON JANUARY 21st, partly cloudy conditions were predicted for the Pas de Calais coast and the Cherbourg Peninsula, providing the Eighth's planners with reasonable prospects for successful strikes on some of the V-1 buzz bomb sites the Germans were still building in these areas. By now these targets had acquired enough importance to merit a code name all their own, "No Ball," and the day was to see a major effort mounted against them. No fewer than 795 heavy bombers, escorted by 628 fighters, were to hit 25 different sites camouflaged in fields and hidden in forests.

The mission plans called for the same type of tactics the Eighth had employed on December 24th and January 14th: The groups would go in at altitudes ranging from 11 to 25 thousand feet in autonomous nine-ship "squadron" formations. All bomb divisions would take part, with the First Division in the lead.

Excerpts from the Division's field order show how the job was supposed to be done: "Target support will be provided by 11 Groups of P-47s, one Group of P-51s, and two Groups of P-38s from Zero Hour minus 5 minutes to Zero Hour plus 55 minutes...2nd Bomb Division will attack targets in same general area from Zero plus 20 minutes [to] Zero plus 35 minutes at 12,000 feet. 3rd Bomb Division will attack targets in same general area from Zero plus 35 minutes to Zero plus 55 minutes...As many runs on targets may be made as necessary, giving consideration to the fact that 2nd Division attacks in the same general area at Zero plus 20 minutes. 1st Division units are responsible for avoiding interference."

The 41st CBW's field order emphasized that "bombs will be salvoed on Squadron Leader's release. Lead Bombardiers will pay particular attention to pinpointing as assurance of locating the immediate target area. Since several runs are authorized, leaders will make sure of accurate sighting operation before actual release."

"Zero Hour," when the 303rd's bombs were supposed to go, was 1400 hours. The first 303rd Fortress actually got aloft at 1151, and 23 ships later at 1203 *Vicious Virgin* climbed into the sky carrying Elmer Brown on his 23rd raid. Brown's diary carries the narrative from this point, capturing him exactly as he was at the time: a confident, seasoned veteran operating at the peak of combat efficiency. His entry shows very clearly how much Brown was responsible for the raid's success:

"January 21, 1944, Friday—a 'No Ball' target in France about 40 degrees 52 minutes N, 01 degrees 40 minutes E. We put up two groups of two nine-plane squadrons. Our group was to hit one target [Beaulieu] and the other group was assigned to another target [Bois Coquerel]. I was in the lead ship of a low squadron, flying with McCormick, Bob Sheets, pilot, and Dubell, copilot. ["Dubell" was Captain Richard P. Dubell, the 427th Squadron Operations Officer.]

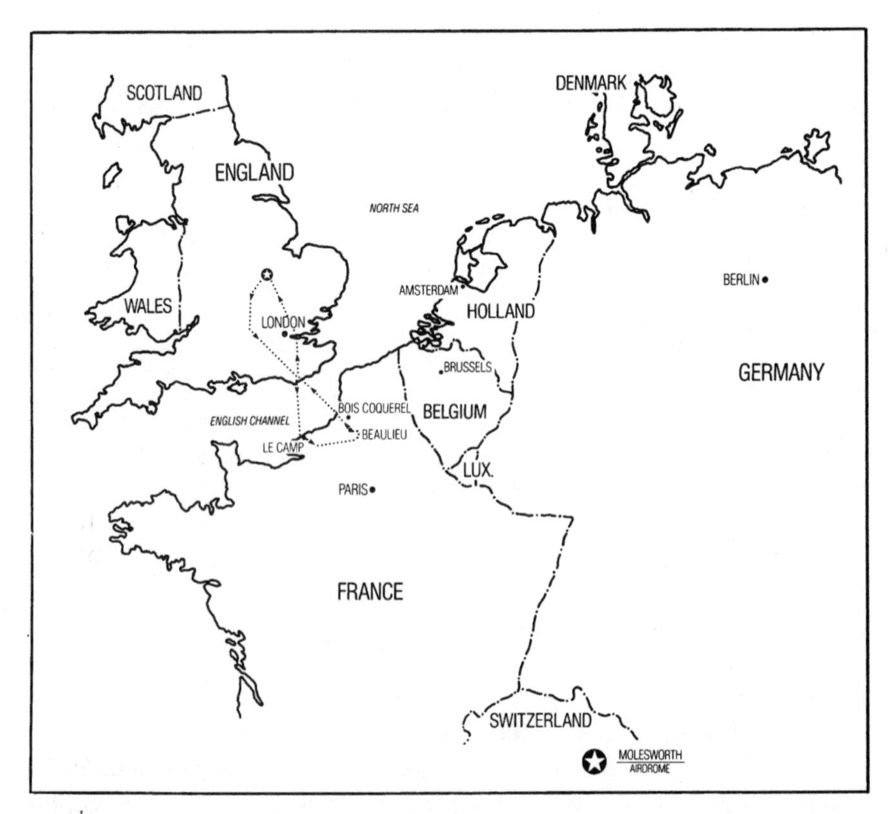

303rd Bomb Group Mission Route(s): Beaulieu, January 21, 1944. (Map courtesy Waters Design Associates, Inc.)

"We were supposed to take intervals crossing the Channel and fly in trail of our lead squadron. We did this, but about 10 miles off the French coast the lead squadron turned into France at the wrong place. We left them and I took my squadron into France at Lecamp, which is where we were supposed to make landfall. I had determined a wind when we crossed the Channel by means of Gee fixes and air plot. Up to the IP I had been doing pilotage, but after the IP we had from 6/10 to 8/10 cloud cover so I had to depend mostly on DR. We turned and came across the IP (Fauville) just as we should have.

"The bomb run was 85 degrees but the pilot started flying about 115 degrees. I caught the error immediately and had him fly off to the left to compensate for it. The bombardier and I had memorized maps and photographs of the target area, as the target was just a tiny woods and required pinpoint precision to pick it up.

"A few minutes before the target there was a break in the clouds and I climbed up in the nose, looking over Mac's shoulder, and pointed out the woods and things leading up to the target. Mac swung his sight on the target but there was a big cloud in the line of sight, so we had to go around.

"I really had to spin that computer in a hurry to figure a rectangular course to the right that would miss all flak-defended areas and that would put us on the same bomb run and give us at least a five-minute run. I used strictly DR and it worked just fine.

"While we were making the rectangular course the copilot, who was the air commander, was complaining because we were taking so long as we were allowed only 20 minutes in the target area. But I was determined to give us a good, long bomb run.

"Mac had taken a few dry runs and had everything preset in his sight. The clouds had cleared a little and we could see the target. I pointed it out to the bombardier and he started working on his sight. We started turning a little to the left, so I hollered, 'Mac, the target is to your right more.' He saw his mistake, unclutched, swung his sight on the target, and as he had things preset, he did a swell job of synchronizing and dropped the bombs right on the target.

"We were not able to get pictures of where the bombs hit, as the clouds closed in. We did get pictures of the approach and some of the ball turret gunners saw the bombs hit. We were carrying 12 500-pound bombs and bombing from 12,000 feet. We were not troubled by flak or fighters. Our fighters had the area completely covered. We had to clear the area in 20 minutes to make room for the other task forces following us. There were probably 700 heavy bombers and almost that many fighters used."

The Group landed between 1533 and 1612, with *Vicious Virgin* touching down at 1547. No aircraft were lost, and Elmer Brown had reason to be pleased.

At interrogation, Lt. Sheets's crew commented that Brown "did excellent work in picking out the target," and the bombing was officially assessed as being "good with a good pattern in a wooded area believed to be the target."

In his diary, Brown noted how the other 303rd formations fared: "Of the other three squadrons in our Group—one made four passes but never could see his target—clouds, I suppose. Another dropped his bombs on what he thought was the target, but it was really 10 miles off. They are not really sure where the other squadron dropped theirs." The Group's mission file confirms his conclusions.

The 303rd's results typified all the bombing missions of this day. Of the 795 B-17s and B-24s dispatched, 394 managed to drop their loads on target. All told, 1070 tons of 500-pound general purpose bombs were dropped on 20 of the 25 assigned objectives, with excellent results on two, good on 11, fair on two, and poor on five. Seventy-two more tons of bombs were unloaded on targets of opportunity with unobserved results.

But the Luftwaffe showed that it still could strike. German fighter pilots were quick to exploit any hole in the fighter escort shield, and one occurred when the Second Division's 44th Group tarried too long in its target area.

The Group's P-47 escort was forced to turn back early due to lack of fuel, and 15 FW-190s and Me-109s rose to intercept the B-24s near Abbeville. They made repeated tail attacks on the bomber formation all the way out to the French coast, shooting down five bombers in exchange for four fighters. Three more Liberators were classified Category E, including one that was abandoned in the air over England.

The Eighth also lost a B-17 to flak, and its fighters and bombers claimed a total of 13 enemy aircraft, but the day's operations really underscored the impact that seasoned combat veterans could have on the outcome of a mission. The skill and determination that Brown and McCormick showed was what made the difference between a successful strike and a failed one on mission after mission that the Eighth flew throughout the air war. These were the kind of men the Eighth's leaders needed to make sure this war would be won.

35

A Mission with Lewis E. Lyle

Frankfurt, January 29, 1944

JUST TWO WEEKS after Oschersleben, the Eighth's planners turned to Germany again.

On January 24th the bombers set out for Frankfurt am Main, but the mission was recalled due to bad weather. A second attempt at Frankfurt was scheduled for January 26th, but it was scrubbed before the crews got to their planes, and a mixture of rain and very strong winds kept the bombers on the ground through January 28th. When January 29th dawned, the crews were ready.

"Finally," wrote Bud Klint, "on our third attempt, we got off for Frankfurt."

The mission plan had the 303rd contributing two separate group formations, "A" and "B," in the high group slots of two parallel wing formations the 41st CBW was putting up. Two groups from the 379th made up the balance of the A wing and two other groups from the 384th completed the B wing.

Hullar's crew was leading the 303rd A group in *Vicious Virgin* on Bob Hullar's first mission as a Captain. The crew included Lt. G.A. Wallen, Jr., as bombardier, Elmer Brown as navigator, and Lt. D. Kendall as tail gunner-observer.

Behind *Vicious Virgin* flying as second element lead for the first time was Lt. James Fowler's crew on their eighth raid. Under them, Klint's crew filled the No. 2 slot in the A group's low squadron aboard *Kraut Killer*, B-17G 42-31423.

On a parallel course to the A group's left in the 41st CBW's B wing was the 303rd B group, led by Major Kirk Mitchell and Captain John Lemmon. It was the 25th raid for both men and the 24th for Lt. Bill McSween, who was aboard as their lead navigator.

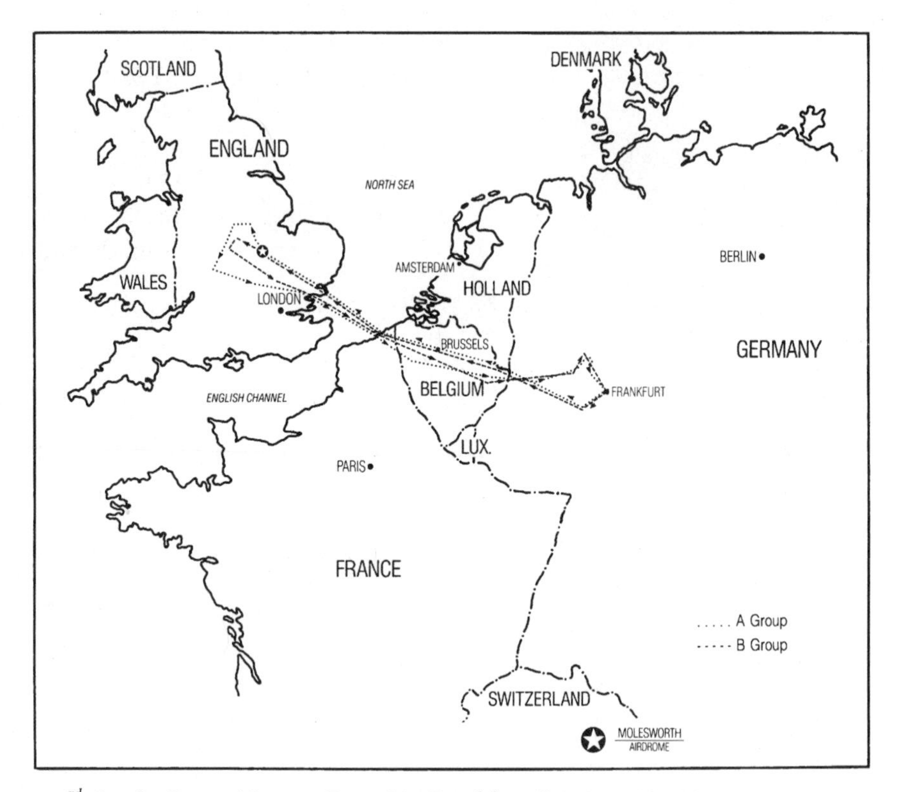

303rd Bomb Group Mission Route(s): Frankfurt, January 29, 1944. (Map courtesy Waters Design Associates, Inc.)

The mission would find the two 303rd groups working closely together in accordance with some special instructions that the field order contained: "CBWs will depart Clacton in columns of 'pairs' leaving four-minute interval between head of lead pair and nose of each succeeding pair. Guide is right. This formation will be maintained until reaching [the IP]. Minimum interval between individual CBWs will be taken at this point to insure unimpeded bomb run. It is preferable for the left CBW in each pair to approach target echeloned slightly to the left of the lead CBW rather than take a larger interval behind the lead CBW. After bomb run, formations will again form pairs of CBWs...and continue throughout withdrawal to enemy coast out."

Elmer Brown noted that the mission was a PFF strike and that "we were the second division to go in and the third wing in our division to cross the target." The force was an awe-inspiring 15 combat wings of 863 heavy bombers—615 B-17s and 188 B-24s—described by news accounts as "the greatest armada of American heavy bombers ever sent into action."

The bombers were escorted by 632 fighters—40 P-51s, 89 P-38s, and 503 P-47s in 16 groups—who provided "corridor" air support. The system allowed different groups of escorts to cover assigned parts of the bomber route in relays, one group relieving another as it reached its fuel limit. Schedules were overlapped to minimize the risks if a group were late.

All this was laid out at the early morning briefing, and after the other mission preliminaries, Hullar's crew waited at *Vicious Virgin*'s hardstand for the raid to start.

Elmer Brown recorded that "Lt. Col. Lewis Lyle on his 32nd mission was the pilot," and George Hoyt clearly remembers him arriving in a jeep "and going right up to the cockpit, where he got into the left seat that Bob usually took."

Merlin Miller recalls Lyle's arrival as well: "We knew that Lyle had flown a lot of missions, and he seemed to me to be a real nice guy. But you could tell before we went on the mission that he was wound real tight. I don't know how to describe it, exactly. He wasn't tense, but he gave the impression that he was ready, willing, and able to do whatever he had to do."

Miller accurately sensed Lyle's state of mind and the leadership role he wanted to project. As Lyle explains:

"I flew with dozens of crews during the war, and I never got close to any of them. I was totally consumed by the need to get on with the war, and by the time I got out to an airplane, I had always worked myself up to a fever pitch of excitement. I would be simply bursting with excitement. I was ready to go!

"Minimizing our losses was important, but improving our bombing was absolutely critical and making our people see that was difficult at times. It used to always frustrate me when people were more interested in getting the mission in instead of getting the bombs on the target. Our learning process was slow, and we made a lot of mistakes. Every time I flew I was concerned about not repeating them.

"I'd start by making it very clear to the crew I was with that I was the one in charge. I'd be real rough with them at first, saying that I didn't want to hear any 'goddamn intercom chatter.' It was important for them to know that I was in control."

Hullar's crew got the message, and Lyle got *Vicious Virgin* off the ground right on time. The Group Leader's Narrative that Bob Hullar prepared describes the trip up to the enemy coast: "Took off at 0750 hours with a 20-ship Group formation. Assembled my Group at 5000 feet over base after climbing as individual ships through the overcast on a magnetic heading of 320 degrees. Broke out on top of the overcast at approximately

3000 feet. Climbed on up to 9000 feet in the vicinity of the base. Rendezvoused with the 379th Groups between base and Eyebrook. Climb to altitude was begun at Eyebrook and was a bit faster than SOP to stay up with the lead (379th) Group. We departed the English Coast at 21,000 feet. Going in, we were almost abreast of the B Wing of the 41st."

Behind the A group formation, the 303rd's B group was having a more difficult time. Major Mitchell described the problems in his Group Leader's Narrative: "Took off at 0804 hours with a 19-ship Group formation. Had trouble assembling and made two circles before picking up any airplanes. We broke out of the overcast at 3200 feet and assembled above it at 4000 feet. We fired 10 sets of flares and left the Base at 8000 feet on time. We were unable to catch the Lead Group and trailed them until about five minutes before the IP. We kept the Group formation together rather than pull excessive power to try and catch the Lead Group. From 24,000 feet we indicated from 155 to 165 to catch the Lead Group. We departed the English Coast at18,000 feet at 0954 hours and crossed the enemy coast on time."

The B group encountered no fighter opposition and Lt. Bill McSween noted the "good fighter cover." Hullar did as well, writing that "The P-47s met us and carried us to the IP, then the P-38s picked us up. The fighter support was good today, only allowing Jerry to make one pass at our Group by five Me-109s from two o'clock."

Bud Klint wrote that "We had 'corridor' fighter support and it was very effective while our 'little friends' were in the area, but their schedules didn't seem to be timed perfectly, and several times we were without fighter cover."

Elmer Brown believed "everything went well until about 10 minutes before the IP. About 15 Jerry Me-109s made one pass at the low group in the wing just ahead of us. They all came in together, flying more or less in formation in a head-on attack. One Fortress went down in flames."

Brown also wrote that "About that time Fowler, who was in the No. 4 position in our squadron, had to drop out of formation and started lagging behind. He had feathered his No. 3 engine just after crossing the enemy coast, but he tried to carry on. We think he had trouble with another engine that forced him to lag. That was the last we saw of Fowler."

Others observed more. Lt. V.A. Wood's crew was in the No. 3 lead squadron position to Hullar's left. They reported Fowler's ship "with one engine feathered. The A/C fell back and was being attacked by enemy fighters at 1046 hours at 25,000 feet. One chute was reported."

Lt. Pharris Brinkley's crew got another apparent glimpse of Lt. Fowler from *Satan's Workshop* in the lead of the A group's high squadron. At

1107 they reported "an unknown B-17 A/C at 18,000 feet going down. This A/C was being attacked by enemy fighters and No. 4 engine caught fire. It headed for France...and an additional report would indicate that this was Lt. Fowler's A/C in distress. Later the A/C exploded."

The "additional report" may well come from Bud Klint, who later wrote: "Just as soon as our fighters were clear of the area, the German jumped the lone B-17 and, after a few runs, sent it down in a sheet of flame."

Lewis Lyle had seen such things countless times, but he never let them interfere with the job at hand:

"I gave no consideration whatever to the possibility of my being lost in action. The only way I figured that would happen was through some fluke and that didn't enter into my calculations. After a while, I could say why somebody got shot down, what should have been done that wasn't, but what happened to individual crews behind me wasn't my concern. I'd get reports about it from my observer in the tail, but my mind was elsewhere. I was constantly thinking about the mission, about what was happening in the wings behind us, what was developing ahead, and what we should do. For me, it was always a very impersonal war."

As Lyle and Hullar's formation neared the IP, getting the bombs on the target suddenly became a real problem. As Bob Hullar described it:

"As we approached the IP, I heard on VHF that our Combat Wing PFF ship's special equipment was out. So the combat Wing S'ed and dropped back to allow the B Combat Wing to bomb first."

By this time, the B wing was finally together. In his own report, Lt. Bill McSween noted that "The actual course made good brought us to a point approximately six miles south of briefed IP. At this point the necessary corrections were made by the Lead Group to bring the Wing into position for the bomb run, which included turning back to the north, then taking a magnetic heading of 145 degrees for the bomb run."

In his notebook, McSween also wrote of "a terrible mix-up of combat wings at the target," and the problem is made clearer by what Major Mitchell reported:

"We were south of course and the two Wings ahead of us bombed on a heading of about 50 degrees magnetic. They turned back into us and we turned on the inside of them. We turned back on the target and bombed on a magnetic heading of 145 degrees at 1123 hours from 25,000 feet."

These delays had a ripple effect on the 303rd's A wing. Elmer Brown wrote that "we had to circle around till we could follow another wing into the target and bomb on their signals," and Bob Hullar reported that "the run on the target had considerable S'ing in it to allow our Wing to stay behind the B Combat Wing. Bombs were released on the Combat Wing Leader from 25,000 feet at 1124 hours on a magnetic heading of 154 degrees."

The 303rd dropped moments later under circumstances Lt. Wallen describes: "The high Group was in back and above the lead group, about one-half mile back and perhaps 1000 feet higher...I dropped my bombs approximately two seconds after the lead group in order to compensate for our position behind them. I saw the smoke from the red flares indicating 'bombs away' and we were very near this smoke when we dropped."

All the while, Elmer Brown recorded that "the wings ahead of us had been throwing out chaff, which worked very well. It made the flak ineffective, although there was quite a bit of it."

Bud Klint agreed. "There were supposed to be about 240 guns in the target area. Quite often, their pattern of fire seemed to be following the bundles of chaff rather than the bomber formations."

Hullar's own formation was dispensing chaff as well, and back in the *Virgin*'s radio room, George Hoyt was busy at the task: "Since we were a lead crew and I had to stay at my radio table, the *Virgin* had been modified with a small flap beside my table through which I had to throw out metal shafts in bundles on the bomb run. As they emerged from this door into the slipstream, the bundles flew apart into paper-thin strips that filled the airspace around our formation. It jammed the Jerry radar sets on the ground that tracked us along with their antiaircraft guns. They threw up volleys of flak that exploded far behind us as the radar tracked the drifting shafts instead of our planes. I threw out in excess of 5000 shafts. It really played hell with the radar flak guns."

The two 303rd formations dropped more than 82 tons of 500-pound M-17 incendiaries over Frankfurt, adding their share of over 1800 tons of bombs that rained on the city through 10/10 clouds.

It was now time to head home, and Bob Hullar wrote that "a right turn was made after bombing and [we] came out almost abreast of the B Combat Wing. Our Low Group (379th) was behind, below and dragging throughout the trip...We let down about 2000 feet after leaving the target."

The letdown was made to tighten up the wing formation, but it was also standard procedure. Beyond the target the need to stay high to avoid flak was diminished, and a gradual loss of altitude allowed stragglers to catch the formation by diving at higher speeds.

The balance of the return flight went without incident until the two 303rd formations got to the Channel at 1302. Hullar's A group was down to 21,000 feet, and Elmer Brown's diary picks up the action:

"We left the wing formation at the Channel and I brought my group home. There was an overcast with a 4000-foot top. We were instructed to make a letdown through the clouds on Splasher No. 16.

"When we got there, two groups were ahead of us waiting to make letdowns. That meant we would have a long wait. Lyle saw small breaks in the clouds that indicated the overcast was thin, so I gave him a heading to the base and he gave it to the group, instructing them to make a letdown at the estimated arrival point as if it were a splasher station."

The course change to Molesworth is the part of the mission Lyle most clearly recalls:

"I do remember the event and the letdown. Some months before, I had taken a B-17 down to Foggia, Italy, to train lead crews from 1st Air Division groups, and I decided to fly back to England at night at low altitude over France. The navigator I was with made a gross error, and we wound up in a heavy flak belt dodging searchlights and antiaircraft fire for over a half hour.

"That was quite an experience, and after it I never accepted a navigator's order to change heading and turn without satisfying myself that he knew where he was. Brown convinced me of our position, so I ordered the letdown."

Brown continues: "We peeled off and dropped through the clouds just over the base. Lyle stood the plane right up on its wing in the traffic pattern. He can really fly the old crate."

George Hoyt also recalls "the spectacular short field landing that Colonel Lyle made at our base that day. This man was already a living legend in the Eighth Air Force, and his landing ended the mission on a perfect note of triumph."

The records show that Lyle and Hullar set *Vicious Virgin* down at 1417 after a flight that lasted six hours and 27 minutes.

Hullar's crew now had only one more mission to fly, but this raid marked less than half the total Lyle would log from the beginning of the air war to its end. His desire to excel as a pilot, his motivations for flying the many missions he did, and his leadership and training role throughout the war are subjects that merit a closer look.

In January 1942 Lewis Lyle was a 25-year-old reserve Second Lieutenant who formed part of the 303rd's original cadre at Boise, Idaho. With seven years prior service in the Army as an Infantry private and service through the ranks at ROTC, in the Reserves, and in the National Guard, Lyle quickly decided that flying was the name of the game.

He aimed high, making it his personal goal to become the world's best B-17 pilot. To hear Lyle tell it, the organizational confusion and lack of control that existed during the 303rd's early days made the task "easy," but his personal log shows that he pursued this goal with single-minded determination.

As a "junior birdman" in a new unit that had precious few planes, Lyle found it difficult to get on the Group's flying schedule; he stood by for every flight and begged to go with the senior flyers, even as a passenger. Most Instructor Pilots let him ride, and in time Lyle was listed in the flight log as an observer. He kept at it, acquiring copilot time, pilot time, and a few landings, none of which was ever scheduled or authorized.

In March 1942 the 303rd received additional B-17Es for training and, to everyone's surprise, Lyle now had 50 hours of B-17 flight time. He was checked out as a pilot and an Instructor Pilot, and his time in the air increased. His flight log for March shows 56 flights; for April, 51 flights; for May, 41 flights, and he continued at the same pace. By September 1942, Lyle had over 500 hours in a B-17, mostly training other pilots and crewmembers.

He accompanied the Group's flying echelon to England in late October of 1942, by which time he was a Captain with his own crew and aircraft, and was a flight commander in the 360th Squadron. Lyle flew many missions as a lead pilot, was given command of the Squadron, and ended his first 25-mission tour in mid-June 1943.

Throughout, Lyle was preoccupied with the challenge of combat and "obsessed with the need for training, sound procedures, and crew discipline." He was promoted to Major and became the 303rd's Deputy Group Commander for Air Operations in August 1943, just in time to take part in the August 17th Schweinfurt mission, and he never stopped flying missions after that.

His interest in operations extended to more than flying, however. He went on many raids in other crew positions to see for himself what conditions were like, and to develop better ways for all combat crewmen to perform. The advice he gave the crew commanders he flew with mirrored his own philosophy: "Know your job, and be the best you can. Know your crew's jobs and demand *their* best, too. Flying combat requires the utmost in discipline. Take command!"

Lewis Lyle was awarded the Distinguished Service Cross for his service to the 303rd as a lead pilot and crew trainer from November 1942 through June 1944. He became acting CO of the Group in August 1944 and remained in that capacity until October, when he was transferred to command of the 379th Bomb Group. He led the 379th until the end of the air war, and during his tenure the 379th established a record as the most accurate heavy bomber group the Eighth had.

"Take command!" A portrait of Lewis E. Lyle as Commanding Officer of the 379th Bomb Group. (Courtesy Lewis E. Lyle.)

In his entire tour of duty with the Eighth Air Force, Lyle flew as a lead pilot with over 55 different crews, being officially credited with 69 missions, and compiling by his own accounting the remarkable total of between 70 and 75 raids.

What awes him, however, is the fact that all those crews followed him to all those targets, and if you ask him today about his own wartime achievements, he will immediately tell you: "*They* are the Real Heroes."

36

Fowler's Fate

Frankfurt, January 29, 1944

WHEN THE 303rd's bombers landed back at Molesworth on their return from the January 29th mission to Frankfurt, few held out any hope for Lt. James Fowler and his crew. There was little reason to. What the other crews reported was a seemingly open and shut situation: a straggling bomber with a feathered prop falling back out of formation, German fighters attacking it like wolves circling a wounded stag, an airplane exploding, and a single chute. Fowler's ship appeared to be just another of the 29 heavy bombers that the Eighth lost on this day.

There is great uncertainty in war, however. What men believe they see in battle is not always what actually occurs. And Fowler's fate was far different from what was reported. His crew's heroic effort to get back home is the story of every straggling bomber crew that fought to return to England, only to fall short of their goal.

The men who made up Fowler's crew this day were, in addition to Jim Fowler himself, Lt. Barney Rawlings, his copilot; Lt. Alvin L. Taylor, the bombardier; Lt. Joseph C. Thompson, navigator; Sgt. Curtis E. Finley, engineer and top turret gunner; Sgt. Donald Dinwiddie, radioman; Sgt. Richard Arrington, ball turret gunner; Sgt. Loren E. Zimmer, right waist gunner; Sgt. Miller O. Jackson, left waist gunner; and Sgt. Jack D. Ferguson, tail gunner. Of all of them, only Ferguson was not an original crew member. He had joined Fowler's crew in early January after having flown two missions for Marson on Hullar's crew; the November 16th trip to Norway and the November 26th attack on Bremen.

The crew's story is told in tandem by Fowler and Rawlings, just as they flew it side-by-side, with a key contribution by Loren Zimmer from the right

Fowler's crew. Bottom row, L-R, Sgt. Curtis E. Finley, engineer; Sgt. Richard Arrington, ball turret gunner; Sgt. Donald Dinwiddie, radioman; Sgt. Loren Zimmer, right waist; Sgt. Miller O. Jackson, left waist; unknown. Top row, L-R, Lt. James F. Fowler, pilot; Lt. Bernard W. Rawlings, copilot; Lt. Alvin Taylor, bombardier; Lt. Joseph C. Thompson, navigator. Not pictured is Sgt. Jack D. Ferguson, the tail gunner. (Photo courtesy James F. Fowler.)

waist. No attempt has been made to resolve the minor factual conflicts; the reader can decide for himself on the details. Fowler begins.

Fowler: "On our first mission, Bob Sheets flew copilot for me and we flew his airplane, the *G.I. Sheets*. On the first mission they always sent a seasoned copilot along with the new crew. We flew it again on our eighth mission, the day we were shot down. The *G.I. Sheets* was slow, as a lot of the old G models were. They were heavy and slow, which meant you had to pull a lot of power. My airplane was in the hanger with a ball turret inoperative, but the *G.I. Sheets* had just come out of the hanger from work being done on No. 2 turbo. It had been test-flown, but it had not been altitude test-flown, and with the old oil regulators you had on some of the early G models you could have trouble with air in the system. This was before they got electronic regulators on the newer G models."

Rawlings: "Fairly early in the day, when we got out to the airplane we were assigned, we had the feeling that we didn't have a really good airplane. I can't tell you precisely why. It was *G.I. Sheets* and we weren't impressed with the airplane. To begin with, it was a G model, and the

B-17G was a dog compared to the F, which was cleaner and flew very nicely. The G had that turret up front that gave you greater firepower, but it was slower than the F and you needed to use more power to stay in formation, which put a greater strain on the engines.

"And there were little things wrong with this one that we discovered when we got in the air. It was out of trim or out of rig. The control pressures weren't right. It was harder to fly in formation than it should have been for a B-17.

"It had just gotten out of the hanger, and had been flight-tested but not altitude-tested. That was the real rub. On the climbout, as we started up we became aware that we were not getting any boost on the No. 2 engine. It would run fine down at low altitude, you could get enough manifold pressure to get the power you needed, but the higher you went, the more the manifold pressure fell off. It meant we were climbing to altitude with an unsupercharged engine. We knew this before we left the English coast."

Fowler: "We were climbing out that day, and it was our first flight as element leader. I was leading the second element of the lead squadron in the No. 4 position. Well, we had no trouble; we climbed out and hit the coast, but right before we hit altitude No. 2 turbo started acting up. It acted up off and on for a while, I don't know how long, and eventually it got uncontrollable. So we had to pull the turbo off, which meant we were pulling more than normal power because we could not get full power at altitude out of No. 2 engine. But we stayed in formation."

Rawlings: "We discussed among our crew whether we should go on the mission, but it was a very brief discussion. Everyone wanted to go. What the hell, a good crew could fly a mission on three-and-a-half engines. Nobody wanted to go back.

"So we were charging along with the throttle of No. 2 full forward and the engine not at full power at 25,000 feet, trying to maintain formation with the other three engines and with the airplane out of rig. I had the feeling that we were torturing this airplane. I didn't like the airplane, so maybe it deserved torture, but I could tell that we were stressing it. We were forcing it to fly, using more power than I would have liked.

"By this time we had crossed the Dutch coast, and we were in Germany. And the engineer, who was in the top turret, said we had a stream of oil coming out of another engine. I recollect it as No. 3, but it could have been No. 4. We saw oil continue to come out, and the RPMs started to fluctuate, and there was no question in our minds that we had to shut the engine down. If we lost all the oil without shutting it down we wouldn't have control of the propeller governor, and we wouldn't be able to feather it. We feathered it despite the hazards, because we had no alternative."

Fowler: "Everything went along fine until right before the IP, when the engineer called over and said we were throwing oil out of No. 4. Well, I looked out at No. 4 and the wing was covered with it. My recollection is that we had started down the bomb run and the engineer said if you're going to feather you had better feather, because we got so much oil gone. And my diagnosis, having had it happen before, was that the scavenger pump at the front of the air sump had gone out, and it would pump the oil out at the breather. So we feathered No. 4 and then we salvoed the bombs."

Rawlings: "Immediately we were unable to stay in formation, and we started to lag behind. We were deep in Germany, vulnerable to attack, but there was never any question of turning back because we were better off trying to get as much protection from the formation as we could get. We got to the vicinity of the IP and by this time we were the last airplane in the formation, and we were still getting some protection from the airplanes that were ahead of us. We were about a hundred yards behind all the rest of the airplanes in the formation.

"In the meantime, other groups in front of us had bombed Frankfurt, and we could see the bomber stream off ahead of us and to the south of us, off to our right. So we figured if we cut across and took up a heading that would take us over to the head of the bomber stream there, even though we were flying slower than everybody else, we could get protection from the rest of the formation for another hundred miles going back towards the Channel. We jettisoned our bombs, figuring that if we were to have any chance at all to survive, we had to get rid of them. We got quite close to the target, but we never got over it.

"We started to turn off to meet with the other bombers, and it was all downhill from there. And as we charged off over towards the south, on pretty much of a southwesterly heading trying to pick up the head of the bomber stream, somebody on the crew, one of the gunners, said, 'Uh-oh, here they come. Four fighters about four o'clock high.' By my recollection it was a 'finger four' of 109s, what the Germans called a *Rotte*, that flew a pattern around us. They made a downwind leg, and a base leg, and then they turned and made a head-on pass at us.

"We thus had a couple of minutes to evaluate what was about to happen and what we were going to do, and we decided to jink the airplane. I can still see those methodical German fighter pilots as they flew past us, then turned on the base leg and came in. We tried to jump around and spoil their aim instead of holding the airplane steady to give our gunners a good firing platform, but they hit us anyway.

"Our gunners were firing and shells were coming at us—you could see that stuff coming to some extent—and the next thing I can recall was a discouraging 'Plump! Plump!' as the cannon shells hit the front of the airplane. They broke the Plexiglas out of the nose, and wounded the bombardier and the navigator. And I could smell gunpowder in my oxygen mask. It puzzled me at the time, but among other damage the cannon shells had destroyed the oxygen system, which was mostly up in the nose. They damaged our left wingtip too."

Fowler: "We didn't make it to the bomber stream. We started over and a fighter hit us from somewhere around one o'clock low. He got a good burst into us, 'cause he hit the instrument panel, he hit the panel by my foot, blew it out, he hit the bombardier in the leg, and he hit the navigator in the small of the back and right up on his helmet, knocking him out.

"Of course we took evasive action, and when we recovered we noted that we were losing our oxygen—and also that we had no vacuum. We dived. We were trying to get to the overcast, which from briefing was supposed to be at around 10,000 feet, but it wasn't. We were going pretty fast, and in the dive No. 4 prop unfeathered.

"The fighters came after us again, and we were scooting along at the top of the overcast. I started to turn, looking back over my shoulder, and we went into the overcast. We got into such a tight spiral, I knew we weren't going to regain control without our instruments, so I rang the bailout bell. But no one could get out of their seat, and when we came out the bottom of the overcast we were in a real tight spiral."

Rawlings: "With our oxygen system gone, we got nothing but this acrid gunpowder stink for a few seconds and then just ambient air. We were at 25,000 feet and we had to hit the deck immediately. We started down and we began to assess our situation.

"We were soon to discover that they had knocked out some vacuum lines behind the instrument panel, so that we had no artificial horizon, or turn-and-bank, but we still had the pitostatic instruments—airspeed, altimeter, and rate of climb. They also knocked out the intercom system.

"We took a turn to the right so that I could check the fighters out my window. We knew they weren't finished with us, and we kept on heading down for the deck on a heading that would make it as difficult as possible for the fighters to get back to us. We were going down as fast as we could, about 330 mph, and since the B-17 redlined at about 300 mph, we were really moving.

"Then we became aware that there was another straggling B-17, and the fighters made a big turn and picked this other poor bastard up first. They milled around him for a couple of minutes, and they must have killed the pilots, because the B-17 peeled up, rolled on its back, and then started spinning down. Then the fighters left that airplane, and resumed their tail chase on us. We hit the top of the clouds before they got within range of us.

"We entered the undercast going full-speed, because one of the gunners in the back said that a fighter was getting close, even though it wasn't yet in firing range of us. Then we had the problem of trying to recover the aircraft. Since we had lost our artificial horizon and turn-and-bank, we used our AFCE equipment—the autopilot—to orient ourselves. You could use it as a straight and level reference if you had to. It had these six axis lights and if you had all the lights out, you had the airplane straight and level. But you could not immediately determine what was going on by looking at those damn lights. So it was a real problem."

Fowler: "We were able to recover above the ground. We were under the overcast and above a river heading northwest, and I remember thinking, 'If we don't get out from over this river, there's a flak gun somewhere that's going to blow us out of the sky.' Eventually we found space above ground underneath the overcast and started west, or northwest back toward England. The overcast was a scut layer anywhere from two to four hundred feet thick laying in about five to six hundred feet above the ground up. And some places were broken. But by holding the aircraft level and pulling back, we were finally able to get up on top."

Rawlings: "We were quite close to the ground when we finally broke out of the clouds and were able to stabilize the airplane in a straight and level attitude. At this time we checked the magnetic compass, because this was the only heading reference we had now. We were below the clouds, and the sun was no longer helping us. The way these magnetic compasses operated, they would sit there and spin interminably until they finally settled down. I was horrified to see that we were heading roughly northeast instead of west when it finally did. So we stabilized the airplane and got it heading pretty much west. We were now in and out of the clouds and quite close to the ground terrain heading west."

Fowler: "Right after we had been hit we had no contact with the nose, and after we had straightened out on top of the overcast, Barney went down to check on the navigator and the bombardier. We had had no communication from up front."

Rawlings: "I went down to the nose to check on the bombardier and the navigator. It was depressing as hell. The navigator had a big scalp wound which involved one of his eyes. There was a mass of blood on his face and I couldn't see the extent of his wounds. And there was a big cannon hole in the bombardier's thigh. The Plexiglas was mostly gone and they were huddled there, among a lot of blood, maps, and papers.

"They had given each other morphine. There is always a question how much morphine you can give a person, but I gave them another shot because I decided they couldn't get out of the plane in their present condition anyway, and they were in obvious pain. I went back up, filled Jim in on what was going on down there, and we kept heading west."

Fowler: "The only compass we had left was the magnetic compass. And the navigator, because he was wounded and had shrapnel in one eye, was having trouble seeing and giving us a course, but he did give us one. Most of the navigating was being done by Barney from the little command chart that they gave the pilot and copilot.

"In the meantime, with that engine windmilling and pumping oil out, we decided to unfeather. And we unfeathered and we used the engine for a short time, about five or ten minutes, and then the engine froze and wrung the prop off the shaft. It was out there dolling, and we were afraid it was going to come off and come back through the fuselage, but it never came off all the way. But *man,* it got so hot you could see pieces dropping off. And it caused a vibration so that you couldn't read the instrument panel.

Rawlings: "I was navigating, and was using a little small-scale map because our navigator was out of commission. And Fowler said we were off course. To be on course and to get to the Channel, we had to be on about 285, and it was just too damn hard to do. The best we could manage what with all that was going on was about 270. Anything as complicated as 285 would have been much too difficult with that magnetic compass.

"We were having a hell of a time flying the airplane on those autopilot lights. If we got up in the clouds, we needed those lights because you very rapidly lose your ability to fly instruments if all you have is your airspeed, altimeter, and rate of climb. So we had a choice of dropping out of the clouds and being visible, or getting up in the clouds and having trouble flying the airplane. So we went from one to the other, all the while trying to fly by that damn magnetic compass.

"The clouds kept getting thinner and thinner, and we saw a couple of German airplanes. We saw a twin-engine airplane, a Ju-88 or a Do-217, that was milling around and that fired at us a couple of times. He was turning in toward us and we pulled up into the clouds and I was afraid he was going to run into us, but he didn't."

Fowler: "How many times we got picked up by fighters I don't know. But it was enough that they wounded everybody on the crew but me and Barney. What we'd do, when they'd start at us and the gunners would call out that they were coming in, we would dump down through the overcast, just set it down and let it pop out through the bottom, taking a chance on hitting a hill."

Rawlings: "At another point the clouds dipped down and we came sliding down out at the bottom, and there was an FW-190 off to the side of us on a parallel course. He was on our left a little bit ahead of us.

"I can still see this guy: He had his helmet pushed back and he had blond hair, and his oxygen mask was hanging down off to the side, and when I first saw him he was looking off in the other direction. He turned around and looked at us, and his mouth flew open, and he flipped up and was gone. It was all over in just a flash. This happened shortly before we were shot down, and this fighter may have had something to do with it. I don't know.

"Fowler didn't see him, and to this day I don't know if he was ever made aware of him. He was quite upset that he couldn't communicate with his crew because of the intercom—I can remember him saying, 'I can't talk to the boys! I can't talk to the boys!'—and he was messing with the intercom while I was flying the plane from the right seat.

"My side was working intermittently. I can remember talking to Arrington, our ball turret gunner. We had called him out of the ball turret as soon as we realized that we were stable and at low altitude, because we knew that there might be a need to get people out of the airplane fast, and it takes a while for the guy in the ball turret to get the doors lined up to get out of there.

"We wore chest packs, and Arrington's chest pack had been hanging on a hook. His pack had become dislodged while we were taking evasive action, and that chute had jammed in the turret's ring gear. And also, we always carried a spare chute and it flew out the window. This was information I got out of my interphone from the back because mine was working part of the time.

"I can remember Arrington calling me on the interphone at least two times, because he was back there with nothing to do, and he said: 'Lootenant, ah'm with you guys whatever you're gonna do, but please don't fergit ah ain't got no parachute.'

"And I said, 'Okay Dick, okay, that's all right, don't worry about it, everything's going to be all right.'"

Fowler: "I didn't know it at the time, but when we were taking all that evasive action a chute fell into the ball turret and jammed the turret, and I'm glad that it did, because the ball turret gunner got out, and we didn't have time at the end and he didn't get trapped in the ball turret when we crashed.

"Well, this business with the fighters went on until we got way back just south of Brussels near a place called Florenne, and a couple of fighters came up to hit us, and we set down through it again. We set down right on top of Florenne air base. It was a German fighter base. It had Focke-Wulf 190s and one of the gunners yelled, 'We're over an airfield and a fighter's taking off.'"

Rawlings: "We popped out of the clouds right over the middle of Florenne, which was a big German fighter field. We blundered right over it out of our sheer good luck. And of course it filled us with some trepidation and despair. The tail gunner yelled, 'Oh God, here they come, they're taking off, here they come, one, two, three, four.' I forget now how many he counted. They were FW-190s.

"So we tried to pull up into the clouds again, but the clouds were getting to be broken. From an overcast that was maybe a 1000 to about 1500 feet thick, it had dwindled down to maybe a 500-foot layer. It was four-eighths coverage perhaps. So maybe three or four minutes after we passed over Florenne there was all of a sudden this horrible battering of machine gun and shell fire in the rear.

"The tail gunner started yelling, 'Kick it, kick it!' and at the same time there was the sound of all this crap hitting the airplane, and an occasional 'Ping!' off the back of the armor plating we had on the rear of our pilots' seats. At the same time, by my recollection, the No. 4 engine blew. Likely it got a cannon shell. I could see it on fire out of the corner of my eye."

Fowler: "Shortly after we got over the airfield we lost all communication with the rear of the aircraft. This fighter that took off was a Focke-Wulf 190, and the other fellows later said he was black, so he must have been a night fighter. I don't think he ever got his gear up. He hung back under us, we couldn't have been at more than three or four hundred feet, and he was just shooting the tar out of us.

"I later learned from the other gunners that before this Jackson, the left waist gunner, had said, 'I'm tired,' and he dropped his flak suit off. Those were quite heavy and he was the oldest man on the crew, 39, and in the extreme cold we had it was quite tiring to have to stand up with all that weight on. And that last fighter just about cut him in half, shot him right in the belly with a 20mm.

"It also hit the right waist gunner, Loren Zimmer, in the small of the back and knocked him just as cold as a cucumber. He was laying on the floor.

"In the meantime the engineer yelled, 'We're on fire on No. 3' and it was shot out and windmilling, and we knew then we had no hope of staying airborne with two shot out and windmilling and the plane on fire, and we were going to have to go in."

Rawlings: "Now we had both engines on the right side out. I was flying the plane at the time, and I kicked the rudder to compensate, and the pedal went all the way to the floor, so we had no rudder. So the options were very small. There was no way we could fly the airplane on two engines with no rudder. We had to pull the power off the good left engines because without the rudder we couldn't keep the plane straight.

"All during this day, when things were going from bad to worse—you know, No. 2 is no damn good, and No. 3 we have to shut down, and now we can't keep up, and we're stragglers and we got attacked, and now we're crippled because the oxygen system's gone, and the vacuum's gone, and we got a couple of wounded guys—I considered all these things sort of trivial handicaps to successfully completing the mission. I was still quite confident we were going to get that damn airplane back to the Channel, and back to England.

"And then, when the No. 4 engine went and I kicked the rudder control and felt that it was gone, I realized instantly that 'No, alligator breath, you're not going to get this airplane back. There is nothing more to do. The mission's over. Sorry."

Fowler: "I rang the alarm bell. Barney dropped the gear as we started in and then he started bringing it back up, and I'll say this for the German fighter pilot, he quit shooting at us. Arrington, the ball turret gunner, had taken over for Loren Zimmer at the right waist gun and Donald Dinwiddie, the radioman, was at the radio gun. Dinwiddie told me that when the alarm bell rang, Loren jumped up and started back for the bailout door. Donald started running toward him and said, 'Don't bail out, we're going to crash.'

"He went out anyway. I thought he hit his hand on the door when he went out and his chute popped and dragged him out, but Loren later told me as he got to the door and looked down, the trees were right in the door. Well, I estimate he was about 150 to 200 feet above the ground. He popped his chute and went out, and Dinwiddie said, 'Oh my God, it's going to wrap around the tail'—he saw it go back and hit the tail—but it didn't wrap around, it went under.

"You talk about Providence or whatever you want to, but what saved Loren Zimmer was that he hit in a field in the only place that was a bog. It was really swampy where he went in; you would sink up to your knees. He hit in the only soft spot in the field and the Belgians who rescued him had quite a time getting him out. They sunk up to their knees trying to get him out."

That bail out, and the events leading up to it, are things that Loren Zimmer will never forget.

Zimmer: "The pilots' maneuvers in and out of the clouds worked good until the clouds diminished as we passed over a large German fighter base. They not only sent up fighters but fired on us with small arm weapons.

"The combat with the fighters became futile. We had lost most of our gunnery protection and were continually receiving damage to the aircraft and more injuries to the crew. I had received several fragment wounds to the abdominal area and one to the left temple that required a compress bandage to stop the bleeding. Arrington, the ball turret gunner, was now out of his position. He applied the bandage and laid a flak suit over me, which probably saved my life. He also took over my gun, which was frustrating to fire—after two or three bursts a short round would have to be ejected. This was probably caused when the machine gun belt flew out of the box during evasive maneuvers.

"In a semiconscious condition I can recall continued fighter attacks until the bailout bell rang. I automatically reacted by snapping on my parachute, jettisoning the waist door, and jumping out. These were actions I had planned over in my mind every night before I went to sleep.

"I remember glancing at the airplane's condition prior to jumping. The number three engine was on fire. There were a few small fires in the fuselage where loose 50 caliber shells had exploded. The fuselage skin at the waist door area had huge holes, and control cables were dangling from the ceiling.

"Not knowing what the cockpit condition was in, I thought the airplane was in a crash situation. Not until I was out of the airplane did I realize how low we were. I pulled the ripcord immediately and then after two very sudden jolts found myself partly buried in a soft spot on the ground. When I stood up, an FW-190 was bearing down on me like he was going to strafe. I dumped my parachute harness and took off running. When he didn't fire on me, I discovered one of my boots was still in the mud. So I went back for it, and that's when a Belgian patriot, Rolin Alberts, directed me by hand signals and his watch to hide in the bushes until 6 p.m. when he would come for me."

Fowler continues from the cockpit.

"We hit about a mile or a mile and a half beyond this. I can remember going in over the trees and everything was happening so fast I can't recall being afraid. I was too busy. As we went in, all I was thinking of was that we were going to hit those trees with what little power and control we had. Well, just before we went in, the radio operator said something did hit and it could have been we hit the trees."

Rawlings: "We had just a little power going, just enough to hold the airplane straight, and we saw a field ahead and took a very slight turn. We put the gear down, and then right back up again, and there was no more gunfire. The German fighter pilots were either out of ammunition or playing the game real nice. We made a good belly landing, and the airplane slid along in pretty good shape."

"After we got on the ground, I opened my side window, got out and ran around to the left side of the airplane where the nose escape hatch was, because I was acutely conscious that I had given the bombardier and navigator morphine. When I got there the hatch was already open, and the bombardier was trying to drag himself out, and the navigator was right behind him. They were attempting to get themselves out of the airplane.

"I grabbed the bombardier and kind of fireman-carried him about 50 yards away. When I started back after the navigator, I heard this noise and glanced up and here came a flight of two FW-190s. One was quite low and the other was in trail behind him, maybe three or four hundred feet up.

I thought, 'Oh hell, they're going to strafe the airplane!' I remember a feeling of terror when I thought they were. But they didn't. The leader did a roll, a victory roll, and I thought, 'Well, you're entitled to that, for God's sake.'

"When I got back to the airplane I didn't see the navigator. I looked in the nose and he was definitely out. I ran around to the waist area of the plane, and somebody had pulled the left waist gunner, Jackson, out of the plane. He was stretched out on the ground. Somebody had also helped the tail gunner out, Ferguson. Both of them were lying on the ground about 50 yards from the airplane. Finley was there, and I ran over to them.

"I glanced at the airplane, and there was a fire going in the right wing by this time. In the top turret there was what appeared to be a man's head. It was Finley's helmet and oxygen mask, but I didn't realize it. I had this feeling of absolute horror and revulsion and terror, and I said, 'Oh, no, Finley's still in there! We gotta go get him! C'mon, let's go get him!' And I was saying this *to* Finley. So although I thought I was coherent and knew what the hell I was doing, I obviously didn't. It was kinda dumb."

Fowler: "We had hit in this little field and slid to a stop. Now, the guns of the top turret were supposed to be forward so that the pilot and copilot could reach out and grab a hold of a gun and pull themselves out. But the guns were not turned towards the front. It wasn't the engineer's fault. He was wounded, too. But it took me a little longer to get out than Barney. I had on so many clothes because we didn't have heated suits and I had on a parachute harness and all and he went before me.

"I got out and I ran around the airplane, circling it from the front and the left wing towards the tail. And when I got back to the tail, Ferguson was out crawling. He was all shot up on his hands and his face and his legs, 'cause they hit that tail pretty good with 20mm. But he had his flak suit on and so did everybody else except Jackson. Everybody who had his flak suit on lived. They all had survivable wounds *even after taking a direct hit from a 20mm*. The one who had it off died. So it was a good piece of equipment.

"Finley had gotten Jackson up to the door and I grabbed his harness and pulled him on out of the door and drug him back beyond the tail of the airplane. In the meantime, Ferguson was still crawling and there was a barbed wire fence that he got tangled up in, and he was too close to the airplane. So I got a hold of Ferguson and got him out of the fence and drug him away back from the airplane. It was on fire."

Rawlings: "Finley was bent over Jackson, and I went over to them. I smoked cigarettes at the time, and Jackson smoked and he wanted a cigarette, so I lit him a cigarette. Finley and I undid his parachute harness and tried to make him comfortable. He had multiple 7.9mm machine gun holes in the chest and abdomen. You look at people, you look at people's eyes, and Jackson's were filled with sorrow. He knew he was in bad shape. He wasn't going to make it. Finley and I were both quite convinced that he was not going to survive."

Fowler: "In the meantime, Barney had come back and he was with Finley, who was kneeling by Jackson, and I yelled to Barney to get Jackson back from the plane more, and he looked up at me and said, 'Jim, it ain't no use.'

"And about then the bombardier, Al Taylor, came up to me and he said, 'Jim, I can't get away, I got a 20mm in my leg.' And he had a hole in his leg and a tourniquet around it and he was still walking, though how I don't know, 'cause you could have stuck your hand practically in the hole he had in his right thigh.

"I said, 'Well, Al, you stay with these two guys and let me see if I can get away.'

"And he said, 'Sure.'"

Rawlings: "Then Finley and I went over to Ferguson. He was shot up pretty bad. He didn't want a cigarette, but we undid his harness and tried to make him comfortable. I had eye contact with him, and it took only an instant to see that he was ticked off at us. He wasn't our regular tail gunner and he had flown only one or two missions with us. I'm sure he felt that he didn't have any business being shot down with a green crew like ours. Anyway, he survived. He got good medical treatment from the Germans.

The burned-out tail of G.I. Sheets, B-17G 42-39786, GN✪R. (Photo courtesy James F. Fowler.)

Fowler: "About that time the Belgian who later hid us, Maurice Lamendin, came running up with some other civilians and the plane was burning and the ammunition was starting to go off and I was scared that plane was going to go at any time. And he told me go—'Parte! Parte!'—get away, and I said 'I got wounded, I got to get a doctor,' and he made me understand 'We'll take them,' and he pointed to a truck he had.

"And all these civilians, God bless them, they came in and they picked them up in their arms and they took all three of them off toward the truck.

And then once I saw they were going to be taken care of, I yelled to the other crewmembers, 'Get going! Get away from here!' 'Cause we had just passed over this German airfield. We weren't too far away from it. And there was a flak observation tower between us and the airfield and the Germans there were looking at us, too.

"So I took off and waded this creek when a guy I later learned was named Emile started guiding me and he took me up to a woods and said, 'Stay here.' And he came back with some civilian clothes, which I put on. And that was the beginning of my stay with the Belgian Resistance.

"They were wonderful people, the most kindhearted folks you could ever hope to meet. They gathered up all but two of my crew, and though I had to care for Thompson's wounds myself before their doctor could get there—Joe had a huge U-shaped flap of scalp which I had to cut off, and a piece of shrapnel in his eye that I had to get out because it was extremely painful and couldn't wait—they took very good care of us, and they were able to hide most of us for many months.

"They had to take Taylor and Ferguson to a civilian hospital because of their wounds, and they fell into the hands of the Germans. Later the Gestapo got real active in our area, and they passed Dinwiddie and me into France, where we evaded until May 13, 1944. We were captured together in Lille, when a doctor and his wife betrayed us by driving us straight to a German roadblock. We ended the war as POWs. Arrington and Zimmer evaded until American ground forces picked them up in September 1944. The Germans got Finley near the French border with Spain, in the Pyrenees. The only one who made it all the way back to England was Rawlings."

Rawlings: "In the field where the airplane was I next remember a group of about five or six civilians who came running down the hill. They were able to communicate the fact—probably in Flemish, I understood some German—anyway, they made me understand there was a doctor and a priest coming. The doctor was going to be there any second. The priest was going to be there any second. I felt like I was in the hands of friends, and I learned later that these people were members of a Belgian resistance cell in the area. They were quite well organized, and they seemed to know exactly what they were doing.

"One of them grabbed me by the arm—he was a youth of maybe 13 or 14—and he pointed across the field and there was a military vehicle coming, sort of a jeeplike vehicle, and it had two or three people with steel helmets. And this fellow said 'Germans coming' and he gestured as if to say, get your rear ends out of here, and pointed to a woods to the north.

"So Finley and I ran north. We ran a couple of miles into the woods, and we finally sat down in a pretty good thicket. I was sitting there with my back against a tree, and Finley started to weep. He was quite close to Jackson, but I was still surprised when he started to weep. It shook me up quite a bit. I was 23 and Finley was 27 or 28. He had been in the Army for quite a while and we sort of looked up to him as the guy who knew all about the airplane, and because he was the old man of the crew, next to Jackson, who was 39. And then I realized it was quite understandable. Christ, even if you were an old guy of 28 you could still cry.

"So I comforted him as best as I could. I pulled some little splinters of glass out of his face that had lodged there when his optical gunsight had been shot up. Already I was feeling guilty that we had left Jackson at the scene. I sat there with my back against the tree and I thought, 'Could we have helped him if we stayed? Probably not. But I feel like a bum.' I know Finley was feeling the same way. We had done what the civilians had told us, but I still felt guilty about it. That guilt has never left me and I'm sure it never left Finley either. So we talked it over.

"I was reviewing my actions of the day and said, 'Well, it really turned out to be a crappy mission. I did the best I could. I feel terrible about the guys getting shot up. I just feel awful. But what the hell. Everything went to hell one piece at a time, what could I do, I did the best that I could. And at least I know that I wasn't scared.'

"And then I realized that I had been chewing gum, and the gum was stuck to the inside of my mouth, and I had to giggle a bit sitting with my back against that tree because I knew that unless you're scared, gum does not stick to the back of your mouth. I thought that was humorous, even at the time. It seemed a humorous sidelight to getting away from the airplane.

"I had intended to go to Europe and be a hero, shoot down a lot of airplanes and all that stuff, and here I was shot down. Finley and I decided it was best to split up, which we did. He was later captured in France, not far from the border with Spain, but I escaped and evaded, making it through France and into Spain. It took me four days to cross over the Pyrenees Mountains.

"And while I was trudging through northern France trying to get to Spain, I would see our bombers high above at 25,000 feet, on their way to bomb Germany. I remember the feeling of pride I had seeing that magnificent Air Force pass by, and the feeling of regret I also had because I was not up there with them."

37

The Last Mission

Bernburg, February 20, 1944

THE DAY AFTER Fowler fell, the Eighth took to the air once more, dispatching 777 heavy bombers and 635 fighters on a deep penetration against enemy aircraft factories located in the Brunswick area. Cloud cover prevented visual bombing, but 742 heavies dropped their loads on the city and targets of opportunity, with a B-24 combat wing hitting Hannover. The Eighth lost 20 bombers and four fighters; the bombers claimed 51 enemy aircraft and the fighters 45.

The 303rd put up 39 aircraft in two groups, A led by Captain "Mel" Schulstad and Lt. W.C. Heller of the 360th Squadron, and B led by Major Richard Cole and Lt. T.J. Quinn of the 359th Squadron. Both bombed the center of Brunswick through 10/10 clouds, and no 303rd ships were lost. The Group's records noted: "Fighter support by P-47s and P-51s was excellent in the target area and on the route out."

A mission for the following day was scrubbed after briefing because of bad weather, and then February arrived. The missions came fast and furious, with the Eighth regularly fielding from 500 to 700 bombers and 600 fighters. On February 3rd the 303rd sent 39 ships in two groups to bomb Wilhelmshaven with PFF equipment through dense haze and 10/10 clouds. It was the last mission of Lt. Bill McSween's tour and the 22nd for Captain Don Gamble. Gamble's diary provides a good description of what went on:

"Up at 0400. Eat at 0430. Briefing at 0530. Sleepy. Briefed twice before for Frankfort. Target today will be Wilhelmshaven. Will lead 41B Combat Wing with Col. Stevens as copilot. Briefed for all kinds of clouds, high, medium, and low. Over to Group Operations after briefing to read orders. Out to ship 574 G. Taxi out late due to plane in our way. Takeoff on time. Carry 12 500-pound HE bombs and 2300 gallons of gas in B-17G.

"Assemble group at 3000 feet and climb through five layer of clouds. Leave base at 8000 feet on time—start climb at 150 mph and 300 ft/min. Keep good airspeed and rate of ascent. We are supposed to fly abreast of 41A but, due to clouds that we have to fly through, are unable to catch 41A.

"Lose all but 13 ships of group in climb through dense clouds. Get well on top at 28,900 feet. AFCE will not keep ship straight and level. Keep resetting it. Friendly fighter support good. See no enemy fighters at all. Very little flak and that is inaccurate.

"Just over target at IP, we flash PFF ship on our right wing a green signal with aldis lamp in tail and they take over lead. Drop PFF, then take lead back and start let-down. Parallel cloud layer then start down through them. Group stays intact till we reach 8000 feet and cloud thickens and splits up ships. Let down to 1000 feet—just below clouds and proceed to base in very rough air. Fair landing. Donuts, coffee, cake and candy at Interrogation taken to Group. Easy raid but for clouds."

It was an "easy raid" but on the way back Lt. McSween recorded that: "Going was slow due to a strong headwind. When nearing English coast you could see planes coming from all directions. White was only crew lost—went out of control over North Sea."

"White" was Captain G.A. White of the 358th Squadron, flying B-17G 42-37927. He and his crew were on their first raid and the cause of their loss was never determined. They were one of four B-17 crews that didn't return that day, along with nine American fighters.

On February 4th the Eighth made another PFF attack on Frankfurt with the 303rd putting up 38 ships in two groups. No Hell's Angels were lost, but this raid was rougher than the one of the preceding day. The Eighth lost 20 bombers and one fighter. The Americans claimed 12 enemy aircraft. On February 5th and 6th the heavies struck back, hitting numerous enemy airfields in France. The Eighth lost eight bombers and six fighters; enemy aircraft claimed were 26.

The 303rd sent 20 bombers against Bricy Airdrome at Orleans on the 5th and lost no aircraft. The Group sent 23 aircraft to Dijon/Lonvic Airdrome on the 6th, and lost the 358th Squadron's *Padded Cell*, B-17G 42-97498, flown by Lt. J.S. Bass on his first raid. Bass was flying with a crew that had more than 10 raids in, among whose missing was Sgt. Cyril Dockendorf, the photographer who had flown with Lt. John Henderson's crew on the Oschersleben raid.

Captain Don Gamble went on the mission and wrote: "Bass goes down, evidently hit by flak. Seven to nine chutes are seen. Dick [Scharch] sees ship hit ground and explode...Hear a copilot was killed by a .50-caliber slug just inside French coast."

Two days later the Eighth returned to Frankfurt for more PFF attacks. The First and Third Division bombed with 195 B-17s. The cost was 13 Fortresses and nine fighters out of 553 dispatched in exchange for 17 enemy aircraft destroyed. The 303rd contributed 22 aircraft, 17 of which bombed and all of which returned. In the meantime, 110 Second Division B-24s hit the V-1 sites at Siracourt and Watten, escorted by 89 P-47s. They suffered no loss.

All during this time Hullar's crew cooled their heels, and Brown kept his diary closed. When he finally picked it up again, he wrote: "It was a long time since my 24th raid. My outfit had sent 12 groups out on raids in that time. They said they were saving us for a DP (deep penetration of Germany). One time we started out for Leipzig but it was recalled just before we left England."

Klint's crew was slated to go on this one too, and he later wrote:

"THIS WAS IT! I had flown 24 'Practice' missions just to get a crack at this one—number 25—the one which would end my operational tour and send me back to the U.S. We started to get this one in on February 9th. We were briefed to go to Leipzig that day—a round trip of seven hours and 51 minutes. In view of that, I was not too disappointed when we were recalled shortly after we left the home field, little thinking that I would have to finish on one just four minutes longer than that.

"While we were in the air on the 9th, two of our ships had a midair collision. That was the third we had had since I had joined the 303rd. The first two had both been during conditions of restricted visibility and in both cases the bomb load of both planes had exploded. We were flying in the second element of the lead squadron, and when the first element went into echelon, the No. 3 ship took the entire rudder and tail gunner's compartment off of the No. 2 ship. Fortunately, no one was hurt. The tail gunner in No. 2 ship had gone forward to check the tailwheel seconds before the crash. Both ships managed to land safely."

The enlisted men in Hullar's plane also witnessed the collision. As Merlin Miller recalls:

"I was up in the waist, looking out the right window when it happened. As we came around in three-plane formations to land, the left wingman would drop down underneath the lead plane, swing off to the right and come up in formation with the right wingman, echeloned off to the right. The left wingman started his turn to the right, and came up with his No. 4 engine right into the tail of the right wingman.

"Metal flew everywhere. He knocked the rudder off the plane and tore into the tail gunner's compartment about three or four feet in. I yelled

One of the 303rd B-17s involved in the aerial collision Hullar's crew witnessed on February 9, 1944. Fortunately, the tail gunner had gone forward to check the tail wheel seconds before the crash. The aircraft is the 358th Squadron's Thru Hel'en High Water, *B-17G 42-39785, VK✪H. It was lost to flak over Hamm, Germany on February 22, 1944.* Photo courtesy National Archives (USAF Photo 51841 AC).

instinctively, 'Look out!' I couldn't help but hope very strongly that the tail gunner was not in there. I learned later that he had just left the tail compartment, and had the holy hell scared out of him when he looked back and saw a lot of jagged metal and open space where his compartment should have been."

George Hoyt remembers that "As we flew over the field to break up our Group, and to form a long straight line of aircraft swinging wide around to the left to land in single file, Dale Rice rushed into the radio room. He was excitedly pointing out the radio room hatch toward two o'clock high. I got up with a bound to look in this direction, and I couldn't believe my eyes! There, flying right next to us, was a B-17 with almost half the vertical stabilizer and rudder chewed off and with shreds flapping in the slip-stream. The wingman had rammed him from behind with his No. 4 engine as he swung over in formation to enter the single file procession. They both made it down safely after firing red flares for an emergency landing, but what a hair-raising way for them to top off a flight!"

On February 11th the Eighth launched a number of attacks against a variety of targets: Frankfurt again, Ludwigshafen, Saarbrücken, and the V-1 site at Siracourt; 424 bombers were dispatched along with 732 fighters. The fighter action was heavy with the escorts claiming 32 German aircraft and the bombers three in exchange for 14 fighters and six missing bombers. The 303rd sent 21 B-17s to Frankfurt, only 14 of which were able to unload. On their return to Molesworth, the 358th Squadron's B-17G 42-39810, flown by Lt. J.R. Worthley's crew, crash-landed, killing the flight engineer and injuring 16 other men.

Meanwhile, Klint recorded: "Our crew went to the 'Flak Home' for a seven-day rest. While I looked forward to that and knew that I would have a wonderful time, I was a little disappointed that it came just then. Bob Hullar and seven other members of my original crew were also 'sweating out' that last raid and I wanted very much to finish at the same time they did. As it turned out, they were still waiting when I returned and we all did finish on the same one—in a full-scale smash on German aircraft production."

The Eighth flew no maximum efforts from February 12th through 19th, but the climactic battles of the Eighth's air war were in the making while Klint's crew took time off and Hullar's men waited for the deep penetration that would mark their final mission. Ever since General "Tooey" Spaatz had taken over as Commander of USSTAF, General Arnold had been emphasizing the need to knock out the Luftwaffe. Even before the formal change of command, Arnold had sent Spaatz a New's Year's message saying: "My personal message to you—this is a MUST—is to 'Destroy the Enemy Air Force wherever you find them, in the air, on the ground and in the factories.'"

With the help of General Fred Anderson and Doolittle, Spaatz had been implementing this directive "in the air" and "on the ground" ever since: In January, VIII Fighter Command began strafing German airfields on the way home from long-range escort missions, and most of the Eighth's targets were chosen in order to draw the Luftwaffe into a continuing war of attrition. But bad weather kept the Eighth from launching the sustained attacks Spaatz wanted "in the factories." His offensive, code named "Argument," remained in limbo as long as clouds shielded the Luftwaffe's vital aircraft "factory complexes" in the central and eastern areas of the Reich.

The picture started to change on February 18th. That day Anderson's chief weather officer at USSTAF Headquarters forecast a meteorological high over central and southern Germany beginning on February 20th, which was expected to provide clear skies over Germany for three days. It was the golden opportunity "Tooey" Spaatz was looking for. Spaatz ordered the decks cleared for Argument, and Anderson put the word out to the Eighth's commanders. Next day the weather officer went even further by *promising* three clear days and possibly a fourth.

Argument actually got underway on the evening of February 19th when, in a rare instance of direct cooperation by RAF Bomber Command, the British night bombers attacked Leipzig, where one of the factory complexes lay. By the morning of February 20th, however, a command crisis arose. Both Doolittle and General Kepner, commander of VIII Fighter Command, were reluctant to order their aircraft off in the cloudy and icy conditions that blanketed the Eighth's bases. They required a direct order from Spaatz, and Anderson took the matter up with him. His reply was, "Let 'em go."

The operation that was about to be presented to the bomber crews and fighter pilots was the most ambitious the Eighth's planners had ever put together. For the first time, more than 1000 heavy bombers were to attack: 417 B-17s from the First Division, 314 from the Third, and 272 B-24s from the Second.

The mission featured another first on the fighter escort side. Among the 17 groups detailed to support the bombers were *two* groups of P-51s: the 354th of IX Fighter Command and VIII Fighter Command's own first Mustang group, the 357th. All in all, 73 Mustangs would safeguard the bombers in the target area, together with 94 P-38s from the 20th and 55th Fighter Groups (one of which would fly two missions). In addition, 668 Thunderbolts would provide penetration and withdrawal support, with three P-47 groups returning to base and flying two missions, a feat made possible by the length of the operation and great improvements in their belly drop tanks.

The main effort, led by the First Division, was against the factory complexes in the Brunswick-Leipzig area. The Eighth was to send 10 wings

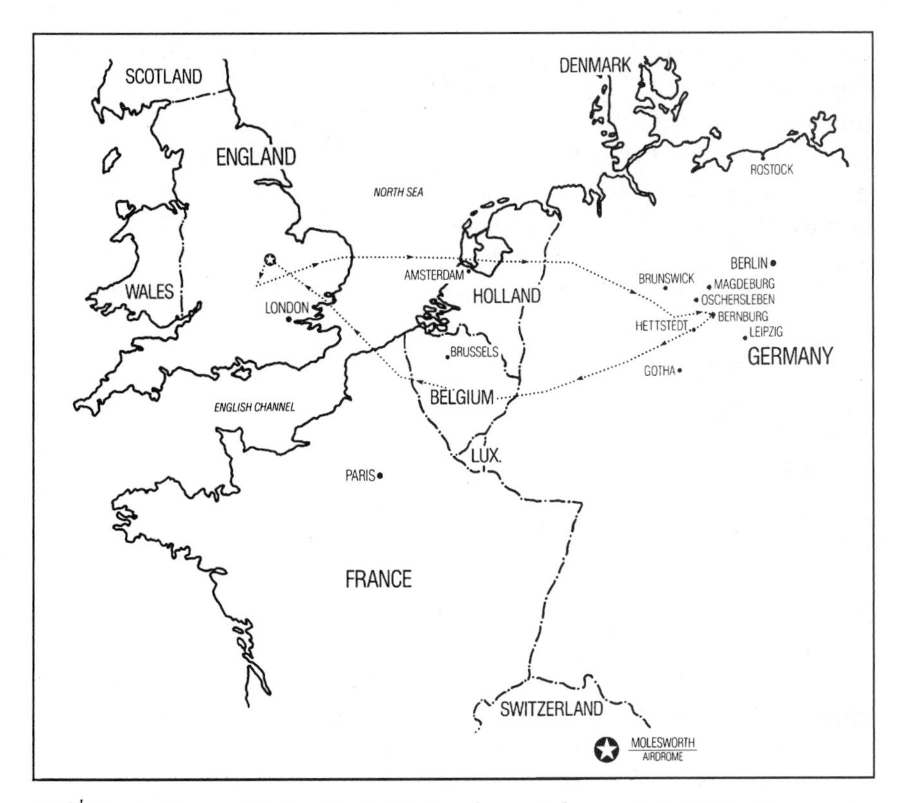

303rd Bomb Group Mission Route(s): Bernburg, February 20, 1944. (Map courtesy Waters Design Associates, Inc.)

there, including those of which the 303rd would be a part. This bomber force would have fighter escort all along the route. Six more wings were to attack aircraft plants at Tutow in eastern Germany and at Posen in Poland. Their routes lay beyond the range of most Luftwaffe fighter units and they would fly unescorted.

For reasons that will soon become evident, the story of the Hullar crew's participation in this historic operation is largely drawn from Bud Klint's experience. He now begins the account of the crew's last mission, February 20, 1944:

"On that Sunday morning we were awakened at 0500 and told that breakfast would be at 0530. The usual GI trucks took us to the mess hall, where we ate the usual powdered eggs and canned bacon and drank several cups of syrupy, black coffee. Then the trucks hauled us back to the briefing room. The CO started his introductory remarks promptly at 0600. He told us how important it was that we destroy the targets which we were to hit that day—talked to us much like a football coach would talk to his players before a big game—and then turned the briefing over to the Intelligence Officers."

"They uncovered the big map to which was pinned a black tape marking our course for the day. They told us that the 303rd was sending two groups of 19 planes each. Group A was to go to Leipzig. Group B, in which we were scheduled to fly, was to hit Bernberg [sic]. We looked at pictures of the target areas, we were assigned our positions in the formations, we were told where, along the course, we could expect to encounter flak and how many guns would be able to reach us at each point, we were told how many enemy fighters were based along our route, and we were given the numbers of the spare airplanes which were serviced and available in case the planes to which we were assigned developed trouble.

"Finally, we were given our time schedule, our bombing altitude, the temperature at that altitude, and were asked to synchronize our watches. The navigators and bombardiers then went into their own detailed briefings while the rest of the crews picked up their equipment and went out to their ships to prepare for takeoff."

Hullar's crew had drawn *Vicious Virgin*, and would be flying with Captain Richard P. Dubell, Operations Officer of the 427th Squadron, who was formation leader. McCormick was the bombardier, and Lt. E.G. Greenwood was the tail observer, with Merlin Miller filling in the right waist. Klint's crew had drawn No. 795, the chin-turreted B-17F named *The Flying Bitch*. Klint now continues:

"In the equipment room, I picked up my parachute, my Mae West, my 'tin hat,' my flying boots and, since we were flying a B-17G [sic], which was like a wind tunnel on the inside, my electrically heated 'Bugs Bunny' flying suit, and also my oxygen mask. Under all this, I wore 'long handled' underwear, a pair of OD pants, an OD shirt (with no insignia), a leather flight jacket, one pair of woolen socks and one pair of silk socks, GI shoes, and my 'escape belt.' This was an innovation of my own—a belt I wore around my waist which, I estimated, contained enough D-ration, cigarettes, Benzadrine tablets, and other essentials to keep me going for at least a week should I happen to be shot down in enemy territory. Besides this we carried escape kits furnished by the Air Corps which contained maps imprinted on silk handkerchiefs, a compass, a hacksaw blade encased in rubber, and a packet of German, Dutch, and Belgian money.

"GI trucks again carried us out to the hardstands, or cement parking strips, where the ground crews were swarming over the ships, making final checks. While the copilot and I gave the ship a thorough going-over, the gunners checked their guns and loaded their ammunition and equipment. Early in my tour, we always carried and often used ten to twelve thousand rounds of ammunition on each raid. Now, with improved fighter support, we seldom carried more than five to seven thousand rounds.

"Once the ship was loaded and checked and the engine run-up completed, we all had a final cup of coffee, which the ground crew always prepared and kept on the stove in their tent. By then it was time to take our places and stand by until 0830—start-engine time.

"Promptly at that time, engines started to roar in the dispersal areas all over the field. I believe the whole crew felt a little extra tension that morning. The fact that I, as pilot, was starting on my 25th mission accounted for part of that, but probably more important were the things the Weather Officer had told us that morning. He had prophesied that the skies over Germany would be clear and that meant that the enemy fighters would be free to operate. We had been promised fighter escort all the way to the target and back, but this was to be an extremely deep penetration and we knew that with the new corridor type escort coverage, there would undoubtedly be times when we would be strictly on our own. To add to our discomfort, our course to the target was planned to feint at Berlin. Undoubtedly, Jerry would be eager to disrupt what would appear to be a major thrust at his capital.

"While such thoughts were running rapidly through my mind—and while similar thoughts were probably passing through the minds of some ten to eleven thousand other American airmen—the copilot and I were going through our preflight check, carefully watching the performance of those 1250-hp Wright engines. The loss of even one of those four powerplants could mean the difference between seeing England again and going down somewhere in enemy territory and I, for one, was determined that everything would be in perfect order before I would signal the crew chief to pull the chocks.

"The engines checked OK. The copilot made a check of each crewmember by interphone and each man replied that he was in position and ready to roll. The ground crew pulled the chocks and gave us their usual good-luck sign—a thumbs-up gesture which was the American counterpart of the English two-finger 'V-for-Victory.'

"We started to move out of the dispersal area at 0840 and as we rolled down the taxi strip to our takeoff position at the end of runway 27, some 'Limey' field workers edged along the sides of the strip to wave their good wishes. We slid into place behind Lt. Melton, who was flying ship No. 081—*Luscious Lady*—and we ran up our engines for the final check.

"At 0850 Capt. Bob Hullar started the lead ship down the runway and, at 30-second intervals the other ships followed. A few minutes later we were airborne and coming about to take up our position in the lead squadron [the No. 6 slot]. As our group began to form, we started to climb and soon fell into place with the other two groups which made up our Combat Wing.

"By 0956, our scheduled time to leave the base, we had reached our bombing altitude of 16,000 feet and our Wing was completely formed. We took up our easterly route which was to take us to the target, and I believe I had more apprehensions then ever before. This was it! I had looked forward to this day for many months, but now I was scared—and plenty. I just couldn't believe that I would come back from this one.

"At 1125 we crossed the enemy coast and the ground batteries let us know that they were aware of our presence. It seemed to me that every burst was aimed for our ship, and with each explosion I shuddered, half expecting the plane to disintegrate around me.

"The lead ship in which Hullar and the majority of our original crew was flying did get hit here. They lost an engine and dropped out of the formation to return to the base. I hated to see them leave the formation and I guess I even felt a little resentment because they were finishing their tour 'the easy way.' In another hour they would be safely back at Molesworth with number 25 under their belts. I still had nearly five hours of flying to go—if we got through OK."

Klint had no way of knowing that things were not quite as easy or clear-cut for Hullar's crew as he imagined. Brown recorded the events as follows:

"We had trouble rendezvousing with the wing and we didn't actually get into wing formation until just before crossing the Dutch coast. Before we aborted, when we crossed the Dutch coast, we got some very accurate moderate flak and it hit Hoeg's No. 3 engine. He was flying in No. 4 position of our high squadron. He dropped out of formation and headed for home about six minutes before we turned around. "The fuel pressure was very low on our No. 2 engine and Hullar was afraid we would lose it and it was too long a trip for three engines, for if we would lose another engine, we would have had it. So we aborted at 05 degrees 35 minutes E, which was just before we crossed the Zuider Zee."

Hoyt also remembers these moments: "During this time I anxiously monitored the intercom conversation of Hullar and Capt. Dubell, our Squadron Adjutant. Bob was concerned about the excessive amount of drop on the fuel pressure on our No. 2 engine to an instrument indication below 12 pounds per square inch. Another engine began to indicate a falloff in critical pressure, so Bob radioed for the deputy lead plane to take over as he dove out of formation, swinging around in a 180 degree turn toward England.

"Below us there was a large layer of clouds ahead at approximately 6,000 feet. It stretched over the Zuider Zee as far as we could see to England. Bob asked Capt. Dubell if he thought it would be a good hiding place to go home in. We were expecting enemy fighters any moment, as we were not far from several fighter bases near the Dutch coast."

From the *Virgin*'s right waist, Miller actually spied enemy aircraft aloft: "I saw what looked to me like a twin-engine German fighter, maybe a Ju-88, milling around with a couple of other planes off in the distance, and I called them out to the cockpit."

Then, Hoyt remembers, "Bob told Dubell, he was going to opt for a high-speed dive for the clouds, and we started down at a sickeningly fast rate of speed. We were all alone in a dangerous piece of airspace. The tail of the plane began to vibrate noticeably, and I watched its shaking with concern. I called Bob to report it, and he asked me to keep a close eye on it, as we were diving at a speed in excess of what was recommended by the manual with the possibility of tail structure failure at high diving speeds. I knew that the airspeed indicator up front was redlined at 305 mph, and Dale later said that we exceeded that point by quite a bit."

To Miller, "the plane was vibrating quite a bit during that dive, and it was kind of exciting, but she held together okay." *Vicious Virgin* plunged into the clouds and Hullar's crew made it back without further incident. They landed at 1305 and an examination of the ship showed that the fuel pressure regulator balance line in No. 2 engine was indeed broken and that the ship's radio compass was out.

This, George Hoyt recalls, raised an interesting question: "We did not know how Base Headquarters would interpret all this. Would they count it as our 25th mission or not?"

"We had to wait several hours to find out," Norman Sampson recollects, and Merlin Miller remembers this as "the only time I'm aware of that some of the enlisted men in our crew ever showed that they were worried."

After *Vicious Virgin* dropped out of formation, command fell to a less-experienced crew. Lt. Arnold Litman of the 358th Squadron dropped down from the high squadron to assume the lead of a badly depleted B group, for there had been two earlier aborts in the low squadron. The ranks were also thinned by Lt. Melton, in *Luscious Lady*, who never found the group (he tagged along with a B-24 formation and bombed with them), and by another high squadron ship flown by Lt. R.W. Snyder, who joined the A group instead. This left the B group with only 12 B-17s, which was not the kind of bomber force to take deep into enemy territory. But Klint and the others pressed on, and his account continues:

"Our escort arrived as we crossed into enemy territory and it gave us a great sense of security to see our little friends flying protectively above. This first wave of friendly fighters was with us for only about 10 minutes and for the next 30 minutes we saw only scattered groups of escort fighters far in the distance. During this interim, two FW-190s came up below our formation and S'ed along a course parallel to our own. Apparently their only interest was observation, for they made no attempt to attack.

"At about 1200 our escort picked us up again and we had intermittent cover from then until we reached the flak zone at the target. Only once more did we see any German fighters. Again it was two 190s and again they did nothing but fly along below us, probably reporting on our progress."

The B group bombers flew on into the target area searching for the Junkers assembly plant at Bernburg. They got to the general area as "breaks in the clouds along the route gave the navigators a chance to check their route." They passed through "moderate and accurate" flak without loss. The crews reported that "the target area was clear with snow on the ground," so that "weather conditions did not interfere with execution of the mission," and Klint's crew "saw 300 planes on ground in target area."

But for all that, the B group bombed a "target of opportunity," which was described as "a factory in a town as yet unidentified somewhere between Bernberg [sic] and Eisleben to the southwest." Later photo analysis showed that it was a copper smelting plant between the small towns of Molmeck and Groborner, and that the bombs all landed in open fields some 17¾ miles from the planned mean point of impact on the Junkers factory. It is not hard to imagine how much better the results might have been had Hullar's crew still led the formation, with the navigator-bombardier team of Elmer Brown and Mac McCormick on the job in *Vicious Virgin*'s nose.

As it was, nothing remained for the bombers except to head back home. The flight over hostile territory passed with the crews having little to occupy their time and even less to report. At 1311 Klint's crew spotted seven barrage balloons over what appeared to be a hanger, and a bit later they saw an observation tower on top of a hill in the Hartz mountains. Small groups of enemy fighters appeared south of Bonn and south of Malmedy, but they made no attempt to attack. The B group encountered some meager and inaccurate flak at Nordhausen and Kassel, and meager but accurate flak at Lille.

Through it all Bud Klint remained nervous and uneasy; there was one final obstacle for him to pass at the enemy coast, northeast of Dunkirk, where the Group hit the Channel. He remembers the moments vividly:

"The flak at the coast scared me more than any I'd experienced in the previous 24 missions. It wasn't too accurate, but I just couldn't believe that I was going to finish. I even closed my eyes once and just waited for the ship to be blown apart. But we came home without a scratch."

The Hell's Angels returned to Molesworth virtually unscathed after a good mission by the A group. Led by Major Richard Cole and Lt. T.J. Quinn's crew of the 359th Squadron, they reached the Junkers Aircraft Motorworks at Leipzig (Mockau) and 18 bombers unloaded on it with excellent results. The mean point of impact was a scant 30 feet short and to the east of the assigned MPI, and many hits on plant buildings were obtained.

The A group got only two tail attacks by Me-109s, and one of these was shot down. The only casualties were two slightly wounded crewmen, and the most excitement of the mission occurred when Lt. J.R. Morrin of the 360th Squadron had to crash-land his bomber near Podington, home of the 92nd Bomb Group. He found the base with two engines out and a third failed while he was in the traffic pattern. He brought B-17F 42-5859 down in an open field. She was a total loss, but his whole crew walked away from the wreck.

Klint's bomber touched down at 1645, and the last Hell's Angel landed at 1723. With the rest of the men Bud Klint went through all the routine postmission procedures: He was interrogated, he handed in his flight gear for the last time, and he headed back to his quarters. But, "I still couldn't believe I had finished. I knelt in the most earnest prayer of my life when I got back to my room. And that night I went to church to thank God again, for it was nothing but His goodness which saw me through 25 missions safely."

There remained only the matter of Hullar's men getting an answer to the big question whether it was all truly over for them. George Hoyt will always remember how they got the word.

"Several hours elapsed as we all waited with suspense in our barracks, and then Bob came by with a smile that said more than the words he spoke: 'That's it boys, we've made it, they've given us credit for our final mission.' As with the others, I was glad. I wanted to go home. I had done my part, I had stuck it out, and I felt triumphant for us all. We had all done our duty, and a damn good job."

Norman Sampson could only marvel, "I am alive, and I am going home. I've never had another feeling like it."

Merlin Miller was surprised: "It was kind of an anticlimax, I guess you'd say. To go out and have an engine go out on us, turn around and come back after we got into enemy territory a little ways—I was really surprised that we got credit for a raid on it. When Hullar came in and told us, I thought at first he was kidding. I guess I was pleased just to get the thing over with, but it was a real anticlimax after all the raids we *did* go through, some of them easy, some of them very difficult."

Elmer Brown duly noted: "We came back to the base and got an abortive sortie out of it," which meant credit for the full mission was awarded. He explained, "They had to give it to us because of Hoeg's battle damage," and he tried to put it in perspective by concluding his diary on the following note: "Eight of our original crew finished their tours on this raid. Hullar, Rice, Hoyt, Sampson, Fullem and Miller and I were in the same plane and Klint finished with his own crew. McCormick got his 18th mission with us that day and Marson got his 22nd with Klint."

But Elmer Brown admits to a sense of incompleteness about it today. "It actually was something of a disappointment. I was very caught up in the war and there was a lot happening. Major Snyder talked to me about a promotion and becoming Squadron Navigator if I would stay on, and part of me wanted to do it. But I was married, and I wanted very much to get back to Peggyann and to start a family, and that made all the difference in the end."

Chuck Marson had to remain to complete his tour and, as things turned out, two of the other men of Bob Hullar's crew decided to stay on to the very end: Hullar himself and Mac McCormick. They were witnesses to what history now calls the "Big Week" and to the final, overwhelming victory the Eighth achieved over the Luftwaffe.

38

The Beginnings of Victory

"Big Week" and Beyond

WHEN THE AMERICAN bombers and fighters returned from the raids flown against the German aircraft factory complexes on February 20, 1944, it was quickly apparent that the Eighth had scored a smashing victory over the Luftwaffe. Some bombing efforts had gone awry, but on mission after mission positive results were reported against negligible enemy opposition.

Describing the day as "the most successful operation to date" by "the largest force of heavy bombers and fighters ever employed on a daylight bombing operation," Eighth Air Force Headquarters reported that "A total of 2218 tons—1668 of H.E. [High Explosive] and 550 of IB [Incendiary Bombs]—were dropped...Targets in the Leipzig area, Brunswick, and Bernberg [sic] were hit with excellent results; those at Gotha and Tutow through the overcast with unobserved results; Oschersleben and Helmstett were bombed as targets of opportunity with fair to good results; and other targets of opportunity were attacked with mainly unobserved results."

There was one passage in the operation summary that said it all: "Considering the depth of the penetration into the Reich, an outstanding feature of the operation is the small loss—21 bombers and four fighters. This was due in a large measure to the excellence of the fighter escort, but it is also apparent that the GAF was surprised and overwhelmed by the large force and its employment, particularly following the large scale RAF attack on Leipzig the previous night. E/A opposition, in view of the location of the targets, was not strong or aggressive and flak was no deterrent. Total tentative claims are 126-40-66, the fighters claiming 61-7-37 and the bombers 65-33-29." (These tallies referred to the totals claimed as destroyed, probably destroyed, and damaged.)

The Eighth's leaders kept the pressure up over the next five days. Spaatz sensed that he was closing in for the kill and took every opportunity to send the Eighth's bombers and fighters—and those of the smaller Fifteenth Air Force—on coordinated attacks against German aircraft factories and other key targets, including Schweinfurt. None of these was as successful as the raid of February 20th, but the net result was a definite turn of the tide.

On February 21st, the Eighth sent 861 bombers and 679 fighters back to the aircraft factory complex at Brunswick and against numerous Luftwaffe air depots and airdromes in central Germany. Bad weather prevented accurate bombing on most of these targets, but the battle of attrition continued. The Eighth lost 16 heavies and five fighters. The bombers claimed 19 enemy aircraft and the fighters 33. There was no letup that night, as RAF Bomber Command launched a large force against Stuttgart, site of other important aircraft plants.

On February 22nd, the Eighth sent 799 bombers and 659 fighters against the factory complexes in central Germany including aircraft plants at Oschersleben, Aschersleben, Halberstadt, and Bernburg. Unfortunately, adverse weather conditions over England forced a recall of the Second and Third Divisions, and in a battle reminiscent of the January 11th Oschersleben raid, the First Division pressed on to bomb its objectives against concentrated Luftwaffe opposition.

This was another bad day for the 303rd, which lost five ships: two from the 358th Squadron and three from the 360th. Thirty-eight First Division Fortresses fell, and 41 of the Eighth's bombers were lost in all, plus 11 fighters. But 99 B-17s got through to their clearly visible targets and bombed them with excellent results. Moreover, they claimed 34 German aircraft and the fighters 59.

That same day, the Fifteenth Air Force attacked a Messerschmitt factory at Obertaubling. Poor weather prevented good bombing, and the attackers lost 14 of 183 heavies as well as 11 fighters, but the latter's claims were impressive: 60 German fighters downed. Total aircraft claimed by the two air forces was a stunning 153.

The Eighth rested on February 23rd, but the Fifteenth Air Force sent a small force of 102 bombers to hit a ball bearing plant at Steyr, Austria, responsible for 15 percent of all German bearings. The plant was 20 percent destroyed.

February 24th saw coordinated efforts by the Eighth and Fifteenth Air Forces and RAF Bomber Command. The Eighth's plan was similar to that of February 20th, with a total of 809 bombers and 767 fighters sent. Five combat wings of the Third Division flew unescorted along the Baltic coast to hit aircraft plant complexes in northeastern Germany and Poland. Three wings of B-24s from the Second Division went after Me-110 plants at Gotha, and the First Division sent its B-17s to Schweinfurt. From the south, the Fifteenth sent a much smaller force back to Steyr to attack the Steyr-Daimler-Puch aircraft factory, and that night the RAF sent a large force of bombers to strike Schweinfurt.

The Fifteenth's formation bombed Steyr at a cost of 17 aircraft. The First Division force hit Schweinfurt at a cost of 11 B-17s while claiming 10 fighters. (The 303rd lost two B-17s on this attack.) The B-24s hit the Gotha plants accurately against strong enemy opposition, losing 33 ships while claiming 50 fighters. The northern force was forced to bomb Rostock through 10/10 clouds, but it lost only five bombers with the B-17s claiming 23 fighters. The Eighth's escorts claimed 38 more at a cost of 10, making this day one of the roughest yet, with 67 bombers lost. On the positive side, German losses were put at 121.

USSTAF and the RAF mounted another series of big raids on the 25th, the last day of "Big Week." Spaatz was briefed that the skies would be clear all over the Reich that day, and he chose to concentrate his forces against aircraft plants located in southern Germany. The Eighth sent 754 bombers and 899 fighters (including 139 P-51s from three Mustang groups) to hit factories in Augsburg, Regensburg, Stuttgart, and Fürth. In addition, the Fifteenth dispatched 176 heavies on an unescorted attack against an Me-109 parts factory at Regensburg.

The Fifteenth's bombers reached their objective despite fierce Luftwaffe opposition. They bombed it with excellent results, but lost 33 bombers. One hour later, 267 B-17s from the Third Division smashed the same Me-109 factory they had hit on August 17, 1943, claiming 13 enemy fighters for 12 bombers. The combined effort drastically reduced Me-109 output at the Regenburg factories for four months. Serious damage was also inflicted by the Eighth at the Messerschmitt facilities in Augsburg, where 196 First Division Fortresses struck, and at Stuttgart, where another 50 B-17s damaged a ball bearing plant. These two bomber forces lost 13 B-17s while claiming eight enemy aircraft.

In the meantime, the Second Division's B-24s bombed an Me-110 parts and assembly plant at Fürth with excellent results, claiming six enemy aircraft for a loss of two bombers. The Eighth's escorts claimed 26 aircraft against three, making a total of 53 German aircraft for 63 bombers and three fighters.

Bad weather brought "Big Week" to a close, ending the most sustained and massive aerial offensive the Eighth had ever launched against the Reich. At the time it was thought that the bombers had fatally crippled

German aircraft output, scoring the kind of knockout blow Eaker had hoped to achieve against the Schweinfurt ball bearing plants in 1943. In both cases, however, many machine tools vital to production were only superficially damaged, and Albert Speer, Nazi Minister of Armaments and Munitions, was quick to implement a program of plant dispersal that returned production to earlier levels after a few hard months of seriously reduced output.

There was, however, one vital war commodity that Speer could not make allowances for: experienced combat pilots. Luftwaffe personnel losses for this week will never be known with certainty because the Germans destroyed many Luftwaffe records at the end of the war. But American claims that 505 German aircraft had been shot down were a real sign that the Luftwaffe was suffering crippling losses in the ranks of its fighter pilots. After "Big Week" the Luftwaffe was never again able to pose a serious *strategic* threat to daylight bombing or to Operation *Overlord*, the invasion of Europe that was to follow in less than four months. The massive scale of daylight bombing operations inaugurated by "Big Week" grew ever larger in the months that followed, and with it, German losses.

March 6, 1944, was a key indicator of the direction that the war was taking. On that day the Americans finally got to Berlin in substantial strength. The Eighth dispatched 730 bombers and 801 fighters to "Big B," drawing forth a furious German response. Sixty nine heavy bombers were shot down, and there were 11 missing escorts, but 672 heavies unloaded on the German capital with the bombers claiming 97 German aircraft and the fighters 82. That it was the most costly mission the Eighth ever flew, and that the total number of German aircraft actually destroyed or damaged did not exceed 90, was beside the point. The Luftwaffe might be able to produce new airplanes, but it could not replace the veteran pilots killed in such fierce combat. The Americans could now replace both.

Not long after this, on March 19, 1944, Captain Don Gamble flew his last mission. Coming home, he "buzzed field while men shot flares from cockpit and radio room." He was "one of the last to land—after dark" and was, he wrote, "Happy to finish."

Unable to challenge the Eighth *en masse*, the Luftwaffe resorted to increasingly desperate tactics. *Sturmgruppen* of heavily armed and armored single-engine interceptors who were "escorted" by conventionally armed fighters were increasingly in evidence. They searched for gaps in the fighter escort cover, and for conditions where local air superiority could be achieved, but their efforts could not alter the outcome. By March 1944, P-51s equipped with twin 108-gallon drop tanks could range 850 miles into the Reich—beyond Germany all the way to places such as Poland, Czechoslovakia, and Austria. And VIII Fighter Command was well on the way to reequipping all its fighter groups with the P-51. By mid-1944, Mustangs made up over half of the Eighth's fighter escort force, and by January 1945 the only unit in VIII Fighter Command that did not fly the P-51 was the famous 56th Fighter Group, which kept the P-47 throughout its wartime service.

May 1944 saw the beginnings of the final blow to any hopes the Luftwaffe had of regaining air superiority through aircraft such as the Me-262, the world's first operational jet fighter. In that month the Eighth flew its first missions against the one real strategic jugular the Germans had: oil. Two days after D-Day, on June 8, 1944, Spaatz; formally made oil USSTAF's top strategic target. ("Tooey wants oil" is how Mel Schulstad remembers the call.) The Fifteenth Air Force was ordered to concentrate on the natural and synthetic oil resources the Nazis had in the Balkans, while the Eighth was to hit the enemy's many synthetic plants in central and eastern Germany. By the time the "oil campaign" ended in March 1945, the Luftwaffe had thousands of aircraft available, but no fuel to

Throughout the air war, the Hell's Angels remained one of the Eighth's premier heavy bomber groups, winning fame as the first bomb group to fly 300 missions and flying a total of 364 by war's end, a number greater than that of any other B-17 group in The Mighty Eighth. Here the Hell's Angels pass in review over the flag at Molesworth in the late stages of the war. (Photo courtesy National Archives (USAF Photo).

train new pilots and precious little fuel to fly and fight. The U.S. Eighth Army Air Force had achieved total victory: The air war had been won.

Throughout the air war, the Hell's Angels remained one of the Eighth's premier heavy bomber groups, building on the record they had achieved in the early days and during the crisis period when Hullar's crew flew. Chuck Marson remained with the Group until he got his last mission in on March 19, 1944, and then went home. Mac McCormick became a Lt. Colonel and stayed with the 303rd to the end of the fighting as Group Bombardier. Hullar stayed on, too, making Major and serving as Group Operations Officer to the very end of the 303rd's existence in the ETO. He flew nine more missions, including at least one trip to "Big B." The 303rd won fame as the first heavy bomber group in the Eighth to complete 300 missions, and by the end, the Group's mission tally was 364, a total greater than any other B-17 group in the Eighth Air Force.

All during the conflict, the 303rd's bomber crews faced many of the risks that Hullar's crew did, since the nature of the air war was always such that individual groups never knew how bad a mission would be for them. For despite a group's experience and its ability to fly good formation, there was never a way to eliminate the role of fate. Too many things

The smile of a survivor. Pete Fullem, left, with two friends relaxing while off duty during the balance of his military service with the Third Air Force in Florida. (Photo courtesy Mrs. Rita Dispoto, Executrix of the estate of Joseph J. Fullem.)

could go wrong—a recall message not received, a missed rendezvous with the fighter escort, location in the bomber stream, navigational error, and a myriad of other factors—any one of which could decimate a group on a mission from which others returned untouched.

The 303rd had two such days after Oschersleben. There was the one immortalized in "Fortresses under Fire," the magnificent mural by Keith Ferris that graces the National Air and Space Museum's World War II gallery in Washington, D.C.—August 15, 1944, when a *Sturmgruppe* snuck up on the Group over Wiesbaden and shot down nine Fortresses in short order. Then there was the bloody trip to Magdeburg on September 28, 1944, when another force of German fighters attacked the Group and sent 11 of its B-17s spinning down out of the sky.

Still, Bob Hullar got it right when he wrote a brief note to Pete Fullem in the copy of *The First 300* that he sent him and the other members of his old bomber crew: "You fellows helped to make the rough part of this story."

Epilogue

Old Soldiers Who Don't Fade Away

WITH THE END of hostilities in Europe on May 7, 1945, the 303rd was soon prepared for other missions. The very next day, in the midst of VE Day celebrations, Bob Hullar was appointed Group Operations Officer, replacing Major Mel Schulstad, and for a while it appeared that the Group would be deployed to the Pacific to take part in the final strategic bombing campaign against Japan. However, this proved unnecessary, and while the Group's bombers flew "Continental Express" missions over Europe that month to show the ground echelon some of the results of its bombing missions against Germany, others in the Group, including Bob Hullar, prepared for a redeployment of the 303rd to the North African Division of the Air Transport Command in Casablanca. The transfer of personnel began on May 30, 1945; its purpose was to have the Group take part in "Green Project" ferry missions transporting ETO veterans back to the United States. Here too, however, the powers that be later determined that the 303rd would not be needed, and the Group soon found itself without a mission. Bob Hullar was present when the Hell's Angels 303rd Bombardment Group (H) was formally deactivated in Casablanca on July 25, 1945.

During the Cold War the Hell's Angels were reactivated as the 303rd Bombardment Wing, Medium, based at Davis Montham Air Force Base, Arizona from 1951 through 1964. The 303rd flew Boeing B-29 bombers and KB-29 aerial refueling tankers from 1951 through 1953, and later transitioned to Boeing B-47 jet bombers and Boeing KC-97 aerial refueling tankers. The Group was deactivated again until, in 1986, the 303rd found a new military incarnation in the 303rd Tactical Missile Wing which was based in Molesworth just like the original Hell's Angels. The new 303rd was responsible for maintenance and readiness of the controversial Ground Launched Cruise Missiles which were deployed to Europe at the end of the Cold War, and Molesworth became the scene of many, well publicized antimilitary protests. The new 303rd soldiered on, however, until the Cold War was won in 1989 when the 303rd was deactivated again.

Today, there is no 303rd in the U.S. Air Force's table of organization, but the military men and women of the Joint Analysis Center, which has been stationed at Molesworth since 1991, are very much aware that the Base they occupy is hallowed ground. The JAC has named its new operations center the "Might in Flight" building, after the 303rd's motto, and the exterior and interior of the building are replete with reminders of what took place there between 1942 and 1945. The old, large main hanger which dates from the war has a large "Triangle C" B-17 tail insignia on its doors. A standing invitation has been issued to every 303rd veteran to visit the facility, and it is very obvious to anyone touring Molesworth today that the JAC is true to one of its own mottoes, "Molesworth Remembers."

The men of the 303rd Association will always remember the comrades-in-arms they served with, and a moving part of each reunion is the reading of a list of those who have passed away since the last gathering. The

Col. Robert J. Hullar while serving as Base Commander at Pleiku Air Force Base, Vietnam, 1966–67. (Photo Courtesy Mrs. Jean J. Hullar.)

deceased on Hullar's crew now include Hullar himself, Elmer Brown, Dale Rice, Pete Fullem, and Chuck Marson.

Bob Hullar stayed in the Air Force after the war, married in 1946, and had three children. Many of his assignments were in staff and quasi-diplomatic posts, and he was also the base commander of a number of Air Force facilities. These included England Air Force Base in Alexandria, Louisiana in 1962,* and Pleiku Air Force Base, Vietnam for 14 months in 1966–77. He retired from the Air Force in 1970, earned a master's degree, and was active in private business until his death of bone cancer in October of 1984. Although the writer never met him, everything I have learned about Bob Hullar from others convinces me that he perfectly embodied the military ideal of "an officer and a gentlemen" until the end of his life.

Although he was tempted to stay in the military, Elmer Brown returned to civilian life and realized his dream of starting a family with Peggyann. They had two children, Larry and Bonnie, in the late 1940s. Elmer returned to the design and construction field that he had worked in as a drafter before the war, earning a BS degree in civil engineering in 1954. He spent his career as a government employee, and was involved in many military construction projects in the Pacific. In 1964 he was working as a government project engineer at a McDonnell Douglas facility in Huntington Beach, California, a position he occupied until his retirement in 1973. He remained in Southern California with Peggyann, where for the last two decades of his life he fought a long and ultimately losing battle against Parkinson's disease, finally succumbing on August 3, 1997. He was very active and involved in the writer's work on the original edition of this book, since he had always wanted a book written about his Eighth Air Force experiences. I take great satisfaction in learning from his son Larry that Elmer's interest in this work did much to sustain him in his last years.

Dale Rice also returned to civilian life after the war, marrying, raising a family, and ultimately settling in Rahway, New Jersey. He made up for the lack of educational opportunity that he felt so keenly before entering the service by obtaining a high school equivalency certificate while working for a Rahway-based pharmaceutical company, and then attending night school at Rutgers University from 1956 to 1963, where he earned a B.S. in Business Administration. Not content with this, Rice then became an educator himself, starting a long career at Rahway Junior High School, during which time he continued to study at night, earning a Master's

---

*Unbeknownst to the writer, I was part of Hullar's command responsibility at the time, since I was living on the same Base as a 13-year-old military dependent!

"The Best Years of Our Lives." Elmer Brown realized his dream of starting a family with Peggyann, and he went on to a successful postwar career as a civil engineer, working all over the world with the U.S. Army Corps of Engineers. This photo shows the Browns with their two children, Larry and Bonnie, circa 1948. (Photo courtesy Elmer L. Brown, Jr.)

Degree in Behavioral Science and accumulating many additional credits in a Master's Plus Program. He loved children and teaching and was, I am told, greatly beloved by his family and students. He died shortly before I first began research for this book in 1985.

Pete Fullem likewise returned to New Jersey after the war, where he periodically got together with Dale Rice. He received training as a watchmaker in 1946, opened his own jewelry store in Jersey City, and died in 1980. Little of his postwar life is known beyond the bare documentary record, which includes some personal photographs and newspaper clippings given to me by one of his relatives when I first began research into Hullar's crew. What I did learn about him convinced me that here was an unassuming man whose deeds should be remembered, which is why I have tried to give his role in Hullar's crew the prominence that it has in this book.

Chuck Marson moved from the backwoods of Maine to settle in Roswell, New Mexico after the war, and he was active in the New Mexico National Guard until he could no longer pass the physical. His widow Sarah has said that he was "one of a kind, a real character. He was tough as nails on the outside and soft as butter on the inside. Kids, especially teenagers, adored him." He died of a heart attack in 1980, during the thirty-fifth year of their marriage.

Norman Sampson lives a quiet but happy retirement in the small town of Ozark, Missouri with Anita, his wife of over fifty years. A deeply religious man, he is quick to say, "I never felt I was any kind of hero," but those knowing the missions he flew can easily draw a very different conclusion.

"Mac" McCormick, Hullar's ace bombardier, shuns the limelight even more. A retired chiropractic physician, he now lives in a small town in Idaho.

What is known of some of the other men who flew with Hullar's crew, some now deceased and others very much alive, follows:

Flamboyant "Woodie" Woddrop was killed in an aircraft accident involving a B-29 while he was in transition training for deployment to the Pacific. The B-29's Wright R-3350 engines were plagued with cooling problems when first mated to the aircraft, and there were frequent engine fires. Woddrop's B-29 experienced one during a takeoff and in exiting the

Crew Reunion. L-R, Merlin Miller, Bud Klint, and Norman Sampson pose for the camera during the 1986 303rd Reunion in Fort Worth, Texas. (Photo Courtesy Wilbur Klint.)

aircraft through the nosewheel door "Woodie" somehow managed to run into the spinning propeller on one of the inboard engines. He suffered massive injuries and his friend, David Shelhamer, learned later that Woddrop died on the operating table of the base hospital.

After the war Shelhamer moved from his native Chicago to Los Angeles, where he raised a family of four children with his first wife and continued his prewar calling as a professional photographer. Always assertive, nothing stopped him when he met his second wife, Lorraine, about a year after the end of his first marriage, and decided instantly to marry her; the two enjoyed many happy years together. When he developed incurable cancer in the late 1980s David elected to forego "heroic measures" and wait out his time in a hospice. The writer exchanged a number of personal tape recordings with him during this period, and I can say with conviction that David Shelhamer faced death in the end as bravely as on any of the missions he flew.

Grover Henderson, who was Woddrop's best friend during the war, returned to his home in Greenwood, South Carolina but maintained a lifelong interest in the Air Force and flying. He was a member of the Air Force Reserve, and in 1955 was recalled to flying status in the Ready Reserve, where he served first with the 77th Troop Carrier Squadron at Donaldson Air Force Base in Greenville, South Carolina and then as operations officer of the Search and Rescue division of the 14th Air Force at Headquarters, Robins Air Force Base, Georgia. He was commander of the 9997th Air Force Reserve Squadron in his hometown of Greenwood until he retired with 20 years of active and reserve duty.

Grover also greatly enjoyed civilian flying. He held a commercial pilot's license and for many years made charter flights for a number of businesses. He also was "on call" as a copilot for a major textile manufacturer. Flying and association with other pilots always gave him pleasure, and he was a most active participant in the 303rd Association as long as his health permitted.

After his death of a heart attack in November 1993, Grover Henderson's name was added to the Memorial to South Carolina's Outstanding Pilots in Orangeburg, South Carolina.

Tail gunner Charlie Baggs, whom Grover Henderson always felt was the second "soldier of fortune" on Woddrop's crew, did not live long after the shooting stopped. Quite by accident George Hoyt ran into Baggs on a street in Atlanta shortly after the war's end, while Hoyt was attending college. The friendly conversation between the two quickly turned grim when Baggs informed Hoyt that a VA doctor had just told him he didn't have long to live because of his alcohol problem. Hoyt was depressed for days afterwards by the memory of his old friend slowly walking down the street in the opposite direction after their discussion. In later years Grover Henderson devoted considerable effort to searching for Baggs in his native Colquitt County, to no avail.

Paul Scoggins, the 427th Sqdn. lead navigator who flew a number of missions with Hullar's crew, returned to the USA in the Spring of 1944 and spent the balance of his service at San Marcos, Texas as a training officer. After the war he earned a B.S. degree in business administration, married, and spent 35 years working for Ralston Purina in a variety of management and sales positions in Texas and Louisiana, followed by five additional years as shop supervisor for a chain of grocery and feed stores. He and his wife, Mary, live in DeRidder, Louisiana, near one of their three daughters. They have 12 grandchildren, and four great grandchildren.

At 77, Scoggins remains active and optimistic about life, as befits a man who has taught Dale Carnegie courses part-time for over thirty years, and who still does so today. Two comments about his combat days nicely sum up both his personal philosophy and his reflections on the war:

"Pop Hamilton, our radio gunner, was always the nervous, chain smoker who worried about everything. He would say, 'They say a 4% loss on each mission is our average. Lieutenant, 25 times 4% is 100%, so we aren't coming back.' My answer was always, 'Pop, every time we go out, we have a 96% chance of coming back!' "

"I've said that we fought a 'Gentleman's war' in the air force. We were in grave danger while we were flying, but we came back to a warm bed (IF we made it, and I did). I was at the right place at the right time. I wouldn't want to relive the experiences, but I have a good feeling for having been through them."

Howard "Gene" Hernan, the experienced 359th Squadron top turret gunner on Claude Campbell's crew, returned to his native Creve Coeur, Illinois after the war, and spent his working life as a loyal "union man." He died before the first edition of this book was published, but his widow, now also deceased, told me that Gene absolutely hated war, and reserved special anger for the politicians who caused young men to have to fight and the "little people" of this world to suffer in it. To the end of his days, however, Gene was proud of his own wartime service, and he was especially proud to have been a member of Campbell's crew.

Eddie Deerfield, the 360th Squadron radioman, likewise has enormous pride in having been a part of Robert Cogswell's crew, so much so that he erected a memorial plaque to his comrades in the gardens outside The

Mighty Eighth Air Force Heritage Museum.* Despite an impressive post-war career as a Foreign Service diplomat who served for more than two decades in the U.S. Information Agency; a parallel career as an officer in the U.S. Army Reserve, including active duty in Korea during the Korean War; and the publication of an historical novel, it is easy to conclude that Deerfield's WWII service as the radioman of *Iza Vailable* ranks higher on his personal scale of accomplishments.

Two other contributors to this book, members of Don Gamble's crew, elected to stay in the service after the war. Bill McSween became a navigator instructor, ending his career as a Lt. Col. and retiring with his wife, Bobbie, to a farm near Shreveport, Louisiana. They had three children. Few things motivated me to see this book through to its initial conclusion more than the opening line in a letter McSween sent me—"I was beginning to wonder if you had forgotten me"—when, early in my research, there was a gap in our communications.

Ralph Coburn decided to make the Air Force a career too, despite the close calls he faced on his missions. After returning to the States he also decided there was a part of his Molesworth experience he could not live without. In 1946 his English girlfriend, Beryl, traveled to the U.S. after being demobilized from the WAAFs. They remained married for fifty years, minus two months, and had five children. After his retirement from the Air Force in 1962, Coburn became an educator. He was Dean of Students at Orlando Junior College in Florida, and later a counselor at an adult technical school until his final retirement.

Together with their wives, both McSween and Coburn were frequent attendees at 303rd reunions until both men passed away earlier this decade. McSween died in 1995 and Coburn six months later, in 1996.

Happily, Don Gamble, their outstanding pilot, continues to enjoy good health, and is invariably at the Group's reunions where he can be seen with Dave Rogan, his old Instructor Pilot, and Charles Schmeltzer, his right waist gunner. Looking back on their collective experience, Don Gamble feels, "We were blessed not to have any of our crew hurt."

Also to be seen at 303rd reunions is Darrell Gust, who still feels blessed to have survived his last mission, to Oschersleben, that terrible day in January 1944. Long retired, after eight and a half years service in the Air Force, and almost 30 years as the Director of Personnel and Labor Relations for a Wisconsin Power Cooperative, Gust keeps himself busy nowadays as a school bus driver for the children in his Wisconsin community.

---

*See pages 415-416.

On John Henderson's crew Ed Ruppel passed away in the early 1990s after a long illness. Ruppel's war experiences had a deep and lasting impact on him, and caused him to refuse interviews with many people whom he felt were only interested in learning, on a superficial level, what he knew about the mission on which Forrest Vosler won the Medal of Honor. From the first of four extraordinarily intense interviews I had with Ruppel, however, it was clear that this man had a much larger story to tell. I will always have a feeling of privilege in the knowledge that he entrusted me to deliver it to a wider audience.

Forrest Vosler died in 1992. After the war he became a member of the original Board of Directors of the Air Force Association, and pursued a long career of service to others as a Veterans Administration counselor. When I interviewed him for the first edition of this book in 1986, 43 years after the fact he still showed the self-confidence and drive which marked him as a man capable of winning the Nation's highest award for military valor. But there was also an indefinable personal quality, and dignity, about him which in my opinion marked him as a genuine hero as much for who he was as for what he did.

Medal of Honor Winner Forrest Vosler and crewmate Bill Simpkins pose together during the Gala B-17 50th Anniversary Celebration in Seattle in 1985. (Photo Courtesy William H. Simpkins.)

Of John Henderson's crew only Bill Simpkins is known to survive today, and it is something of a miracle that he is here at all. On February 24, 1944 his B-17 was shot down by flak over Schweinfurt, and while everyone parachuted safely, Simpkins faced many dire threats to his life while in captivity.

Simpkins' ordeal began on the way to interrogation in Frankfurt. While being shepherded by a single German guard through the main city Bahnhof (railroad station), his group of POWs were pursued by a mob of enraged German civilians who had just lynched some other U.S. airmen. Simpkins saw their bodies hanging from the main overhead girders in the station. He and the other Americans owed their lives to the solitary German soldier, who rushed them to a hiding place underneath the station to escape the mob.

Simpkins was shipped, together with Ed Miller, the crew's new ball turret gunner, to Stalag 6 on the border of Lithuania in East Prussia. When the Russians began to cut off this part of Germany late in the war, Simpkins and the other Stalag 6 POWs were forced to stand in the hold of a ship during a perilous journey from Königsberg to Stettin, where they arrived in the middle of an Allied air attack. Angry civilians took their belongings and made the POWs run gantlet of bayonets and dogs. The POWs were marched to Stalag 4, but this was only an interim stopping place. When the Russians again threatened, Simpkins and the others were forced, in an atmosphere rife with rumors that orders had gone out to kill all POWs, to make another long and difficult march in the snow. Their journey temporarily ended in Nuremberg, where the forlorn group was strafed by P-47s who killed and wounded some of their number. The POWs were then marched across the Danube River at Weiner Neustadt and put on the road once again until they ended up at Moosberg, Germany where the men were finally liberated by a U.S. Army armored division.

During their long and terrible trek Simpkins and his comrades traded cigarettes for raw meat, and Ed Miller credits Bill Simpkins with saving his life when, one evening, Simpkins literally carried Miller on his back to their resting place for the night. (Those POWs who could not keep up died in the snow.)

After the war Bill Simpkins returned to his native New Jersey where he held a variety of jobs until settling on a career with Lenox China in the early 1950s. He worked as a supervisor of decorating, was active in research and development methodology, and helped establish five factories. He and his wife, Edie, live in the small town of Cologne, New Jersey and are longtime members of the 303rd Association. When, during one of my interviews, I asked how he was able to fly all those missions and then endure the POW experience, he reminded me that he was only 19 years old at the time and said simply, "You had to be young."

POW experiences like those of Bill Simpkins cry out for a realistic portrayal in writing by someone who was there, and Carl J. Fyler has done just that in his book, *Staying Alive*. It should be required reading for anyone who harbors "Hogan's Heroes" type illusions about what being an American POW in the ETO was really like. After the war, Fyler returned to his native Topeka, Kansas where he married, became a Doctor of Dentistry, and practiced his profession until his retirement. A past President of the 303rd Association, he has never stopped being a fierce advocate of veterans' rights and benefits, and has given selflessly not only to the goal of obtaining a posthumous Medal of Honor for Joseph Sawicki, but to many other deserving cases as well.

The Reverend Charles W. Spencer, the other Kansas veteran who figures large in this book, lived a life characterized by service to others right up to the date of death on April 16, 1998. After graduating from the Baptist Theological Seminary, Kansas City, Kansas, in 1951, he served as Chaplain to the Kansas Soldiers' Home in Fort Dodge from 1953 to 1982, and continued to assist in Christian ministry at the First Baptist Church in Dodge City as long as his health permitted. According to Jeanne, his wife of 57 years, he always preferred devotion to his ministry to dwelling on his war injuries. If you ask her today what kind of man he was, she will tell you that being with him "was like living a miracle every day. He touched many lives."

Bill Fort, Spencer's pilot on the fateful Bremen mission of November 26, 1943, still lives with his wife, Rachael, in a Jacksonville, Florida suburb. Like Charles Spencer, Fort is also a man of few words when it comes to the war. He and Spencer met only once after the November 26 mission and that briefly, and privately, during the 1990 303rd Reunion in Norfolk, Virginia. It is enough to know that their paths crossed again after the war, without trying to delve into what was discussed.

Until his death in 1990, Jack Hendry lived in the same neighborhood as Bill Fort, just a block or two away. Though they flew in the 358th Squadron at the same time, neither knew the other until after the war, when they got acquainted as neighbors and became friends. Jack Hendry married in 1947 and remained in the service until 1950, when he was honorably discharged. He graduated from the University of Florida in 1954 with a Bachelor's degree in Building Construction. Like Elmer Brown he had two children, and also like Brown he traveled extensively overseas doing construction work for the U.S. Army Corps of Engineers. However, while working in Europe he refused a grand tour of Berlin at Corps' expense. His wife, Gloria, believes he did this not because of any lingering animosity towards the Germans, but because "he did not want to see some of the bombing damage still visible there." A great amateur golfer in retirement, he was clearly an even greater father. When I contacted Mrs. Hendry for information about Jack for this Epilogue, I also received a letter from his son, Michael.

Of Jack Hendry's brave bomber crew, only Howard Abney remains alive today. John Doherty, who replaced Abney in the *War Bride's* tail gunner's position, died shortly before the publication of the first edition of this book in 1989. Jim Brown, the crew's radioman and by common consensus one of the crew's real leaders, is buried in his native Georgia but definitely not forgotten. Brown's niece, Mrs. Henrietta S. Duke, traveled to Savannah for the most recent 303rd Reunion, where she met Abney to learn more about her uncle.

Ed Snyder, the Hullar crew's squadron commander, remained in the Air Force after the war, retiring in 1970 as a full Colonel. Like Hullar his career included a number of significant headquarters assignments, in the Far East, Europe, and the Pentagon. The one big surprise to the writer when I interviewed Ed at his home in Tacoma, Washington, was discovering that he was also a combat qualified *fighter* pilot in F-101s, and that he had logged time in F-106 interceptors as well. Somehow I had expected that he would stick with bombers, but on the other hand...Snyder and his wife, Dot, are frequent attendees at 303rd reunions.

Mel Schulstad, who also lives in the greater Seattle area, is another who stayed in the Air Force after the war. Ironically, in the early 1950s he was assigned to Germany as a staff officer implementing a Military Assistance Plan to rebuild the Luftwaffe and West Germany's military-industrial infrastructure. He retired as a full Colonel in the 1960s, but his son, Jon M. Schulstad, and his grandson, Jon Jeffrey Schulstad, have both followed in his footsteps as Air Force officers. His son is retired but his grandson is on active duty as a Captain, flying the closest thing to the B-17 in the Air Force inventory—the four-engined, straight-winged, turboprop C-130 "Hercules."

Today Mel Schulstad is still active in a significant second career. Twenty-six years ago he co-founded the National Association of Alcoholism and Drug Abuse Counselors, a professional organization which now has over 17,000 members. He is presently helping to write a history of the organization.

George Hoyt, Hullar's radioman, also helped Germany after the war, and he concluded a separate peace with the Germans in a very special way. During the Berlin Airlift Hoyt returned to active duty on a volunteer basis, traveling to Celle, Germany in the British Sector. There he assisted in the establishment of an air traffic control system to guide cargo flights going to and from Berlin in one of the three air corridors to the City. While in Celle he met Elfriede, a young German woman who was a refugee from the East. They were married by the Mayor of Frankfurt am Main while Hoyt was still on active duty.

Upon his return to Georgia Hoyt pursued a successful banking career, and he has a second career as part of the management team at the Stone Mountain Park outside Atlanta. He and Elfriede have two sons and live in the Atlanta suburb of Lilburn.

Merlin Miller, Hullar's indomitable tail gunner, returned to the United States after his tour was over and spent a number of months in a B-25 training unit. However, he soon tired of this and volunteered for combat in the Pacific. He flew 13 combat missions as the tail gunner with one of the most aggressive, low level B-25 "skip-bombing" and strafing units in the whole Pacific theatre—the "Air Apaches" of the 345th Bomb Group. Hepatitis, not hostile fire, finally knocked Miller out of action, and he returned home to Sullivan County, Indiana, where he married a high school classmate named Marjorie. He and his wife spent many years in the south Chicago suburb of Chicago Heights, where Miller worked as a lumber salesman. Following his retirement they returned to their roots in Sullivan County, where the Millers now reside in the tiny town of Dugger, Indiana. They have one son who works as a corporate attorney in Denver, Colorado.

Merlin wields an oil paintbrush almost as effectively as the twin fifties he used during the war, and he has produced a number of excellent B-17 paintings, though his talents are by no means confined to aviation art. He is also an excellent public speaker, and frequently gives talks about the Eighth Air Force to students. Those wishing to hear him tell what the air war was like in both the ETO and the Pacific should make arrangements to borrow the tape recording of a joint speech—entitled "Eighth Air Force History: Living It and Writing It"—that he and the writer gave on October 18, 1995 before a large audience at The U.S. Air Force Museum, Wright-Paterson Air Force Base near Dayton, Ohio.

The need to create tangible reminders of what was accomplished in the aerial battlefield over Occupied Europe during World War II *and* what the victory cost in human terms, is one which many Eighth Air Force veterans still work daily to achieve. No one has worked harder at this task than Lewis E. Lyle, the 303rd's outstanding combat leader. Following his retirement as a Major General in the Air Force, Lyle criss-crossed the country by auto countless times to create interest in and raise financial support for a "people's museum" to tell the story of the Eighth Air Force in human terms. His efforts have borne fruit in the last two years with the opening and impressive growth of The Mighty Eighth Air Force Heritage Museum, located off Interstate 95 in Pooler, Georgia near Savannah. From its startling "Mission Experience" audio-visual exhibit to its touching exhibits of personal memorabilia, and its sometimes shocking photographs, the Museum succeeds admirably in showing "the way it was."

It may come as a surprise to some to learn that there are many smaller but no less significant "people's memorials" located in out-of-the-way fields and odd pieces of real estate all over Europe, from Poland and the Czech Republic to France, the Netherlands, Belgium, Britain, and even Germany. They mark the spots where Allied bombers and fighters crashed, and men were killed. In some cases they are erected by Europeans who have rediscovered the crash sites, and have spared no effort to find out whose life it was that ended amidst the bits of metal, broken Plexiglas, rubber, and even personal effects that may be found with a careful raking of the ground. In others they reflect the wishes of old resistance fighters to mark the spot where they arrived on the scene and helped Allied airmen before the Germans arrived. In all cases they show

Crew Memorial. Donald Dinwiddie, radioman, (Left) and James F. Fowler pilot (Right) pose on each side of Maurice Lamendin, Belgian Resistance Fighter, before the monument the Belgians dedicated on June 27, 1987 to honor Fowler's crew near the January 29, 1944 crash site of their B-17, G.I. Sheets, *in Soire-St-Géry, Belgium.* *(Photo Courtesy James F. Fowler.)*

the strong desire of people now free to honor the memory of those who died helping to make them so.

There are many such memorials honoring 303rd Bomb Group crews, including one erected and dedicated by the townspeople of Soire-St-Géry, Belgium on June 27, 1987 to commemorate the place where Lt. James Fowler's *G.I. Sheets* crashed on January 29, 1944 and Sgt. Miller O. Jackson died. Surviving members of the crew and their families attended, including Jim Fowler and his copilot, Barney Rawlings. After a lifelong career as an airline pilot after the war, Rawlings was moved to write his own book about their crew, entitled *Off We Went (Into the Wild Blue Yonder)*.

Bud Klint has been active in placing a number of memorials in memory of the 303rd at various locations in the United States, but these activities occurred long after his return to civilian life. Back in America after the war he met and married his wife, Mary, resumed his job at the candy company in 1946, and followed its fortunes through a number of corporate mergers and position changes until he moved to Fort Worth, Texas, where he was Director of Marketing and Sales for the King Candy Company. Klint left the candy business in 1972 in order to join Tandy Corporation, where he worked in the advertising department as Director of Newspaper Media until his retirement in 1981.

All during these years, Bud Klint's priorities were the same as those of the other veterans and fathers of his generation—earn an income, raise his children (the Klints have three and now have eight grandchildren), get ahead in his career—but in all this time Klint's World War II experiences were "always there, in the background." So when Bud Klint "finally started to live" after his retirement, his interests returned to that period in his life when he wondered if he would live at all. He became an active member of the 303rd Association, of which he is a past president;* of The Eighth Air Force Historical Society; and last but not least, of a special veterans group whose membership is necessarily limited: The Second Schweinfurt Memorial Association, or SSMA.

Bud Klint's efforts to ensure that those who died in the war are remembered found its ultimate expression in June of 1998, when he headed up a small contingent of SSMA veterans and family members on his last mission to Schweinfurt. They were there at the invitation of Georg Schäfer, who heads up a German veterans' organization of the *Luftwaffenhelfer* who manned the City's flak guns against the Americans throughout the war. There, the two groups of former enemies jointly dedicated an extraordinary memorial, crafted by one of the *Luftwaffenhelfer,* to remember the

*"This Bud's for you" was his slogan at the time.

Old enemies united in friendship. Bud Klint, (Left) and George Schäfer pose before the German-American Memorial in Schweinfurt during dedication ceremonies on June 16, 1998. The monument was created by G. Hubert Neidhard, one of the Luftwaffenhefler *who manned the flak guns against the American bombers. The inscription reads, Left, (Author's Translation from the German): TO THE MEMORY OF THE MEN, WOMEN AND CHILDREN, AND OF THE MEMBERS OF THE U.S. EIGHTH AIR FORCE, AND OF THE GERMAN LUFTWAFFE WHO DIED DURING THE AERIAL ATTACKS AGAINST SCHWEINFURT IN THE YEARS 1943 THROUGH 1945. The inscription reads, Right, IN MEMORY OF CITIZENS OF SCHWEINFURT AND AIRMEN OF THE 8TH U.S. AIR FORCE AND THE GERMAN LUFTWAFFE WHO LOST THEIR LIVES IN MISSION 115, OCTOBER 14, 1943, KNOWN TO THOSE WHO WERE THERE AS BLACK THURSDAY. An inscription in both German and English at the bottom of the Memorial reads, DEDICATED BY SOME WHO WITNESSED THE TRAGEDY OF WAR, NOW UNITED IN FRIENDSHIP AND THE HOPE FOR LASTING PEACE AMONG ALL PEOPLE. 16 JUNE 1998. (Photo Courtesy Wilbur Klint.)*

fallen on both sides during the many bombing raids against the City, and on "Black Thursday" in particular.

The occasion was one of utmost solemnity and celebration, with wonderful speeches delivered by the Mayor of Schweinfurt, the American Counsul from Munich, Georg Schäfer, and Bud Klint. Hundreds attended and the event received extensive coverage in both the electronic and print media. This was as it should be, for all men and women should take note whenever old enemies reconcile with the wish that there be no more war.

Appendix I

Honor Roll of 303rd Bomb Group (H) Personnel Killed, Wounded Or Missing In Action During The Hullar Crew's Tour August 12, 1943 - February 20, 1944

Key to Abbreviations

P = Pilot
CP = Copilot
NAV = Navigator
BOM = Bombardier
ENG = Engineer
RAD = Radioman
BTG = Ball Turret Gunner
LWG = Left Waist Gunner
RWG = Right Waist Gunner
TG = Tail Gunner
PHOT = Photographer
PASS = Passenger
EVD = Evaded Capture
KIA = Killed in Action
K.I. Cr. = Killed in Crash
POW = Prisoner of War
REP = Repatriated
MACR = Missing Air Crew Report

8/12/43 Gelsenkirchen, Ger.
B-17F 42-29640 - *Old Ironsides* - A/C Mis. 16 - 359th BS
MACR 256 - Crashed Wuppertal, Ger.

Rank	Name	Position	Fate	Mission #
1Lt	A.H. Pentz	P	POW	16
2Lt	T.E. Mulligan	CP	POW	20
1Lt	B.B. Street	NAV	KIA	21
2Lt	R.C. Philpit	BOM	POW/REPR	15
T/Sgt	J.A. Dougherty	ENG	POW	13
S/Sgt	R.L. Murray	RAD	POW	16
T/Sgt	H.L. Edwards	BTG	POW	11
S/Sgt	R.A. Horvath	LWG	POW	15
S/Sgt	G.E. Howard	RWG	POW	16
S/Sgt	E.T. Knudsen	TG	POW	14

8/19/43 Gilze-Rijen, NL
B-17F 42-3192 - *G for George* - A/C Mis. 12 - 358th BS
MACR 284 - Crashed near Odsterhout, NL

Rank	Name	Position	Fate	Mission #
1Lt	J.S. Nix*	P	KIA	24
2Lt	D.A. Shebeck*	CP	KIA	5
2Lt	D.M.Curo	NAV	POW	3
1Lt	R.K. Solverson	BOM	KIA	23
T/Sgt	F.G. Krajacic*	ENG	KIA	5
T/Sgt	C.O. Brooke	RAD	POW	8
S/Sgt.	F.F.Perez	BTG	POW	3
S/Sgt	G.W. Buck	LWG	POW	3
S/Sgt	J. Gross	RWG	POW	3
S/Sgt	F.C. Boyd	TG	POW	2
1Lt	L.T. Moffatt**	PASS	POW	1

* Buried in Netherlands American Cemetery, Martgratten, Netherlands.
** from 359th Sqdn.

8/19/43 Gilze-Rijen, NL
B-17F 42-5392 - *Stric Nine* - A/C Mis. 21 - 427 BS
MACR 285 - Ditched near Dutch Coast

Rank	Name	Position	Fate	Mission #
2Lt	L.H. Quillen	P	KIA	1
2Lt	J.R. Homan	CP	KIA	1
2Lt	B.W. Colby, Sr.	NAV	KIA	2
2Lt	W.N. Irish	BOM	KIA	1
T/Sgt	E.F. Richter	ENG	KIA	1
S/Sgt	DiCosmo	RAD	KIA	1
S/Sgt.	E.O. Price	BTG	POW	2
S/Sgt	J.H. Brown	LWG	POW	1
S/Sgt	P.W. Abernathy	RWG	POW	3
S/Sgt	S.K. Sauer	TG	POW	1

8/27/43 Watten, Fr.
B-17F 42-29754 - *Shangrila Lil* - A/C Mis. 12 - 360 BS
MACR 405 - Crashed St. Omer, Fr.

Rank	Name	Position	Fate	Mission #
2Lt	C.W. Crockett	P	POW	6
2Lt	M.A. Barsam, Jr*	CP	KIA	5
2Lt	I. Millne	NAV	POW	7
2Lt	W.J. Cramsie	BOM	KIA	6
T/Sgt	M. Tepper	ENG	KIA	6
T/Sgt	A.D. Jaynes	RAD	POW	7
S/Sgt	E.F. Williams	BTG	POW	6
S/Sgt	E.D. Homer	LWG	POW	6
S/Sgt	B. Clarke, Jr.	RWG	POW	6
S/Sgt	J.C. Burke	TG	POW	6

*Buried in Normandy American Cemetery near Coleville-sur-Mer, France.

8/31/43 Amiens, Fr.
B-17F 42-29635 - *Augerhead* - A/C Mis. 18 - 358 BS
MACR 470 - Crashed near Abbeville, Fr.

Rank	Name	Position	Fate	Mission #
1Lt	W.J. Monahan	P	POW	13
1Lt	L.M. Benepe	CP	POW	6
2Lt	W.P. Maher	NAV	EVD	12
2Lt	W. Hargrove	BOM	EVD	13
S/Sgt	W. Gasser	ENG	POW	13
S/Sgt	F. Kimotek	RAD	EVD	12
S/Sgt	A.R. Buinicky	BTG	POW	13
Sgt	V.E. Olson	RWG	POW	2
S/Sgt	J.H. Comer, Jr.	LWG	POW	13
S/Sgt	D. Miller*	TG	KIA	13
Sgt	V.B. Pryor	PHOT	EVD	2

*Buried in Normandy American Cemetery near Coleville-sur-Mer, France.

10/2/43 Emden, Ger.
B-17F 42-5260 - *Yardbird II* - A/C Mis. 43 - 360 BS
MACR 738 - Exploded over North Sea

Rank	Name	Position	Fate	Mission #
1Lt	P.S. Tippet	P	KIA	18
2Lt	L.H. Nemitz	CP	KIA	3
2Lt	E.E. Mims	NAV	KIA	3
1Lt	B. Rice	BOM	KIA	24
T/Sgt	O.H. Smith	ENG	KIA	3
T/Sgt	T.L. Richardson	RAD	KIA	3
S/Sgt	A.J. Rasch	RTG	KIA	3
S/Sgt	H.F. Gibney, Jr.*	LWG	KIA	3
S/Sgt	W.R. Greason	RWG	KIA	3
S/Sgt	J.H. Randall	TG	KIA	3
Sgt	C.C. Rhodes	PHOT	KIA	1

*Buried in Ardennes American Cemetery near Liege, Belgium.

10/4/43 Frankfurt am Main, Ger.
B-17F 42-29846 - (No Name) - A/C Mis. 16 - 359 BS
MACR 780 - Crew Parachuted SW of Aachen, Ger.

Rank	Name	Position	Fate	Mission #
1Lt	V.J. Loughnan	P	POW	20
2Lt	H.W.Gredvig	CP	POW	16
1Lt	J.L. Maxwell	NAV	POW	13
1Lt	E.J. Pullman	BOM	POW	12
T/Sgt	E.W. High	ENG	POW	13
T/Sgt	R.M. Daley Jr.	RAD	POW	19
S/Sgt	K.L. McGee	BTG	POW	19
S/Sgt	G.E. Barr	LWG	POW	13
S/Sgt	P.C. Robillard	RWG	POW	5
S/Sgt	B.K. Knorpp	TG	POW	13

10/9/43 Anklam, Ger.
B-17F 42-5221 - *Son* - A/C Mis. 25 - 427 BS
MACR 874 - Crashed in Baltic Sea Near Lolland, Den.

Rank	Name	Position	Fate	Mission #
2Lt	B.J. Clifford	P	KIA	4
2Lt	C.O. Jahn	CP	KIA	4
2Lt	C.M. State	NAV	KIA	4
2Lt	R.V. Bruce*	BOM	KIA	4
S/Sgt	A.E. Horning	ENG	KIA	4
T/Sgt	F.G. Hartzog	RAD	KIA	4
S/Sgt	C.L. Gale	BTG	KIA	4
S/Sgt	J.L. McLarty	LWG	KIA	4
T/Sgt	A.A. Dyke*	RWG	KIA	4
S/Sgt	W.O. Heller	TG	KIA	4

*Buried in Ardennes American Cemetery near Liege, Belgium.

10/14/43 Schweinfurt, Ger.
B-17F 42-29477 - *Joan of Arc* - A/C Mis. 4 - 358th BS
MACR 907 - Crashed near Bamberg, Ger.

Rank	Name	Position	Fate	Mission #
1Lt	R.C. Sanders	P	POW	8
2Lt	H.F. Nogash	CP	POW	6
2Lt	G.W. Wood	NAV	POW	8
2Lt	P.A. Peed	BOM	POW	6
T/Sgt	J.A. Rozek	ENG	POW	7
T/Sgt	J.G. Trest	RAD	POW	6
S/Sgt	M.J. Kremer	BTG	POW	6
S/Sgt	R.K. Smith	LWG	POW	7
Sgt	M.V. Carll*	RWG	KIA	7
S/Sgt	A.E. Cockrum*	TG	KIA	7
Sgt.	W.G. Martin	PHOT	POW	5

*Buried in Lorraine American Cemetery near Moselle, France.

10/20/43 Duren, Ger.
B-17F 41-24629 - (No Name) - A/C Mis. 4 - 358 BS
MACR 1032 - Crashed near Valenciennes, Fr.

Rank	Name	Position	Fate	Mission #
1Lt	J.W. Hendry, Jr.	P	POW	24
2Lt	W.B. Harper*	CP	KIA	12
2Lt	B.T. McNamara	NAV	POW	25
2Lt	R.E. Webster	BOM	POW	25
T/Sgt	L.C. Biddle	ENG	POW	24
T/Sgt	J.J. Brown	RAD	KIA	25
S/Sgt	A.J. Hargrave	BTG	POW	18
S/Sgt	D.E. Guhr	LWG	POW	3
S/Sgt	W.G. Raesly	RWG	POW	15
S/Sgt	J.J. Doherty	TG	POW	24

*Buried in Normandy American Cemetery near Coleville-Sur-Mer, France.

10/20/43 Duren, Ger.
B-17F 42-29571 - *Charley Horse* - A/C Mis. 32 - 358 BS
MACR 1033 - Crashed near Mons, Bel.

Rank	Name	Position	Fate	Mission #
2Lt	W.R. Hartigan	P	EVD	7
2Lt	E.N. Goddard	CP	POW	6
2Lt	L.F. Douthett	NAV	EVD	7
2Lt	B.F. Dorsey	BOM	POW	6
T/Sgt	C. Resto	ENG	POW	7
T/Sgt	R.L. Ward	RAD	EVD	7
S/Sgt	V.F. Stoddard	BTG	POW	7
S/Sgt	C.J. Dove	LWG	KIA	7
S/Sgt	J.W. Lowther	RWG	EVD	5
S/Sgt	J.T. Ince	TG	POW	7
Sgt	R.P. Moffett	PHOT	POW	7

10/23/43 Local Night Training Flight
B-17F 42-29930 - *Miss Patricia* - A/C Mis. 2 - 360 BS
No MACR - Crashed Shortly After Takeoff at Keyston, Eng.

Rank	Name	Position	Fate	Mission #
1Lt	L.E. Jokerst	P	K.I. CR.	15
2Lt	T.M. Jackson*	CP	K.I. CR.	6
S/Sgt	W.H. Stephen*	ENG	K.I. CR.	15
Cpl	S.B. Morse	RAD	K.I. CR.	0
Sgt	E.A. Chuhran	GUN	K.I. CR.	3
Sgt	R.L. Long*	GUN	K.I. CR.	0
Sgt	H.R. Sherman	GUN	K.I. CR.	0
Pvt	R.V. Morgan	PASS	K.I. CR.	0

* Buried at Cambridge, England

11/5/43 Gelsenkirchen, Ger.
B-17F 41-24565 - *Rambling Wreck - A/C Mis. 28 - 359 BS**
MACR 1157- Crashed near Rosenthal, Germany

Rank	Name	Position	Fate	Mission #
2Lt	A.G. Grant	P	POW	8
2Lt	F.C. Hall	CP	POW	8
2Lt	J.F. Berger	NAV	POW	8
2Lt	M.D. Blackburn	BOM	POW	8
S/Sgt	T.T. Kujawa	ENG	POW	7
T/Sgt	E.J. Sexton	RAD	POW	8
S/Sgt	C. Petroksy	BTG	POW	8
T/Sgt	H.A. Kraft	LWG	POW	23
Sgt	J.J. Hauer	RWG	KIA	3
S/Sgt	F.D. Andersen	TG	POW	8

*Also known by the name *Idaho Potato Peeler*

11/26/43 Bremen, Ger.
B-17F 42-29955 - *Mr. Five by Five* - A/C Mis. 24 - 427 BS
MACR 1324 - Crashed at Den Helder, NL

Rank	Name	Position	Fate	Mission #
Capt	A.A. Cote	P	KIA	24
2Lt	C.C. Bixler	CP	KIA	1
1Lt	W.R. Barnhill	NAV	KIA	18
1Lt	J.W. Hull*	BOM	KIA	24
T/Sgt	J.R. Arter	ENG	KIA	23
T/Sgt	V. Reaves	RAD	KIA	24
S/Sgt	C.M. May	BTG	KIA	19
S/Sgt	T. Gomes	LWG	KIA	4
S/Sgt	J.M. Micek	RWG	KIA	23
S/Sgt	P. Gunsauls	TG	KIA	13

*Body washed up on Terschelling Island, Holland on 13 February 1944.
Buried in Netherlands American Cemetery, Martgratten, Netherlands.

11/29/43 Bremen, Ger.
B-17F 42-29498* - *Dark Horse* - A/C Mis. 7 - 360 BS
MACR 1657 - Crashed 25 km SW of Bremen, Ger.

Rank	Name	Position	Fate	Mission #
1Lt	C.J. Fyler	P	POW	25
2Lt	R.C. Ward	CP	POW	13
1Lt	G. Molnar	NAV	POW	24
2Lt	J.S. Petrolino	BOM	POW	11
T/Sgt	B.J. Addison	ENG	POW	25
Sgt	R.B. O'Connell	RAD	KIA	2
S/Sgt	R.D. Ford**	BTG	KIA	24
S/Sgt	M.G. Stachowiak	LWG	POW	21
S/Sgt	G.C. Fisher	RWG	POW/REP	23
S/Sgt	J.R. Sawicki	TG	KIA	15
Sgt	N.P.S. Egge	PHOT	KIA	5

*Reportedly equipped with a chin turret.
**Buried in Ardennes American Cemetery near Liege, Belgium.

11/29/43 Bremen, Ger.
B-17F 42-5483 - *Red Ass* - A/C Mis. 5* - 360 BS
MACR 1656 - Crashed at Renslage, Ger.

Rank	Name	Position	Fate	Mission #
2Lt	F.A. Brumbelow**	P	POW	3
2Lt	D.H. Marsh	CP	POW	3
2Lt	J.R. Groves	NAV	POW	3
2LT	C.D. Garneau	BOM	KIA	3
S/Sgt	P.J. Kiebish	ENG	POW	3
S/Sgt	I.A. Johnson	RAD	POW	3
Sgt	H.S. Payton***	BTG	KIA	3
Sgt	W.C. Steele	LWG	KIA	3
Sgt	D.H. Hoffman***	RWG	KIA	1
Sgt	W.R. Perryman	TG	POW	3

*Official mission number, believed by the author to be significantly understated.
**Lost a heel - used a cedar post as a prothesis while a POW.
***Buried in Ardennes American Cemetery near Liege, Belgium.

12/1/43 Solingen, Ger.
B-17G 42-39781 - (No Name) - A/C Mis. 4 - 360 BS
MACR 1325 - Crashed at Radingham, 10 km W of Lille, Fr.

Rank	Name	Position	Fate	Mission #
2Lt	G.W. Luke, Jr.	P	POW	3
2Lt	F. Mitchell	CP	EVD	3
2Lt	G.A. Ballagh	NAV	POW	3
2Lt	L.L. Dahnke	BOM	POW	3
S/Sgt	M.A. Boreen	ENG	POW	3
S/Sgt	B.T. Day	RAD	POW	3
Sgt	R.H. Washburn	BTG	KIA	3
Sgt	E.D. Yaekel	LWG	POW	3
Sgt	S.G. Wright	RWG	POW	3
Sgt	I.J. Walter	TG	EVD/POW	3

12/20/43 Bremen, Ger.
B-17G 42-39764 - *Santa Anna* - A/C Mis. 6 - 360 BS (427 BS crew)
MACR 1706 - Crashed near Bremen, Ger.

Rank	Name	Position	Fate	Mission #
2Lt	A. Alex	P	POW	1
2Lt	L.E. Jackman, Jr.*	CP	KIA	6
2Lt	N.J. Goldschmidt, Jr.	NAV	POW	2
2Lt	A.L. Farrah	BOM	POW	1
S/Sgt	J. Adamczyk	ENG	POW	1
S/Sgt	W.R. McCarren	RAD	POW	1
Sgt	H.P. Micheles	BTG	POW	1
Sgt	R.J. Newcomb**	LWG	POW/REP	1
Sgt	H.A. Brown**	RWG	POW/REP	1
Sgt	S.G. Hall**	TG	POW/REP	1

*Buried in Netherlands American Cemetery, Martgratten, Netherlands
**Badly wounded and repatriated after capture.

12/20/43 Bremen, Ger.
B-17G 42-31233 - (No Name) - A/C Mis. 5 - 427 BS
MACR 1707 - Crashed near Bremen, Ger.

Rank	Name	Position	Fate	Mission #
2Lt	F. Leve*	P	KIA	6
2Lt	D.L. Libbee*	CP	KIA	5
2Lt	R.D. Morehead	NAV	KIA	6
2Lt	D.J. Murphy	BOM	KIA	6
T/Sgt	J.C. Spross	ENG	POW	6
S/Sgt	P.D. Craig*	RAD	KIA	6
S/Sgt	E.W. Drees*	BTG	KIA	6
S/Sgt	O.L. Keefer	LWG	KIA	6
S/Sgt	F.L. Midkiff*	RWG	KIA	6
S/Sgt	A.O. VanLandingham	TG	POW	6

*Buried in Netherlands American Cemetery, Martgratten, Netherlands.

12/30/43 Ludwigshafen, Ger.
B-17G 42-39795 - *Women's Home Companion* - A/C Mis. 7 - 359 BS
MACR 1674 - Crashed near Cerfontaine, Bel.

Rank	Name	Position	Fate	Mission #
1Lt	W.C. Osborn	P	EVD/POW	7
2Lt	J. Jernigan, Jr.	CP	EVD/POW	7
2Lt	E.L. Cobb	NAV	POW	13
2Lt	N. Campbell	BOM	EVD	7
S/Sgt	W.E. Wolff	ENG	EVD	7
S/Sgt	G.L. Daniel	RAD	POW	7
Sgt	L.W. Fitzgerald	BTG	KIA	7
Sgt	E.D. Wolfe	LWG	POW	6
Sgt	V.J. Reese*	RWG	EVD/POW/KIA	2
Sgt	L.B. Evans	TG	KIA	7

*Buried in Netherlands American Cemetery, Martgratten, Netherlands.

1/4/44 Kiel, Ger.
B-17G 42-31526 - *Sweet Anna* - A/C Mis. 1 - 427 BS
MACR 1682 - Ditched in North Sea

Rank	Name	Position	Fate	Mission #
1Lt	F.C. Humphreys	P	KIA	16
2Lt	J.H. Clemons	CP	KIA	15
2Lt	H. Hladun	NAV	KIA	1
2Lt	B.N. Mire	BOM	KIA	5
T/Sgt	F.J. Janisch	ENG	KIA	15
T/Sgt	A.H. Woods	RAD	KIA	14
S/Sgt	M.H. Ross	BTG	KIA	15
S/Sgt	M.M. Dare	LWG	KIA	15
S/Sgt	E.P. Madak	RWG	KIA*	10
S/Sgt	W.C. Sparks	TG	KIA	13

*Body recovered on Oct. 4, 1944 on Ameland Island, NL

1/5/44 Kiel, Ger.

B-17G 42-31441 - (No Name) - A/C Mis. Unk. - 360 BS

Destroyed in Collision with 379 BG A/C During Assembly

Rank	Name	Position	Fate	Mission #
2Lt	B.G. Burkitt	P	K.I. CR.	5
2Lt	H.J. Kuhn*	CP	K.I. CR.	4
2Lt	H.A. Foote, Jr.	NAV	K.I. CR.	4
2Lt	F.J. Reith+	BOM	K.I. CR.	4
S/Sgt	P.H. Gatewood	ENG	K.I. CR.	3
S/Sgt	A.D. Cantrell	RAD	K.I. CR.	4
Sgt	A.E. Berntzen**	BTG	K.I. CR.	4
Sgt	W.E. Stoffregen	LWG	K.I. CR.	4
Sgt	L.E. Brown*	RWG	K.I. CR.	4
Sgt	C.C. Rush*	TG	K.I. CR.	4

*Cambridge American Cemetery, England (Wall of Missing).
**Brittany American Cemetery near Manche, France (Wall of Missing) next to his brother, F/O S. Berntzen, KIA 29 May 1943.

1/11/44 Oschersleben, Ger.

B-17F 42-24587- *Bad Check - A/C Mis. 45 - 427 BS**

MACR 1922 - Crashed near Lienen, Ger.

Rank	Name	Position	Fate	Mission #
1Lt	G.S. McClellan, Jr.	P	KIA	18
2Lt	W.A. Fisher	CP	KIA	1
2Lt	J.C. Kaliher	NAV	POW	18
2Lt	M.L. Cornish	BOM	POW	17
T/Sgt	D. Tempesta	ENG	KIA	17
T/Sgt	G.A. Callihan**	RAD	KIA	17
S/Sgt	R.G. Yarlan	BTG	POW	16
S/Sgt	A.B. Chiles, Jr.	RWG	POW	14
S/Sgt	B.S. Heaton	LWG	POW	17
S/Sgt	C.E. Dugan	TG	POW	15

*Original 303rd BG B-17F
**Buried in Ardennes American Cemetery, Liege, Belgium

1/11/44 Oschersleben, Ger.

B-17F 42-3131 - *Flak Wolf* - A/C Mis. 40 - 427 BS

MACR 1966 - Crashed near Kloster Oesede, Ger.

Rank	Name	Position	Fate	Mission #
2Lt	J.W. Carothers*	P	KIA	4
2Lt	C.E. Frost	CP	KIA	3
2Lt	A. Linnehan	NAV	POW	7
2Lt	H.W. Barriscale	BOM	POW	4
S/Sgt	H.E. Scott	ENG	POW	4
S/Sgt	G.S. Rajcula	RAD	KIA	4
Sgt	R.T. Peavy	BTG	KIA	4
T/Sgt	W.A. Roer	LWG	KIA	4
Sgt	F.J. Morneau*	RWG	KIA	4
Sgt	R.R. Ziegler	TG	KIA	5

*Buried in Netherlands American Cemetery, Martgratten, Netherlands

1/11/44 Oschersleben, Ger.
B-17G 42-39896 - (No Name) - A/C Mis. 3 - 360 BS
MACR 9554 - Crashed near Kirchlengern, Ger.

Rank	Name	Position	Fate	Mission #
2Lt	R.H. Hallden*	P	KIA	3
2Lt	R.L. Gentry	CP	KIA	2
2LT	G.N. Limon	NAV	POW	4
F/O	J. Hubenschmidt	BOM	POW	4
S/Sgt	H.M. Beben	ENG	POW	4
Sgt	D.R. Hutchins	RAD	POW	4
Sgt	C.E. Moore	BTG	POW	4
Sgt	R.B. Robinson	LWG	POW	4
Sgt	H.G. Hays	RWG	POW	4
Sgt	C.H. Chatola	TG	POW	4

*Buried in Netherlands American Cemetery, Martgratten, Netherlands.

1/11/44 Oschersleben, Ger.
B-17F 42-5360 - *War Bride* - A/C Mis. 35 -358 BS
MACR 1926 - Crashed near Detmold, Ger.

Rank	Name	Position	Fate	Mission #
2Lt	H.A. Schwaebe	P	POW	7
2Lt	H.F. Dumse	CP	POW	4
2Lt	P.T. Degnan	NAV	POW	7
2Lt	W.W. Wiley	BOM	POW	7
S/Sgt	R. Foreman	ENG	POW	7
T/Sgt	M.E. Tudor	RAD	POW	5
Sgt	B.F. Harvey	BTG	POW	6
Sgt	J.F. Malcolm	LWG	POW	7
Sgt	R.O. Whitesell	RWG	KIA	3
Sgt	F.G. Iott	TG	POW	8

1/11/44 Oschersleben, Ger.
B-17F 42-29894 - *Baltimore Bounce* - A/C Mis. 15 - 359 BS
MACR 1928 - Crashed near Laube/Lippe, Ger.

Rank	Name	Position	Fate	Mission #
2Lt	W.A. Purcell	P	KIA	5
2Lt	F.D. Krohn	CP	KIA	4
2Lt	M.H. Mussett*	NAV	KIA	5
2Lt	J.B. Kyne	BOM	KIA	5
S/Sgt	P.C. Castriciano*	ENG	KIA	5
S/Sgt	J.C. Beeny	RAD	KIA	5
Sgt	L.N. Faner*	BTG	KIA	5
Sgt	K.W. Nye	LWG	KIA	5
Sgt	H.R. Eastburn*	RWG	KIA	4
Sgt	J.W. Swanson	TG	KIA	4

*Buried in Ardennes American Cemetery, Liege, Belgium.

1/11/44 Oschersleben, Ger.
B-17F 41-24619 - *S for Sugar - A/C Mis. 52 - 427 BS**
MACR 1923 - Crashed near Oberhaus, Ger.

Rank	Name	Position	Fate	Mission #
2Lt	T.L. Simmons	P	POW	4
2Lt	F.E. Reichel	CP	POW	3
2Lt	W.L. Clyatt, Jr.	NAV	POW	5
2Lt	R.W. Vaughn	BOM	POW	5
S/Sgt	W.S. Elliott	ENG	POW	4
Sgt	J.A. Bennett	RAD	POW	4
Sgt	D.C. DiPietra	BTG	POW	4
Sgt	R.F. Livingston**	LWG	POW/Died	4
Sgt	R.D. Stewart	RWG	POW	4
Sgt	W.L. Hasty	TG	POW	4

*Original 303rd BG B-17F.
**Died of pneumonia on March 27, 1944 in Stalag 17B, Krems, Austria.

1/11/44 Oschersleben, Ger.
B-17G 42-30865 - (No Name) - A/C Mis. 9 - 358 BS
MACR 1927 - Crashed near Nordhausen, Ger.

Rank	Name	Position	Fate	Mission #
1LT	P.W. Campbell	P	POW	16
2Lt	J.C. Doty	CP	POW	3
1Lt	J.P.D. Nothstein	NAV	POW/REP	16
2Lt	W.J. Millner*	BOM	KIA	7
T/Sgt	S.J. Backiel	ENG	POW	19
Sgt	D. Di Martino*	RAD	KIA	4
S/Sgt	J.W. Brooks	BTG	KIA	16
S/Sgt	J.F. Hoy*	LWG	KIA	14
S/Sgt	A. Wisniewski	RWG	KIA	16
S/Sgt	E.J. Cassidy*	TG	KIA	18

*Buried in Ardennes American Cemetery, Liege, Belgium

1/11/44 Oschersleben, Ger.
B-17F 41-24562 - *Sky Wolf - A/C Mis. 60 - 358 BS**
MACR 1925 - Crashed at Wolsdorf, Ger.

Rank	Name	Position	Fate	Mission #
1Lt	A.L. Emerson	P	POW	20
2Lt	M.L. Riddick	CP	POW	8
1Lt	J. Halliburton	NAV	POW	14
2Lt	D.J. De Laura	BOM	POW	8
T/Sgt	G.C. Mullins	ENG	POW	13
S/Sgt	J.C. Supple	RAD	POW	14
S/Sgt	H.H. Zeitner	BTG	KIA	12
S/Sgt	J.G. Visneki	LWG	POW	13
S/Sgt	J.H. Pleasant	RWG	POW	13
S/Sgt	B.J. Sutton	TG	POW	12

*Original 303rd BG B-17F

1/11/44 Oschersleben, Ger.
B-17G 42-39794 - (No Name) - A/C Mis. 10 - 358 BS
MACR 1929 - Crashed near Oschersleben, Ger.

Rank	Name	Position	Fate	Mission #
2Lt	W.C. Dashiell*	P	KIA	8
2Lt	H.C. Mable*	CP	KIA	5
2Lt	T.A. Sutherland*	NAV	KIA	7
S/Sgt	G.H. Fee**	BOM	KIA	8
S/Sgt	R.L. Stevenson*	ENG	KIA	8
S/Sgt	B.J. Radebaugh**	RAD	KIA	8
Sgt	A.H. Robinson*	BTG	KIA	8
Sgt	R.A. Parker	LWG	KIA	4
Sgt	R.J. Owen	RWG	KIA	8
Sgt	C.M. McKinney	TG	KIA	8

*Group burial in Jefferson Barracks National Cemetery, St. Louis, Missouri
**Buried in Netherlands American Cemetery, Martgratten, Netherlands.

1/11/44 Oschersleben, Ger.
B-17F 42-3448 - (No Name) - A/C Mis. 13 - 359 BS
MACR 1924 - Crashed in Steinhuder Lake, Ger.

Rank	Name	Position	Fate	Mission #
2Lt	H.J. Eich, Jr.	P	POW	8
2Lt	W.E. Woodside	CP	POW	8
2Lt	J.E. Carroll	NAV	POW	8
2Lt	W.G. Stein	BOM	POW	8
S/Sgt	H. Lenson	ENG	POW/REP	8
T/Sgt	D.S. Harvey*	RAD	KIA	8
S/Sgt	J.P. Celoni	BTG	POW	8
S/Sgt	E.A. Maggia	LWG	POW	8
S/Sgt	D.C. Erdmann	RWG	POW	8
S/Sgt	R.M. Gilstrap	TG	POW	8

*Body Washed up near Neinberg, Germany on March 6, 1944.

1/11/44 Oschersleben, Ger.
B-17F 42-29524 - *Meat Hound* - A/C Mis. 25 - 358 BS
No MACR - Landed at Metfield, Eng. After 9 Crew Bailed Out

Rank	Name	Position	Fate	Mission #
2Lt	J.W. Watson	P	RET	6
2Lt	C.C. David	CP	EVD	3
2Lt	J.G. Leverton	NAV	POW	8
2Lt	V.R. Calvin* **	BOM	KIA	6
S/Sgt	S.J. Rowland**	ENG	KIA	6
S/Sgt	H. Romaniec	RAD	POW	6
Sgt	F.H. Booth**	BTG	KIA	6
Sgt	E.H. Stewart	LWG	POW	6
Sgt	W.H. Fussner**	RWG	KIA	6
Sgt	K.P. Kosinski	TG	POW	6

*Buried in Netherlands American Cemetery, Martgratten, Netherlands.
**Drowned in Zuider Zee.

1/14/44 Le Meillard & Gueschart, Fr.
B-17F 42-3029 - *Wallaroo* - A/C Mis. 35 - 359 BS A/C 358 BS Crew
MACR 1965 - Crashed at Foret de Crecy, 10 mi. N. of Abbeville, Fr.

Rank	Name	Position	Fate	Mission #
Capt.	M.R. Hungerford*	P	POW	13
2Lt	K.B. Arundale*	CP	POW	2
2Lt	J.B. Vogel*	NAV	POW	3
2Lt	J.F. Barlow*	BOM	POW	2
S/Sgt	C.C. Finch	ENG	POW	2
S/Sgt	R.A. Davis, Jr.	RAD	POW	2
Sgt	R.L. Clink	BTG	POW	6
Sgt	A.P. Petiz	LWG	POW	2
Sgt	J.T. Elovich	RWG	POW	2
Sgt	J.F. Fertitta	TG	POW	2
Sgt	A.C. Wilson, Jr.**	PHOT	POW	6

*Imprisoned at Stalag Luft 1.
**Possibly "Nilson"

1/29/44 Frankfurt, Ger.
B-17G 42-39786 - *G.I. Sheets* - A/C Mis. 11 - 427 BS
MACR 2260 - Crashed at Solre - St. Gery, Belgium

Rank	Name	Position	Fate	Mission #
2Lt	J.F. Fowler*	P	EVD/POW	8
2Lt	B.W. Rawlings**	CP	EVD	7
2Lt	A.L. Taylor***	NAV	POW	8
2Lt	J.C. Thompson, Jr.***	BOM	POW	3
T/Sgt	C.E. Finley****	ENG	POW	8
T/Sgt	D.J. Dinwiddle*	RAD	EVD/POW	7
S/Sgt	R. Arrington*****	BTG	EVD	8
Sgt	M.O. Jackson	LWG	KIA	7
S/Sgt	L.E. Zimmer*****	RWG	EVD	7
S/Sgt	J.D. Ferguson***	TG	POW	12

*Evaded capture until May 13, 1944.
**Returned to England after crossing the Pyrenees into Spain.
***Badly wounded. Taken to a civilian hospital by Belgians where captured by Germans.
****Evaded with Rawlings, captured near Spanish border after they split up.
***** Evaded capture. Picked up by American Troops in September, 1944.

2/3/44 Wilhelmshaven, Ger.
B-17G 42-39727 - (No Name) - A/C Mis. 11 - 358 BS
MACR 2238 - Ditched in North Sea

Rank	Name	Position	Fate	Mission #
Capt.	G.A. White	P	KIA	1
2Lt	W.W. Newsom	CP	KIA	1
2Lt	E.A. Kruse	NAV	KIA	1
2Lt	W. Stafford, Jr.	BOM	KIA	1
S/Sgt	S.L. Bottomley	ENG	KIA	1
S/Sgt	L.E. Znidersich	RAD	KIA	1
Sgt	R.W. Slate	BTG	KIA	1
Sgt	A.J. Quinlan	LWG	KIA	1
Sgt	W.E. Becknell	RWG	KIA	1
Sgt	R.E. McCoy	TG	KIA	1

2/6/44 Dijon, Fr.
B-17G 42-97498 - *Padded Cell* - A/C Mis. 11 - 358 BS
MACR 2384 - Crashed near Bricy in Paris, France Area

Rank	Name	Position	Fate	Mission #
1Lt	J.S. Bass	P	POW	1
2Lt	M.M. Goldman	CP	KIA	2
2Lt	M.B. Abernathy	NAV	POW	12
2Lt	M.S. Zientar	BOM	POW	10
S/Sgt	A. Quevedo	ENG	POW	11
S/Sgt	J.C. Hensley	RAD	EVD/POW	11
Sgt	H.J. Brown	BTG	POW	10
Sgt	M.J. Canale	LWG	POW	11
Sgt	J.P. Grsetic	RWG	POW	11
Sgt	C.J. Dockendorf	TG	POW	7

Appendix II

Suggestions for Further Reading and Research

Publications

Those interested in learning more about the Hell's Angels 303rd Bombardment Group (H) or the individuals in this book are urged to consult the following works. A ✪ listed next to the work indicates that the individual or work is specifically mentioned or referenced in this book.

Clark, Peter, *Where the Hills Meet the Sky* (Glen Graphics, Wooler, Northumberland, England, 1995). A guide to wartime crashes in the Cheviot Hills, Northumberland, England, including the 12/16/44 crash of the 303rd BG/360th BS B-17G 44-6504, PU✪M.

Crawford, William, *Angels Over the Reich—Combat with a B-17 Flight Crew* (Marietta, Georgia, Private Printing, 1996).

✪ David, Clayton C., *They Helped Me to Escape: From Amsterdam to Gibraltar in 1944*, (Sunflower University Press, Manhattan, Kansas, 1988). Clayton David was a one of the surviving members of Lt. Jack Watson's crew who bailed out of the badly damaged *Meat Hound* on its return from the 1/11/44 mission to Oschersleben, Germany. This is the story of his successful escape and evasion (E&E) and ultimate return to England.

✪ Deerfield, Eddie, *The PsyWarriors* (Northwest Publishing, Inc., Salt Lake City, Utah, 1994). An historical novel inspired by the author's service with a U.S. psychological warfare unit during the Korean War.

✪ DeJong, Ivo M., *Mission 85, a Milk Run That Turned Sour: The United States Eighth Air Force and its mission on August 19, 1943 over south-*

west Holland. (Liberation Museum, 1944, Groesbeek, The Netherlands, 1998.) Strongly recommended for its in-depth treatment of this mission from the perspectives of the combatants on both sides and the Dutch civilians on the ground. Contains much information about the 303rd's involvement in the day's operations, including personal accounts, photographs, and detailed information about 303rd aircraft and personnel lost on the raid.

✪ Ethell, Jeffrey, *Cowboys and Indians*, (Graphically Speaking, Alexandria, Virginia, 1977). Excellent monograph about the research behind and the painting by noted aviation artist Keith Ferris of the huge mural, "Fortresses Under Fire" in the National Air and Space Museum, Washington, D.C. Explains why the 303rd's *Thunderbird* was selected and provides substantial information about the aircraft's operational history and crews and the ill-fated August 15, 1944 303rd mission to Wiesbaden, Germany.

✪ Feeney, William, Sgt. (Editor), *The First 300—Hell's Angels* (Public Relations Office, 303rd BG(H), Molesworth, England, under the supervision of Captain Walter R. Donnelly, Group Public Relations Officer, 1945). This is the Group's wartime history, which is mentioned in a number of places in the main body of this book. An excellent source of basic information about the 303rd, which is all the more valuable for its 1940s flavor.

Fleming, Samuel P. as told to Ed Y. Hall, *Flying with the Hell's Angels* (The Honoribus Press, Spartenburg, South Carolina, 1991).

✪ Flyer, Carl, J., *Staying Alive* (J. H. Johnson III, Leavenworth, Kansas, 1995). Carl J. Fyler's detailed personal account of his pilot training, deployment overseas, combat tour with the 303rd, and POW experiences. Strongly recommended.

Giering, Edward J., *B-17 Bomber Crew Diary* (Sunflower University Press, Manhattan, Kansas, 1985).

✪ Gobrecht, Harry D., *Might in Flight, Daily Diary of the Eighth Air Force's Hell's Angels 303rd Bombardment Group (H)* (303rd Bomb Group Association, Inc., 2d ed., 1997). The most comprehensive and detailed single source of information about the Hell's Angels Bomb Group ever written. Strongly recommended.

✪ Hand, Robert A., Sr., *"Last Raid," A Personal Account of the 303rd Bomb Group Participation in the Thousand Plane Raid on Berlin, 3 February, 1945...and Other Stories*. (Privately Printed, Boynton

Beach, Florida, 1995). The author's personal memoir of service in the 303rd during the latter part of the war, organized around the account he wrote in 1945 of the mission referred to in the title. With personal, color illustrations by the author. Strongly recommended for the picture it paints of life in the 303rd during the late war period.

✪ Heller, William, *Airline Safety, A View from the Cockpit,* (Rulorca Press, Half Moon Bay, California, rev. ed., 1986). Though mentioned only in passing in this book, William Heller (distinguished in the 303rd during the war years by the moniker "Heller not Hullar"), was a significant participant in the missions the Hullar's crew flew (including both of the 1943 Schweinfurt missions). He ultimately became 360th Squadron CO and served with Bob Hullar through the end of the war. A postwar airline pilot with an incredible personal log of over 32,000 cockpit hours at the time of his retirement in 1980, the author's book is considered by many as a standard reference on its subject.

Howell, Forrest W., *Whispers of Death—Yankee Kriegies* (Rainbow Books, Moore Haven, Florida, 1985). A 303rd radio operator's account of his late war tour, shot down on his seventeenth mission, and POW experiences.

✪ Rawlings, Barney, *Off We Went (Into the Wild Blue Yonder)* (Morgan Printers, Inc., Washington, N. Carolina, 1994). Privately published memoir of James F. Fowler's copilot, which provides a detailed history of their crew, its last mission on January 29, 1944, the author's successful escape and evasion, their postwar reunion, and the dedication of the Belgian monument to their crew in 1987. Strongly recommended.

Smith, Ben, Jr., *Chick's Crew—A Tale of the Eighth Air Force* (Rose Printing Co., Inc., Tallahassee, Florida, 2d ed., 1983). Another well-written crew memoir by a 303rd radioman whose tour occurred in 1944, after the Hullar crew's missions.

303rd Bomb Group and Related Web Sites

Those interested in the 303rd Bomb Group, The Mighty Eighth Air Force Heritage Museum in Pooler, Georgia, and The U.S. Air Force Museum near Dayton, Ohio are encouraged to visit the following Internet sites on the World Wide Web, where many additional links may be found.

http://www.303rdbga.com/ The official Web site of the 303rd Bomb Group Association (H).

http://www.molesworth.af.mil/303bg/303bg.htm Historical 303rd Bomb Group page maintained by the 423rd Air Base Squadron, the military unit that provides base facility services to the Joint Analysis Center located at RAF Molesworth.

http://www.imall.com/stores/mighty8th museum/The official Web site of The Mighty Eighth Air Force Heritage Museum in Pooler, Georgia.

http://www.nasm.edu/GALLERIES/GAL205/gal205.html Web site page from the National Air and Space Museum's World War II Aviation Gallery containing partial views of the Keith Ferris mural, "Fortresses Under Fire."

http://www.fairmont.wvnet.edu/www/webteam/bob/Mural.html Web site page from "Bob's B-17 Page" providing a full view of the Keith Ferris mural.

http://www.wpafb.af.mil/museum/ Web site of the United States Air Force Museum, Wright-Paterson Air Force Base, Dayton, Ohio.

Index

Note: Boldface numbers indicate illustrations.

Index of B-17s by Serial Numbers

General Terms Index

Named Aircraft Index

Place Names Index

Military Units

Names Index

About the Author

Brian D. O'Neill, is the General Counsel of Curtiss-Wright Corporation, the famous aviation company that manufactured the Curtiss P-40 fighter and the B-17's Wright Cyclone engines during World War II. Before becoming an attorney, he served with the U.S. Navy as a destroyer gunner officer and shipyard repair officer. An avid student of Eighth Air Force history, he resides in East Windsor, New Jersey.